AF307789

Springer Series in Information Sciences 32

Editor: Thomas S. Huang

Springer
Berlin
Heidelberg
New York
Barcelona
Budapest
Hong Kong
London
Milan
Paris
Santa Clara
Singapore
Tokyo

Springer Series in Information Sciences

Editors: Thomas S. Huang   Teuvo Kohonen   Manfred R. Schroeder
Managing Editor: H. K. V. Lotsch

Volumes 1–29 are listed at the end of the book.

Nailong Wu

The Maximum Entropy Method

With 53 Figures

Springer

Dr. Nailong Wu

Department of Physics and Astronomy
York University
Petrie Science Building
4700 Keele Street
Toronto, Ontario, M3J 1P3, Canada

Series Editors:

Professor Thomas S. Huang

Department of Electrical Engineering and Coordinated Science Laboratory,
University of Illinois, Urbana, IL 61801, USA

Professor Teuvo Kohonen

Helsinki University of Technology, Neural Networks Research Centre,
Rakentajanaukio 2C, FIN-02150 Espoo, Finland

Professor Dr. Manfred R. Schroeder

Drittes Physikalisches Institut, Universität Göttingen, Bürgerstrasse 42-44,
D-37073 Göttingen, Germany

Managing Editor:

Dr.-Ing. Helmut K. V. Lotsch

Springer-Verlag, Tiergartenstrasse 17,
D-69121 Heidelberg, Germany

ISSN: 0720-678X
ISBN-13:978-3-642-64484-9   e-ISBN-13:978-3-642-60629-8
DOI: 10.1007/978-3-642-60629-8

Springer-Verlag Berlin Heidelberg New York

Library of Congress Cataloging-in-Publication Data.

Wu, Nailong, 1946– . The maximum entropy method / Nailong Wu. p. cm. – (Springer series in information sciences; 32) Includes bibliographical references and index. ISBN 3-540-61965-8 (hardcover: alk. paper) 1. Maximum entropy method. 2. Spectral theory (Mathematics) 3. Signal processing. 4. Mathematical physics. I. Title. II. Series.   Q370.W8   1997   003'.54–dc21   96-54843

This work is subject to copyright. All rights are reserved, whether the whole or part of the material is concerned, specifically the rights of translation, reprinting, reuse of illustrations, recitation, broadcasting, reproduction on microfilm or in any other way, and storage in data banks. Duplication of this publication or parts thereof is permitted only under the provisions of the German Copyright Law of September 9, 1965, in its current version, and permission for use must always be obtained from Springer-Verlag. Violations are liable for prosecution under the German Copyright Law.

© Springer-Verlag Berlin Heidelberg 1997
Softcover reprint of the hardcover 1st edition 1997

The use of general descriptive names, registered names, trademarks, etc. in this publication does not imply, even in the absence of a specific statement, that such names are exempt from the relevant protective laws and regulations and therefore free for general use.

Typesetting: Camera ready by author
Cover design: *design & production* GmbH, Heidelberg
Production editor: P. Treiber, Heidelberg

SPIN: 10130564   54/3144 - 5 4 3 2 1 0 - Printed on acid-free paper

Preface

Forty years ago, in 1957, the Principle of Maximum Entropy was first introduced by Jaynes into the field of statistical mechanics. Since that seminal publication, this principle has been adopted in many areas of science and technology beyond its initial application. It is now found in spectral analysis, image restoration and a number of branches of mathematics and physics, and has become better known as the Maximum Entropy Method (MEM). Today MEM is a powerful means to deal with ill-posed problems, and much research work is devoted to it.

My own research in the area of MEM started in 1980, when I was a graduate student in the Department of Electrical Engineering at the University of Sydney, Australia. This research work was the basis of my Ph.D. thesis, *The Maximum Entropy Method and Its Application in Radio Astronomy*, completed in 1985.

As well as continuing my research in MEM after graduation, I taught a course of the same name at the Graduate School, Chinese Academy of Sciences, Beijing from 1987 to 1990. Delivering the course was the impetus for developing a structured approach to the understanding of MEM and writing hundreds of pages of lecture notes.

Between 1991 and 1994, I was a member of the Image Restoration Project at the Space Telescope Science Institute in the United States of America, where I applied MEM as well as other methods to restore images from the Hubble Space Telescope. The research results, including image restoration software, were presented in a number of published papers and internal technical reports.

It was these three activities: my Ph.D. thesis, the lecture notes, and the recently published papers and reports, that formed the basis for this monograph.

Before actually rolling up my sleeves to start writing the manuscript, I thought for a while about the areas to be covered, the style of writing, the prospective readers and so on. Based on my learning, teaching and researching experience, I decided to select spectral analysis, image restoration, some branches in mathematics, and statistical mechanics for the purposes of elaborating upon the principle and demonstrating applications of MEM. A step-by-step style of writing is used for enhanced readability, so that this monograph can be used by both beginners and more experienced researchers in MEM. Relevant discussions and comments are included after, rather than interleaved with, the presentation of the main results. I have tried my best to follow the above guidelines throughout the book.

This monograph consists of five chapters, two appendices, a list of references and an index. Chapter 1 is an overview of MEM. This was written especially with the needs of beginners in mind. Chapters 2, 3 are devoted to MEM and its applications, respectively, in spectral analysis and image restoration, including many algorithms in practical use. Chapter 4 is concerned, both experimentally and theoretically, with the analysis and comparison of three schools of thought on MEM in their basic ideas, properties and applications. Chapter 5 presents MEM and its applications to mathematics (including the solution of moment problems, integral equations and partial differential equations) and statistical mechanics. Appendices A, B provide the minimum relevant knowledge of cepstral analysis and image restoration, respectively, for understanding Chap. 3.

This monograph will be found to be very useful as a reference for researchers, as a textbook for graduate students in a one-semester course, or as lecture notes for MEM disciples in an intensive course of some forty hours.

Writing a monograph on MEM has been a somewhat challenging task for me. Much effort has been made to maximize the readability and to bring the contents up to date. My greatest satisfaction from publication of this monograph will come if it contributes to the further development of MEM theory and applications.

Acknowledgments. My first thanks are due to Professor Trevor Cole, Department of Electrical Engineering at the University of Sydney, Australia, for his supervision of the research on MEM in my graduate studies, and for his continuous encouragement.

I would like to acknowledge the assistance received from my colleagues at the following institutions during the preparation of the manuscript: the Space Telescope Science Institute, the United States of America; Dominion Radio Astrophysical Observatory, Canada; and York University, Canada. In particular, I would like to express my appreciation to Professor John Caldwell, Department of Physics and Astronomy at York University, for his generous support.

I am deeply grateful to Dr. Clifford Maldonado, for the time he took to read over the manuscript, and for the corrections and suggestions he made to improve the presentation of this monograph.

Finally, I would like to thank the Series Editor, Professor Thomas Huang, for his recommendation of the publication of this monograph, and the Managing Editor, Dr. Helmut Lotsch and his editorial staff at Springer-Verlag, for their kind assistance and friendly cooperation.

Toronto, Canada   *Nailong Wu*
February, 1997

Table of Contents

1. Introduction

This chapter is an introduction to the Maximum Entropy Method (MEM). Many aspects of MEM are covered. General ideas are presented with few technical details being concerned. Having gone through this introduction, the reader can selectively read in the following chapters what is considered to be of interest and applicable to their research areas. On the other hand, general ideas presented in this chapter can be well understood only after reading the relevant parts in the remaining chapters.

This chapter is organized as follows: In Sect. 1.1 a general statement of MEM is presented and the method is exemplified by the problem of power spectral analysis. This is followed in Sect. 1.2 by a brief introduction to the information-theoretic entropy which is used in MEM. Section 1.3 is devoted to the justification of the use of MEM. Finally, in Sect. 1.4 the three schools of thought on MEM research are investigated; the applications of MEM, including spectral analysis, image restoration, and mathematics and physics, to be covered in this book are outlined, and prospective development in MEM research is pointed out.

1.1 What is the Maximum Entropy Method

In mathematics, physics, engineering and other areas, problems are often solved on the basis of measured data, given conditions or assumptions. (They are generally called data in the following.) Usually, we are concerned with the three properties of solution: existence, uniqueness and stability. A problem for which at least one of the three above requirements is not met is called an *ill-posed* (*ill-conditioned*) *problem*. This kind of problem is caused by incomplete and/or noisy data. (Here the term "noise" is used in a broad sense. Any discrepancy between the measured and "true" data is called noise.)

Many methods, one of which is the *Maximum Entropy Method*, have been invented in order to deal with such ill-posed problems. MEM states that *of all the feasible (possible) solutions, we should choose the one that has the maximum entropy.* The statement of MEM is so simple. However, several points must be clarified.

1. Among the three requirements to the solution, stability is usually put aside for the moment. This property would be investigated, if one wants to do so, after determining the MEM solution.

2. If the solution does not exist, it makes no sense to say "choose the solution that has the maximum entropy." In this case the way out is usually to relax the constraints imposed by the data so that the solution exists. This relaxation often leads to multiple or even infinitely many solutions. This is good news because we can then choose the solution according to the criterion "maximum entropy".

3. "Maximum" here is in the global, but not local sense. This guarantees the uniqueness of the maximum entropy solution.

MEM described above may be exemplified by the following one-dimensional (1-D) power spectral analysis (Fig. 1.1), where three cases are illustrated.

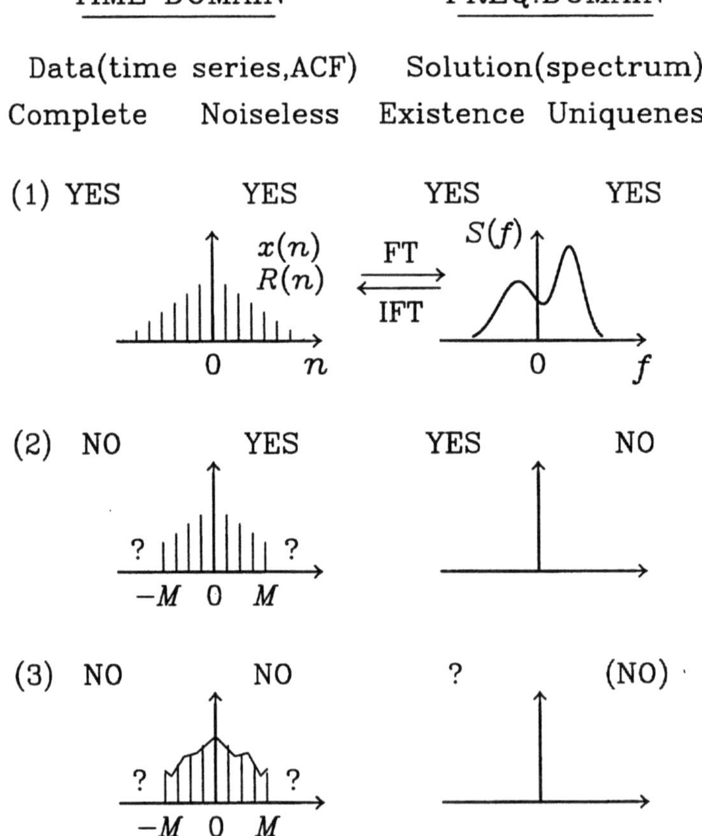

Fig. 1.1. Ill-posed problems in 1-D spectral analysis

(1) The data in the time domain, a time series or its autocorrelation function (ACF), are complete and noiseless (exact). In this case the solution (spectrum) in the frequency domain exists and is unique. The data and solution are related by the Fourier transform (FT). (IFT stands for the inverse FT.)

(2) The data are incomplete, but noiseless. The data are only partially given on the interval $[-M, M]$. In this case the solution (spectrum) exists, but is not unique. Different data extensions outside the interval $[-M, M]$ will lead to different solutions. Generally speaking, infinitely many solutions exist.

(3) The data are both incomplete and noisy (inexact). The partially given data on $[-M, M]$ are contaminated by noise. In this case the solution may not exist. For instance, when the noise is sufficiently strong, the solution formally calculated by FT does not meet the essential requirement of nonnegativity (i.e., the spectrum must not be negative) for any extension of the ACF. Then, the uniqueness is, of course, out of the question. The situation here is no better than in Case (2), even if the noise is so weak that positive definite solutions exist. To sum up, in this case the existence of solution is in doubt and the uniqueness requirement is definitely not met.

The problem in Case (1) is well-posed; therefore, it is not the subject of MEM research. In Cases (2, 3) the problems are ill-posed; hence MEM is of great use.

Let us consider Case (2) first, in which the solution exists, but is not unique. As mentioned before, different data extensions will lead to different solutions. According to the criterion of maximum entropy, a straightforward way of choosing the unique solution is to calculate the entropy for every data extension or solution, compare all the entropy values, and then single out the data extension or solution having the maximum entropy. This is a primitive version of MEM in spectral analysis. However, this simple way of finding the MEM solution cannot be put into use in practice because the number of calculations would be infinite. Thus, it is needed to properly formulate MEM in mathematics. In short, the solution is determined by maximizing the entropy subject to the data constraints. It should be pointed out that the data extension with the maximum entropy in the time domain does not correspond to the solution with the maximum entropy in the frequency domain. This results in the two schools of thought on the Maximum Entropy Method, namely, MEM1 and MEM2. In this respect, the details will be discussed in Sect. 1.4.

Let us now turn our attention to Case (3), in which the solution may not exist, and the requirement of uniqueness cannot be met. Our strategy in this case was mentioned before. Specifically, in maximizing the entropy, it is required not to strictly meet the data constraint at each datum point, but to fit data within a certain error limit on the whole interval $[-M, M]$. In this way the existence of solution is guaranteed.

As stated above, in MEM the solution having the maximum entropy is to be chosen of all the feasible ones. Then, one may raise questions: What is the definition of entropy? Why should we choose the solution according to the criterion of maximum entropy? They will be answered in the following two sections.

1.2 Definition of Entropy

Entropy is an old concept. Its first introduction into thermodynamics, by Clausius, dates back to the middle of the nineteenth century. Later Boltzmann derived the entropy expression in statistical mechanics, i.e., the well-known Boltzmann relation: $S = k \ln W$. The entropy in thermodynamics or statistical mechanics is a measure of the degree of disorder existing in a thermodynamical system. (See any standard textbook of thermodynamics and statistical mechanics, e.g., [1.1].)

Entropy is also a modern concept. In establishing information theory in 1948, Shannon found a unique quantity H that measured the uncertainty of an information source. Because of the similarity of the mathematical form and physical significance, the quantity H is also named *entropy*, as is the quantity S in thermodynamics and statistical mechanics. However, the entropy H is sometimes explicitly called "information-theoretic entropy" or "Shannon's entropy" for clarity. It is the information-theoretic entropy that is in use in MEM. Therefore, a more detailed discussion about the entropy H is necessary. (See any standard textbook of information theory, e.g., [1.2].)

A discrete information source may be represented by

$$
X : \begin{pmatrix} x_1 & x_2 & \cdots & x_n \\ p_1 & p_2 & \cdots & p_n \end{pmatrix},
$$

where p_i is the prior probability that the random variable X assumes the value (symbol) x_i, $i = 1, \ldots, n$; $P(X = x_i \cap X = x_j) = 0$, $i \neq j$, that is, $X = x_i$ and $X = x_j$, $i \neq j$, are mutually exclusive events; $\sum_{i=1}^{n} p_i = 1$. Our partial knowledge about the information source or the uncertainty of the information source is represented by the prior probability distribution $(p_1, \ldots, p_n)$. If we require that the measure of the uncertainty H satisfy the following three conditions, then H can be determined to within a constant factor. The three conditions are:

(1) $H(p_1, \ldots, p_n)$ is a continuous function of p_i, $i = 1, \ldots, n$.

(2) If all p_i are equal, the quantity $H(1/n, \ldots, 1/n)$ is a monotonically increasing function of n.

(3) The *composition law*. If the values are arbitrarily divided into m groups $(x_{11}, \ldots, x_{1k_1})$, $(x_{21}, \ldots, x_{2k_2})$, $\ldots$, $(x_{m1}, \ldots, x_{mk_m})$, the corresponding probabilities are $w_1 = p_{11} + \cdots + p_{1k_1}$, $\ldots$, $w_m = p_{m1} + \cdots + p_{mk_m}$, then we must have

$$
\begin{aligned}
H(p_1,\ldots,p_n) \;=\; & H(w_1,\ldots,w_m) \\
& + w_1 H(p_{11}|w_1,\ldots,p_{1k_1}|w_1) \\
& + \cdots \\
& + w_m H(p_{m1}|w_m,\ldots,p_{mk_m}|w_m) \;,
\end{aligned}
$$

where the vertical bar denotes conditional probability. The composition law is also called *consistency* or *additivity*, which means that the uncertainty measure H must be independent of the form of the probabilistic test, or of the grouping of the values.

The measure H is of the unique form:

$$
H(p_1,\ldots,p_n) = -k \sum_{i=1}^{n} p_i \log p_i \;, \tag{1.1}
$$

where the constant k depends on the unit to be used, and is usually set to unity. Different logarithmic bases result in different entropy units. The unit is of no significance in maximizing entropy. For the sake of the convenience of mathematical operation, the logarithmic base is always assumed to be e (the base of natural logarithm) in MEM. Therefore, unless otherwise specified, we have $\log \equiv \log_e \equiv \ln$.

The quantity H in (1.1) is called the (information-theoretic) *entropy*, which measures the uncertainty of the information source. Another interpretation of H is that the uncertainty is eliminated after receiving a message (a stream of values) from the information source; entropy H is equal, on average, to the uncertainty eliminated, namely the information obtained, by receiving one value.

We have discussed the case of discrete information source. For a continuous information source, the distribution of the random variable X of the continuous type is expressed in terms of probability density function (p.d.f.), which is assumed to be a continuous one for simplicity. If we take the continuous case as the limit form of the discrete case, i.e., let $p_i = p(x_i)\delta x_i$, $n \to \infty$, $\delta x_i \to 0$, $\sum \to \int$ in (1.1), then the integral will diverge. A common way to cope with this situation is to write out the following entropy expression similar to (1.1):

$$
H(p(x)) = - \int_{-\infty}^{+\infty} p(x) \log p(x) \mathrm{d}x \;. \tag{1.2}
$$

However, it should be borne in mind that H here does not represent the absolute value of the uncertainty of the information source or the information obtained. It can only be used to calculate the change (increase or decrease) of the uncertainty or information. Another difficulty encountered here is that unlike H in (1.1) in the discrete case, H in (1.2) in the continuous case is not invariant in a transform of the coordinate system. That is to say, $H(p(x))$ may change its value in different coordinate systems. In order to overcome this difficulty, we introduce a *measure* $m(x)$ in the entropy expression. Then, (1.2) becomes

$$H(p(x)) = -\int_{-\infty}^{+\infty} p(x) \log[p(x)/m(x)]\mathrm{d}x \ . \tag{1.3}$$

(Here $m(x)$ may or may not be normalized.) As a consequence, if $p(x)$ changes, $m(x)$ also changes in the same manner so that the value of H remains unchanged. In MEM $m(x)$ is the so-called *prior* (*estimate*) of the solution or *model*. Equation (1.1) can also be reformed to

$$H(p_1,\ldots,p_n) = -\sum_{i=1}^{n} p_i \log(p_i/m_i) \ . \tag{1.4}$$

The invariance of H in (1.3) is proved as follows:

Suppose $X = f(Y)$ is a one-to-one transform between X and Y. The p.d.f. of Y is $q(y) = p(x)|f_y'|$, the measure becomes $n(y) = m(x)|f_y'|$. Thus we have

$$
\begin{aligned}
H(p(x)) &= -\int_{-\infty}^{+\infty} q(y) \log\{[p(x)|f_y'|] \,/\, [m(x)|f_y'|]\}\mathrm{d}y \\
&= -\int_{-\infty}^{+\infty} q(y) \log[q(y)/n(y)]\mathrm{d}y \\
&= H(q(y)) \ .
\end{aligned}
$$

A similar proof can be carried out in the case of a random vector. The transform is then $\boldsymbol{X} = \boldsymbol{f}(\boldsymbol{Y})$. To show $H(p(\boldsymbol{x})) = H(q(\boldsymbol{y}))$, we have only to replace $|f_y'|$ by the modulus of the Jacobian of the transform, $|J|$, in the above.

The entropy of a random variable X is said to be the probabilistic entropy. In MEM, X may represent a random field; p_i may represent frequency, proportion, or so-called degree of belief. In formulating MEM, a variety of derived entropy expressions may be used instead of the original definitions (1.1–1.4). We will see them in due course.

1.3 Rationale of the Maximum Entropy Method

Why is MEM used? In other words, why do we choose the solution that has the maximum entropy among the feasible ones? As pointed out in Sect. 1.1, various methods exist to cope with ill-posed problems. What are the merits of taking "maximum entropy" as the criterion for choosing the solution? There are answers of various kinds to these questions. They may be summarized in the following five points:

1. Principle of the Increase of Entropy. The *Principle of the increase of entropy* in thermodynamics states that the (thermodynamic) entropy in any isolated system never decreases, i.e., spontaneously tends towards the maximum value possible. Noting that the thermodynamic entropy is proportional to the information-theoretic entropy (this relationship is to be established in Chap. 5), we can easily see that the choice of solution in MEM is "natural". On the other hand, if we know beforehand that the entropy is decreasing in the evolution of a particular system (definitely not isolated), we should, of course, choose the solution having the minimum entropy.

2. First Principle of Data Reduction. This principle states that when we solve a problem with incomplete data, the solution shall incorporate and be consistent with all relevant data and be maximally noncommittal with regard to unavailable data. The solution of a problem may be viewed as a procedure of extracting information from data. The information comes from the two sources: the measured data, and the inevitable assumption about the unavailable ones because of the incompleteness of the data. Making the assumption means artificially "adding" information, which may be true or false. Maximum entropy means that the total information obtained is minimum. That is to say, the "added" information is minimum. Therefore, the MEM solution is having "least assumption" and "maximally noncommittal".

3. Principle of Maximum Multiplicity. In statistical mechanics, image processing and other areas, the entropy H, multiplicity W of a state, and the number of particles or pixels N are related by

$$H \sim N^{-1} \log W \; ,$$

($a \sim b$ means $a/b \to 1$ as $N \to \infty$). The *multiplicity* of a state is defined as the number of possible microscopic states in the (macroscopic) state, or the number of possible ways in which the system finally reaches that (macroscopic) state in evolution. The greater the multiplicity of a state is, the greater is the possibility that the system is found to be finally in that state. Obviously, the most probable state has the maximum multiplicity. This is the *principle of maximum multiplicity*. From the expression above, we have

$$W \sim \mathrm{e}^{NH} \; .$$

It is evident that the maximum H corresponds to the maximum W. Therefore, MEM is a consequence of the principle of maximum multiplicity. If N is very large, a small deviation of H from H_{max} will lead to a big drop in W. So we can conclude that absolute majority of the probable states will concentrate in the vicinity of the maximum entropy state. This is the *principle of entropy concentration*. On the basis of this principle, we may say that the prediction using MEM is quite accurate.

4. Consistency. Consistency requires that the final solution should depend only on the total information (data) available. In other words, the final solution should be independent of the steps of determining the solution. Let us take a simplest example. Suppose the given data are divided into two groups, D_1 and D_2. Then we have the three ways of determining the final solution.

(1) Use D_1 to determine the intermediate solution S_1, then modify S_1 by D_2 to obtain the final solution $S_{1,2}$. (2) Use D_2 to determine S_2, then modify S_2 by D_1 to obtain $S_{2,1}$. (3) Use D_1 and D_2 at once to obtain the final solution S_{1+2}. Consistency requires $S_{1,2} \equiv S_{2,1} \equiv S_{1+2}$. It should be noticed that the data are divided in an arbitrary way. The number of data groups is also arbitrary in more complex cases.

One of the strongest arguments in favor of MEM is that the MEM solution is the only one that meets the requirement of consistency. This valuable property of MEM stems from the composition law imposed in deriving the entropy function (1.1). In Sect. 4.3.3 we will show through specific examples that both MEM1 and MEM2 solutions indeed meet the requirement of consistency.

At this point one may raise a question: Why should we impose the requirement of consistency on the solution? The reason is the same as the one for imposing the composition law on the entropy function. Now that this requirement or condition is by its nature an axiom, nothing serious would happen to you if you dare to ignore it. This is what the school of so-called Generalized MEM has done. But you might be expelled from the MEM community by the fundamentalists.

5. The Proof of the Pudding is in the Eating. The study of the properties of solution and applications in practice have shown that MEM is indeed a good method to deal with ill-posed problems. Specifically, in spectral analysis and image restoration, the most remarkable merits of MEM are the enhancement of resolution and the suppression of noise; both can be achieved at the same time. In this regard many examples will be given in the following three chapters. In statistical mechanics, we can use the principle of maximum entropy (synonym of MEM) instead of assumptions such as ergodicity, equiprobabilities to infer the probability distribution of a system, and then calculate all the thermodynamical quantities. This method is applicable to both classical and quantum, and to both equilibrium and nonequilibrium statistical mechanics. And yet one may well be faced with difficulties in practice due to the complexity of the problem. In mathematics MEM is employed as a new method to solve some old problems, and is becoming a new tool for tackling some thorny problems. In one word, this MEM pudding is greatly appreciated for it is not only colorful and attractive in theory, but also, more importantly, rich in nutrition and appealing to many researchers' taste in practice.

We conclude this section by solving a problem of probability distribution which seems to be very simple. Suppose that a random variable X may assume any one of the values $x_1, \ldots, x_n$. We do not know anything about X otherwise. Our task is to determine the probability distribution of X.

In the orthodox probability theory, the solution is $p_i = 1/n$, $i = 1, \ldots, n$. The *Laplacian principle of indifference* states that two events are to be assigned equal probabilities unless there is a reason to think otherwise. Consequently, all the p_i are equal; assigning the value $1/n$ to p_i is just to meet the normalization condition. It is unfair to say that this way of determining the probability distribution is completely unreasonable. But we feel, after thinking for a while, that this way of solving the problem is a bit passive and the statement of the Laplacian principle is somewhat pessimistic.

MEM can be used to solve the above problem, which may be formulated as follows:

Determine the MEM solution $p_1, \ldots, p_n$ by

$$\text{maximizing } H(p_1, \ldots, p_n) = -\sum_{i=1}^{n} p_i \log p_i \ ,$$

subject to $\sum_{i=1}^{n} p_i = 1$ (normalization condition).

It is easy to find that the MEM solution is also $p_i = 1/n$, $i = 1, \ldots, n$, using the Lagrange multiplier method. So we see that the Laplacian principle of indifference is "derived" by MEM. By determining the solution through maximizing entropy, we show activeness and optimism instead of passiveness and pessimism. But one may argue that this is no more than a matter of wording: a pessimist says that a bottle is *half empty* while an optimist says that the bottle is *half full* – they refer to the same thing in fact!

1.4 Present and Future Research

It is forty years already since the introduction of MEM into statistical mechanics by *Jaynes* [1.3] in 1957. During these years MEM developed rapidly. The introduction of MEM into spectral analysis by *Burg* [1.4] in 1967, and into image restoration by *Frieden* [1.5] in 1972, is considered to be two milestones. Today papers on MEM frequently appear in many reputable international publications. Every year in summer, researchers from all over the world gather together, exchanging new information on MEM research at an MaxEnt meeting. Many researchers' unremitting efforts have helped make MEM become a method of firm foundation and wide applications. It is not an easy job to describe the present state of MEM research. The prospect of its future is even more difficult.

In the following we present the two approaches of describing the current situation of the MEM research: investigating the three schools of thought on MEM first, and then showing its main applications [1.6].

I. Three Schools of Thought on MEM

The three schools are divided on the basis of entropy definition and expression. They are MEM1, MEM2 and the Generalized MEM (GMEM).

1. MEM1. In spectral analysis of a time series, the entropy is defined in the time domain:

$$H1 \stackrel{\triangle}{=} - \int p(\boldsymbol{x}) \log p(\boldsymbol{x}) \mathrm{d}\boldsymbol{x} \;, \tag{1.5}$$

where $p(\boldsymbol{x})$ denotes the joint p.d.f. of the time series $\boldsymbol{x}$, $\int$ denotes an infinitely dimensional integral, and the integral limits are $-\infty$ and $+\infty$. (A more sophisticated definition will be given in Chap. 2.) The derived expression of $H1$ in the frequency domain is

$$H1 = \int \log S(f) \mathrm{d}f \;, \tag{1.6}$$

where $S(f)$ denotes the power spectrum density of the time series, f denotes frequency, and the integral interval is the whole frequency band. Equation (1.6) is the expression used in mathematical operations in MEM1.

The problem to be solved in MEM1 is constrained maximization: The quantity to be maximized is the entropy $H1$, and the constraints come from the given partial time series or its ACF. Usually, this problem is solved using the Lagrange multiplier method. Only in the case of 1-D and noiseless data, does the explicit solution exist. In Chap. 2 the derivation of the entropy expression and the procedure of finding a solution will be elaborated upon. Here we only present the formulae concerning the explicit solution.

The MEM1 spectrum

$$S_1(f) = \frac{p_M}{\left| \sum_{k=0}^{M} a_k \exp(-2\pi \mathrm{i} f k) \right|^2} \;,$$

where $\mathrm{i} = \sqrt{-1}$, p_M is the power of prediction error, and a_k are the coefficients of the prediction-error filter with $a_0 = 1$. In the case where the partial ACF, $R(k)$, $|k| \le M$, is given, p_M and a_k are the solution of the following normal equations:

$$\sum_{k=0}^{M} a_k R(n-k) = p_M \delta_n \;, \quad (a_0 = 1) \;, \; n = 0, \ldots, M \;.$$

When a segment of the time series, $x(k), 1 \le k \le N$, is given, p_M and a_k can be estimated directly from $x(k)$. The order of the filter, M, must be less than N.

MEM1 is mainly used in areas related to 1-D spectral analysis such as the processing of seismic, radar and speech signals.

2. MEM2. Although basically MEM2 and GMEM are not used for 1-D spectral analysis, the technical terms and symbols in MEM1 are also used in the following for uniformity. Here entropy is defined directly in the frequency domain:

$$H2 \stackrel{\triangle}{=} - \int S(f) \log S(f) \mathrm{d}f \;. \tag{1.7}$$

A more sophisticated definition contains a measure $m(f)$ as shown in (1.3). The only difference between the formulations of MEM1 and MEM2 is in the entropy expression. In general, no explicit solution exists. Even in the case of 1-D and noiseless data, very restrictive conditions, i.e., real, causal and minimum-phase, must be imposed on the time series in order to find an explicit solution. This problem will be discussed in considerable detail in Chap. 3. Here we only present the form of the solution.

The MEM2 spectrum

$$S_2(f) = \exp\left[\sum_{l=-M}^{M} \lambda_l^* \exp(-2\pi i f l) \right] ,$$

where λ_l^* denote the complex conjugate of the Lagrange multipliers. λ_l^* are the complex cepstrum of the ACF sequence $R(k)$:

$$\lambda_l^* = \mathrm{IFT}[\log \mathrm{FT}[R(k)]] ,$$

and are also (double) the real cepstrum of the time series $x(n)$:

$$\lambda_l^* = 2\,\mathrm{IFT}[\log |\mathrm{FT}[x(n)]|] .$$

MEM2 is used mainly for image restoration (equivalent to two-dimensional (2-D) spectral analysis) such as astronomical image restoration, medical imaging, crystallography, and photo processing. The Principle of Maximum Entropy in statistical mechanics also falls into the category of MEM2 (Chap. 5).

3. GMEM. "Entropy" is also defined directly in the frequency domain:

$$H_G \triangleq \int E(S)\mathrm{d}f . \tag{1.8}$$

Here the function $E(S)$ satisfies the following conditions: $E' \to \infty$ as $S \to +0$ (to ensure the positivity of the solution); $E'' < 0$ (to ensure the uniqueness); and $E''' > 0$ (so that the solution has sharp peaks and flat background). For example, a family of nonlinear functions $E(S)$ for which

$$E''(S) \propto -S^{-n}, \qquad n \geq 1$$

can be used in the "entropy" functions. The generalized "entropy" functions of this kind include $H1$ ($n = 2$) and $H2$ ($n = 1$). The basic ideas underlining GMEM and various forms of H_G will be presented in detail in Sect. 4.1.

Basically, GMEM is used only in imaging for aperture synthesis telescopes of radio astronomy. It is not so widely used as MEM1 and MEM2.

In MEM research, MEM1 and MEM2 are dominating while GMEM is not generally accepted. In Sect. 4.8 we will show the main points in the argument between the three schools and explain the current situation.

II. Applications of MEM

We briefly describe the applications and point out possible development of MEM in the three areas, namely, spectral analysis, image restoration, and physics and mathematics. They constitute the subjects of Chaps. 2, 3, 5, respectively.

1. Spectral Analysis: MEM1. Spectral analysis mostly uses MEM1. 1-D MEM1 is mature in theory and application. Computer algorithms and programs can be found in books and journals. Gratis and commercial Software is available. Some fast algorithms can be implemented by hardware and used for real time processing. Although quite a number of researchers have been working on the theory and application of MEM1 in 2-D spectral analysis, results are not widely used as in the 1-D case due to the complexity. From recently published papers, it seems that the research work is concentrated mainly on fast algorithms in 1-D MEM1 and the problem of order selection when the prediction-error filter coefficients are estimated directly from the time series.

2. Image Restoration: MEM2. In many cases image restoration is equivalent to 2-D spectral analysis. Image restoration mostly uses MEM2. MEM2 is not so mature as MEM1 in theory. No generally applicable explicit solution has been found even in the simple case in which 1-D noiseless data are given. Iterative algorithms must be used in practice without exception. The properties of solution are investigated mainly by means of inspecting the solution structure and numerical calculation. Nevertheless, MEM2 has won popularity in areas related to image restoration.

As mentioned before, its most remarkable merits are the enhancement of resolution and the suppression of noise; both can be achieved at the same time. And yet we should not be reticent about its drawbacks. Because of the high degree of nonlinearity of the method, it is hard to quantitatively analyze the MEM2 results, the mathematical operations are complicated and the computation is time consuming. The last problem was very serious especially in (2-D) image restoration. However, because of the advent of new efficient algorithms and the ever increasing power of computers, this problem now is not as serious as it used to be.

In recent years MEM2 has been put in the framework of the Bayesian method, and the performance of the solution is greatly improved.

3. Physics and Mathematics: MEM2. Statistical mechanics is the birthplace of maximum entropy. In physics it is a branch in which MEM research is the most active. Jaynes' Principle of Maximum Entropy has provided a new means by which old results can be obtained more easily. Some researchers also claimed that new results had been derived which could not be obtained using the conventional statistical mechanics method.

MEM intrinsically has superiority in dealing with irreversible processes and nonequilibrium states because statistical mechanics is viewed as statistical inference based on incomplete knowledge instead of physical theory and hence various physical assumptions are irrelevant. Schemes of various kinds have been worked out, but much more effort is needed before significant advances can be made. Other physics' branches in which the principle of maximum entropy has recently found its applications are quantum mechanics, condensed matter physics, etc.

In mathematics MEM is used to solve moment problems, partial differential equations and integral equations, which are quite common in mathematics and physics.

From the description above, we can see not only the current situation of MEM research, but also its trend of development. Topics of interest may be summarized as follows:

1. Order selection, fast algorithms and hardware implementation in MEM1. In the last respect, neural network is most promising.

2. Relaxation of the conditions for the explicit solution in 1-D MEM2 so that it can be used in the general case. Estimation of the Lagrange multipliers directly from the time series, and a related problem of filter order selection.

3. The integration of MEM with the Bayesian method.

4. Image restoration by MEM2 with variable resolution.

5. Nonstationary processes in stochastic signal processing. The definition of entropy. The applicability of MEM.

6. A vast variety of MEM applications in areas such as radar, astronomy, physics and economy.

2. Maximum Entropy Method MEM1 and Its Application in Spectral Analysis

This chapter is devoted to MEM1, namely, MEM with the entropy expression

$$H1 = \int_{-1/2}^{1/2} \left(\log \frac{S(f)}{m(f)} - \frac{S(f)}{m(f)} + 1 \right) df \, .$$

In most cases $m(f) = 1$ so that the above expression degenerates to

$$H1 = \int_{-1/2}^{1/2} \log S(f) df \, .$$

Here $S(f)$ denotes power spectrum density, or spectrum for short.

This chapter is organized as follows: In Sect. 2.1 the entropy expression $H1$ in the frequency domain is derived from its original definition in the time domain by two approaches. Then, in Sect. 2.2 the problem of spectral analysis by MEM is formulated and the solution is worked out through a lengthy mathematical derivation. In Sect. 2.3 six equivalents to MEM1 and a signal model are described in considerable detail. Algorithms and numerical examples of MEM spectral analysis are presented in Sects. 2.4, 2.5, respectively, for the cases of given ACF and time series as the data. Section 2.6 concludes this chapter by dealing with the problem of filter order selection, an important issue in MEM spectral analysis.

Understanding the mathematical manipulation in this chapter is essential and important for readers who are concerned with the MEM theory. On the other hand, readers who are interested mainly in the MEM application should concentrate on algorithms.

2.1 Definition and Expressions of Entropy $H1$

Our aim in this section is to derive the entropy expression in the frequency domain:

$$H1 = \int \log S(f) df \, . \tag{2.1.1}$$

Two approaches are presented. The first one uses the time series which is explicitly assumed to be a Gaussian (normal) stochastic process. The second one uses the ACF (autocorrelation function), and no explicit assumption of the Gaussian process is made. Finally, we will point out the relationship between the two approaches.

2.1.1 Approach 1

The derivation procedure consists of two steps [2.1].

Step 1. Entropy definition and expression in the time domain

Given a time series $x = \{\ldots, x_{-1}, x_0, x_1, \ldots\}^T$, where T denotes transposition and x is a column vector. Each x_k is a real or complex random variable and x is a stochastic (random) process (Fig. 2.1). We are concerned only with (weakly) stationary processes, i.e., the first and second moments of x_k are independent of the index k.

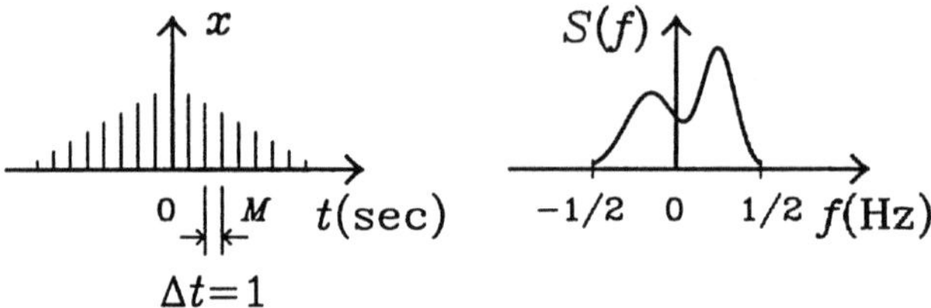

Fig. 2.1. A band-limited time series and its spectrum

Actually, x can be viewed as a sampled signal. The sampling interval is assumed to be $\Delta t = 1$ second for simplicity, and then the bandwidth of the signal is $[-1/2, 1/2]$ Hz (Shannon's sampling theorem).

The entropy of this stochastic process is of a form similar to (1.2) with the joint probability density function (p.d.f.) $p(x)$ replacing $p(x)$. In order to avoid difficulty of divergence of an infinitely dimensional integral, we first consider a finite segment $x_{M+1} = \{x_0, x_1, \ldots, x_M\}^T$ contained in the time series of infinite length. Thus we have

$$H1 \overset{\triangle}{=} - \int_{-\infty}^{+\infty} p(x) \log p(x) \mathrm{d}x , \qquad (2.1.2)$$

where $\int$ is a simplified notation of a multidimensional integral.

Equation (2.1.2) is the definition of the entropy $H1$ and an expression of $H1$ in the time domain. It is difficult to use this expression in maximization. Therefore, we must derive an expression which is easy to use in practice.

Suppose that the process under consideration is Gaussian and its joint p.d.f. is (omitting the subscript $M + 1$)

$$p(x) = [(2\pi)^{M+1} \det(C_x)]^{-1/2} \exp\left[-\frac{1}{2}(x - \mu_x)^H C_x^{-1}(x - \mu_x)\right] , \quad (2.1.3)$$

where C_x is the autovariance matrix of x_{M+1}, whose elements are

$$c_{ij} = c_{(i-j)} = \mathrm{E}[(x_i - \mu_x)(x_j^* - \mu_x^*)] \ .$$

Here E denotes expectation and $*$ complex conjugation. Since we consider only stationary processes, the expectation μ_x and variance σ_x^2 of x_i are independent of the index i. The components of $\boldsymbol{\mu}_x$ are all equal to μ_x. The superscript H denotes the Hermitian operation. Performing a Hermitian operation on a matrix means taking its transposition and then the complex conjugation.

Substituting (2.1.3) into (2.1.2), we have

$$\begin{aligned}
H1 &= -\int p(\boldsymbol{x}) \log \left\{ \left[(2\pi)^{M+1} \det(C_x) \right]^{-1/2} \right. \\
&\qquad \left. \times \exp\left[-\frac{1}{2}(\boldsymbol{x} - \boldsymbol{\mu}_x)^{\mathrm{H}} C_x^{-1}(\boldsymbol{x} - \boldsymbol{\mu}_x) \right] \right\} \mathrm{d}\boldsymbol{x} \\
&= \frac{1}{2}\log[\det(C_x)] + \frac{1}{2}(M+1)\log(2\pi) \\
&\quad + \frac{1}{2}\int p(\boldsymbol{x})(\boldsymbol{x} - \boldsymbol{\mu}_x)^{\mathrm{H}} C_x^{-1}(\boldsymbol{x} - \boldsymbol{\mu}_x)\mathrm{d}\boldsymbol{x} \ .
\end{aligned}$$

The calculation of the last term needs coordinate transformation. Because C_x is a positive definite matrix, a linear transformation of full rank must exist (its specific expression is of no concern for our purpose) by which C_x is diagonalized and consequently the quadratic form $(\boldsymbol{x} - \boldsymbol{\mu}_x)^{\mathrm{H}} C_x^{-1}(\boldsymbol{x} - \boldsymbol{\mu}_x)$ becomes the sum of squares

$$\boldsymbol{y}_{M+1}^{\mathrm{H}} \Lambda \boldsymbol{y}_{M+1} = \sum_{k=0}^{M} \frac{1}{\sigma_k^2}|y_k|^2, \quad (\Lambda \text{ is a diagonal matrix}) \ ,$$

and the modulus of the determinant of the transformation matrix is equal to one. That is to say, the p.d.f. of the $(M+1)$-dimensional Gaussian distribution is transformed to the product of the p.d.f's of $(M+1)$ independent and identical Gaussian distributions with a zero mean and variance σ_k^2 by the translation and rotation of the coordinate system. As a result, the joint p.d.f. of $\boldsymbol{y}_{M+1}$ is

$$q(\boldsymbol{y}) = \prod_{k=0}^{M} q_k(y_k) \ .$$

It follows that the last term in the above expression of $H1$ is simplified to

$$\begin{aligned}
\frac{1}{2}\int q(\boldsymbol{y}) &\sum_{k=0}^{M} \frac{1}{\sigma_k^2}|y_k|^2 \mathrm{d}\boldsymbol{y} \\
&= \frac{1}{2}\sum_{k=0}^{M} \frac{1}{\sigma_k^2}\int q_k(y_k)|y_k|^2 \mathrm{d}y_k \prod_{\substack{l=0 \\ l \neq k}}^{M} \int q_l(y_l)\mathrm{d}y_l
\end{aligned}$$

$$= \frac{1}{2} \sum_{k=0}^{M} \frac{1}{\sigma_k^2} \cdot \sigma_k^2 \cdot 1^M$$

$$= \frac{1}{2}(M+1) \,.$$

Therefore, we have at last

$$H1 = \frac{1}{2} \log[\det(C_x)] + \frac{1}{2}(M+1) \log(2\pi e) \,. \tag{2.1.4}$$

With no loss of generality, we assume further that the time series x_{M+1} has a zero mean, $\mu_x = 0$. Then, its autovariance matrix C_x becomes ACF matrix R_x. Hence, the entropy of a zero-mean Gaussian process of finite length is

$$H1 = \frac{1}{2} \log[\det(R_x)] + \frac{1}{2}(M+1) \log(2\pi e) \,. \tag{2.1.5}$$

The last term may be discarded for a fixed M and then we have

$$H1 = \frac{1}{2} \log[\det(R_x)] \,. \tag{2.1.6}$$

Now we revert to the case of the time series x of infinite length. If we simply let $M \to \infty$, then the first term on the right-hand side of (2.1.5) will diverge and the second will also approach infinity. To resolve this problem, we use the *entropy rate* h (entropy per sampling point) instead of the entropy $H1$ in (2.1.5):

$$
\begin{aligned}
h \ &\overset{\triangle}{=} \ \lim_{M \to \infty} \frac{H1}{M+1} \\
&= \ \lim_{M \to \infty} \frac{1}{2} \log[\det(R_x)]^{1/(M+1)} + \frac{1}{2} \log(2\pi e) \,.
\end{aligned}
\tag{2.1.7}
$$

Noting that the last term is a constant, we may write

$$h = \lim_{M \to \infty} \frac{1}{2} \log[\det(R_x)]^{1/(M+1)} \,. \tag{2.1.8}$$

We will see that h in the above indeed converges.

Equation (2.1.6) or (2.1.8) is another expression of the entropy (rate) $H1$ (h) in the time domain. They are still difficult to use in maximization because of the determinant of the ACF matrix (and the limit operation) included. In the following we continue the derivation so that an entropy expression like (2.1.1) is finally obtained in the frequency domain.

Step 2. Entropy expression in the frequency domain

Because the properties of the ACF matrix R_x will play an important role in our derivation, we first briefly introduce some of those which are relevant. The subscript x is omitted since there is only one time series involved.

The expanded form of R is

$$R = \begin{bmatrix} R(0) & R(-1) & \cdots & R(-M) \\ R(1) & R(0) & \cdots & R(1-M) \\ \cdots & \cdots & \cdots & \cdots \\ \cdots & \cdots & \cdots & \cdots \\ R(M-1) & R(M-2) & \cdots & R(-1) \\ R(M) & R(M-1) & \cdots & R(0) \end{bmatrix} . \tag{2.1.9}$$

This is a Toeplitz (equidiagonal) and Hermitian ($R^{\mathrm{H}} = R$) matrix, for its elements satisfy the relationships

$$r_{ij} = R(i-j) = R^*(j-i) = r_{ji}^* , \quad R(i-j) = \mathrm{E}[x_i x_j^*] .$$

Property 1. The eigenvalues of R, λ_m, are all nonnegative and real, and

$$\det(R) = \prod_{m=0}^{M} \lambda_m \geq 0 . \tag{2.1.10}$$

That is, R is a nonnegative definite matrix.

Property 2.

$$\lim_{m \to \infty} [\det(R)]^{1/(M+1)} = 2B \exp\left[\frac{1}{2B} \int_{-B}^{B} \log S(f)\mathrm{d}f \right] , \tag{2.1.11}$$

where B is the bandwidth of the signal, i.e., $-B \leq f \leq B$. In our case, $B = 1/2$.

Proof. Szegö's theorem states that if $g(\cdot)$ is a continuous function, then

$$\lim_{M \to \infty} \frac{g(\lambda_0) + g(\lambda_1) + \cdots + g(\lambda_M)}{M+1} = \frac{1}{2B} \int_{-B}^{B} g(\, 2BS(f)\,)\mathrm{d}f .$$

Assuming $g(\cdot)$ is a logarithmic function, we have

$$\lim_{M \to \infty} \frac{\log \lambda_0 + \log \lambda_1 + \cdots + \log \lambda_M}{M+1}$$
$$= \lim_{M \to \infty} \log[(\lambda_0 \lambda_1 \cdots \lambda_M)^{1/(M+1)}]$$
$$= \frac{1}{2B} \int_{-B}^{B} \log[2BS(f)]\mathrm{d}f .$$

Utilizing (2.1.10) and the above expression, we get

$$\lim_{M \to \infty} \log[\det(R)]^{1/(M+1)} = \frac{1}{2B} \int_{-B}^{B} \log[2BS(f)]\mathrm{d}f . \tag{2.1.12}$$

Changing the order of the limit and logarithm operations on the left-hand side and then taking the exponentiations of the two sides, we can prove (2.1.11). But for our purpose, (2.1.12) will suffice.   Q.E.D.

Now we stop investigating the properties of R and revert to the entropy rate h in (2.1.8). From (2.1.12) with $B = 1/2$, we have

$$
\begin{aligned}
h &= \lim_{M \to \infty} \frac{1}{2} \log[\det(R)]^{1/(M+1)} \\
&= \frac{1}{2} \int_{-1/2}^{1/2} \log S(f) \mathrm{d}f \ .
\end{aligned}
$$

Usually, we also call h entropy, use the notation $H1$ and omit the constant factor $1/2$. Thus, we finally obtain the usual entropy expression in the frequency domain

$$
H1 = \int_{-1/2}^{1/2} \log S(f) \mathrm{d}f \ . \tag{2.1.13}
$$

It is $H1$ in this form that will be used later in the mathematical operations in MEM1. Quite often the integral limits are not explicitly specified.

2.1.2 Approach 2

To begin with, we give two notes preparatory to deriving the entropy expression $H1$ in the frequency domain [2.2].

(1) For the convenience of mathematical operations, quantities in the frequency domain are also sampled. Thus, both the time and frequency are discrete and therefore the time series, ACF and spectrum are all periodic. Their common period is, say, N. Results in the continuous frequency case are obtained from those in the discrete frequency case by letting $N \to \infty$.

We denote a period of the zero-mean time series by $x = \{x_0, \ldots, x_{N-1}\}^{\mathrm{T}}$ and its FT by $X = \{X_0, \ldots, X_{N-1}\}^{\mathrm{T}}$, i.e.,

$$
X = \mathrm{FT}[x] = Wx, \quad X_k = N^{-1} \sum_{n=0}^{N-1} x_n \mathrm{e}^{-2\pi \mathrm{i}kn/N} \ , \tag{2.1.14a}
$$

where W is the transformation matrix. The inverse transform is

$$
x = \mathrm{IFT}[X] = W^{-1}X, \quad x_n = \sum_{k=0}^{N-1} X_k \mathrm{e}^{+2\pi \mathrm{i}kn/N} \ . \tag{2.1.14b}
$$

The ACF is

$$
R(j) = \mathrm{E}[x_n x_{n-j}^*] \ .
$$

The (power) spectrum is

$$
S(k) = \mathrm{E}[|X_k|^2] = \int q_k(X_k)|X_k|^2 \mathrm{d}X_k = \int q(X)|X_k|^2 \mathrm{d}X \ ,
$$

where $q(X)$ and $q_k(X_k)$ are joint p.d.f. of X and the marginal p.d.f. of X_k, respectively. $R(j)$ and $S(k)$ form an FT pair, $S(k) = \mathrm{FT}[R(j)]$.

From

$$|\det(W)| \cdot |\det(W^{-1})| = 1, \quad |\det(W)| = N^{-1}|\det(W^{-1})|$$

we get that the moduli of the Jacobians for the FT and IFT are $N^{-\frac{1}{2}}$ and $N^{\frac{1}{2}}$, respectively, i.e.,

$$|\det(W)| = N^{-\frac{1}{2}}, \quad |\det(W^{-1})| = N^{\frac{1}{2}} .$$

(2) That $|\det(W^{-1})| \neq 1$ should be taken into consideration in calculating the entropy in the frequency domain by FT. Specifically

$$
\begin{aligned}
H1 &= -\int p(\boldsymbol{x}) \log p(\boldsymbol{x}) \mathrm{d}\boldsymbol{x} \\
&= -\int q(X) \log[\, q(X)/|\det(W^{-1})| \,] \mathrm{d}X \\
&= -\int q(X) \log q(X) \mathrm{d}X + \frac{1}{2} \log N .
\end{aligned}
\tag{2.1.15}
$$

Now we derive the expression of $H1$. The partially given ACF $R(j), j \in D \subset \{0, 1, \ldots, N-1\}$ together with the normalization condition does not suffice for uniquely determining $p(\boldsymbol{x})$ or $q(X)$. Hence we invoke MEM.

The problem is stated directly in the frequency domain:

$$\text{Maximization}: \quad H1 = -\int q(X) \log q(X) \mathrm{d}X \ \text{(omitting the constant)},$$

$$\text{Constraints}: \quad \int q(X) \mathrm{d}X = 1 , \tag{2.1.16}$$

$$\sum_{k=0}^{N-1} \int q(X)|X_k|^2 \mathrm{d}X \cdot \mathrm{e}^{-2\pi \mathrm{i}jk/N} = R^*(j) , \quad j \in D . \tag{2.1.17}$$

Our main task is to derive the expression of $H1$ in terms of $S(k)$.

Introducing the Lagrange multipliers μ and λ_j for the normalization condition (2.1.16) and ACF constraints (2.1.17), respectively, we form the objective function to be maximized:

$$
\begin{aligned}
G &= H1 + \mu \left[\int q(X) \mathrm{d}X - 1 \right] \\
&\quad + \sum_{j \in D} \lambda_j \left[\sum_{k=0}^{N-1} \int q(X)|X_k|^2 \mathrm{d}X \cdot \mathrm{e}^{-2\pi \mathrm{i}jk/N} - R^*(j) \right] .
\end{aligned}
$$

By the variational method letting $\delta G = 0$, it is easily shown that

$$q(X) = C \exp \left(-\sum_{j \in D} \lambda_j \sum_{k=0}^{N-1} |X_k|^2 \mathrm{e}^{-2\pi \mathrm{i}jk/N} \right) ,$$

where C is a constant related to μ. Supplementing $\lambda_j = 0$, $j \notin D$ and denoting the FT of $\lambda = \{\lambda_0, \ldots, \lambda_{N-1}\}^{\mathrm{T}}$ by $\Lambda = \{\Lambda_0, \ldots, \Lambda_{N-1}\}^{\mathrm{T}}$, we get

$$
\begin{aligned}
q(X) &= C \exp\left(-N \sum_{k=0}^{N-1} \Lambda_k |X_k|^2\right) & (2.1.18) \\
&= \prod_{k=0}^{N-1} q_k(X_k) ,
\end{aligned}
$$

$$
q_k(X_k) = C_k \exp(-N\Lambda_k |X_k|^2) . \tag{2.1.19}
$$

It can be seen that X, as well as x, has the Gaussian distribution. From the normalization condition we have

$$
\begin{aligned}
C_k &= N^{\frac{1}{2}} \sqrt{\Lambda_k/\pi} , \\
C &= \prod_{k=0}^{N-1} C_k = \prod_{k=0}^{N-1} N^{\frac{1}{2}} \sqrt{\Lambda_k/\pi} . & (2.1.20)
\end{aligned}
$$

The variance is

$$
\sigma_k^2 = \frac{1}{2} N^{-1}/\Lambda_k .
$$

It follows that

$$
\begin{aligned}
S(k) &= \int q_k(X_k) |X_k|^2 \mathrm{d}X_k \\
&= \sigma_k^2 = \frac{1}{2} N^{-1}/\Lambda_k , & (2.1.21)
\end{aligned}
$$

$$
\log S(k) = -\log \Lambda_k - \log N - \log 2 .
$$

From (2.1.20) and the above equation, we get

$$
\begin{aligned}
\log C &= \frac{1}{2} \sum_{k=0}^{N-1} \log \Lambda_k + \frac{1}{2} N \log N - \frac{1}{2} N \log \pi \\
&= -\frac{1}{2} \sum_{k=0}^{N-1} \log S(k) - \frac{1}{2} N \log(2\pi) . & (2.1.22)
\end{aligned}
$$

Now we can calculate the entropy, which concerns us most. From (2.1.15, 2.1.18), we have

$$
\begin{aligned}
H1 &= -\int q(X) \log q(X) \mathrm{d}X + \frac{1}{2} \log N \\
&= -\int q(X) \left(\log C - N \sum_{k=0}^{N-1} \Lambda_k |X_k|^2\right) \mathrm{d}X + \frac{1}{2} \log N \\
&= -\log C + N \sum_k \Lambda_k \int q(X) |X_k|^2 \mathrm{d}X + \frac{1}{2} \log N
\end{aligned}
$$

$$\begin{aligned}
&= -\log C + N \sum_k \Lambda_k S(k) + \frac{1}{2}\log N \\
&= -\log C + N \sum_{k=0}^{N-1} \Lambda_k \cdot \frac{1}{2} N^{-1}/\Lambda_k + \frac{1}{2}\log N \quad (\text{ by } (2.1.21)) \\
&= \frac{1}{2} \sum_{k=0}^{N-1} \log S(k) + \frac{1}{2} N \log(2\pi e) + \frac{1}{2}\log N \quad (\text{ by } (2.1.22)) .
\end{aligned}$$

$$(2.1.23)$$

Ignoring the constant terms and factor, the above expression becomes

$$H1 = \sum_{k=0}^{N-1} \log S(k) . \tag{2.1.24}$$

This is the entropy expression of discrete form that we want. $H1$ here is the entropy in one period.

Expression (2.1.23) is similar to (2.1.5). The first two terms on their right-hand sides correspond to each other. The third term in (2.1.23) is the result of the coordinate transformation. If $N \to \infty$, (2.1.23) will diverge. Therefore, we must use the entropy rate instead of the entropy. Specifically, we have

$$h = \lim_{N\to\infty} \frac{H1}{N} = \lim_{N\to\infty} \frac{1}{2} \cdot \frac{1}{N} \sum_{k=0}^{N-1} \log S(k) + \frac{1}{2}\log(2\pi e) + 0 .$$

Noting that if $N \to \infty$, then

$$\frac{1}{N} \to \mathrm{d}f , \quad S(k) \to S(f) , \quad \sum \to \int ,$$

and ignoring the constant independent of N, we obtain

$$h = \frac{1}{2} \int_{-1/2}^{1/2} \log S(f)\mathrm{d}f .$$

This expression is exactly the same as the result in the first approach.

2.1.3 Discussion

In the following we discuss the relationship between the two approaches and related issues.

(1) Approach 2 seems to be simpler than Approach 1. This is because in Approach 2 we stated the problem in the frequency domain right at the beginning while in Approach 1 we started off with the problem in the time domain and related the quantities in the two domains by Szegö's theorem.

(2) In Approach 1 we explicitly assumed that x was a zero-mean Gaussian process so that we could write out the joint p.d.f. $p(x)$ and then calculate

the integral. In contrast, in Approach 2 we did not explicitly make the assumption of a Gaussian process. What we did was to maximize the entropy with the partially given ACF (second moments) as constraints. The resulting expression (2.1.18) showed that the time series x under consideration was a zero-mean Gaussian process. That is to say, the assumption of a Gaussian process was hidden in Approach 2. The consistency of the two approaches stems from a well-known fact: A stochastic process (with the state space $(-\infty, +\infty)$) having the maximum entropy under the constraints of the first and second moments must be Gaussian.

(3) In the two approaches, the measure $m(x)$ in the entropy definition was left out. This was because we did not consider the prior knowledge about $p(x)$. The entropy expression giving consideration to $m(x)$ is the negative of cross-entropy. (For details, see Sect. 2.3.2.)

(4) The entropy is defined in the time domain, so the requirement of invariance with regard to the coordinate transformation should be normally posed in the time domain. If one is so peculiar as to impose this requirement in the frequency domain, then $H1$ might take the following forms:

$$H1 = \int_{-1/2}^{1/2} m(f) \log[S(f)/m(f)] \mathrm{d}f \qquad (2.1.25)$$

and

$$H1 = \sum_{k=0}^{N-1} m(k) \log[S(k)/m(k)] , \qquad (2.1.26)$$

where $m(f)$ and $m(k)$ are the prior estimates of the spectra playing the role of measure. Indeed, with the above $H1$, the MEM1 solution incorporates the prior estimate of the spectrum (see the next section). But the solution does not meet the requirement of consistency (see Sect. 4.3 for details).

(5) We should emphasize that either the continuous form (2.1.13) or discrete form (2.1.24) is a derived expression of the entropy $H1$ in the frequency domain, but not the original definition, which is in the time domain. This is why the commonly used $H1$ expression looks unlike the information-theoretic entropy.

(6) In the 2-D case although the entropy $H1$ in the time domain may be defined like (2.1.2), the entropy expression like (2.1.13) in the frequency domain cannot be derived from this definition. Thus, we can only directly write out the 2-D entropy expression in the frequency domain by analogy with (2.1.13):

$$H1 = \int \int \log S(f_1, f_2) \mathrm{d}f_1 \mathrm{d}f_2 .$$

2.2 Formulation and Solution

This section is concerned with the solution of MEM1. To begin with, we formulate MEM1, that is, state the problem in mathematical language. Then the solution and data extension formula are derived. Finally, discussion about some issues concludes this section.

2.2.1 Formulation

In the preceding section we have derived the expression of the entropy $H1$ (2.1.13) in the frequency domain. The given data are partial ACF $R(k)$, $|k| \leq M$. The spectrum $S(f)$ and ACF form an FT pair:

$$
\begin{aligned}
S(f) &= \mathrm{FT}[R(k)] = \sum_{k=-\infty}^{\infty} R(k)\exp(-2\pi ifk)\,, \\
R(k) &= \mathrm{IFT}[S(f)] = \int_{-1/2}^{1/2} S(f)\exp(+2\pi ifk)\mathrm{d}f\,.
\end{aligned}
\tag{2.2.1}
$$

MEM1 is the following constrained maximization problem:

$$
\text{Maximization}: \quad H1 = \int_{-1/2}^{1/2} \log S(f)\mathrm{d}f\,,
\tag{2.2.2}
$$

$$
\text{Constraints}: \quad R(k) = \int_{-1/2}^{1/2} S(f)\exp(2\pi ifk)\mathrm{d}f\,, \quad |k| \leq M\,.
\tag{2.2.3}
$$

It can be shown that the solution to this problem is unique, i.e., there exists only one maximum point (Sect. 4.3). As usual we convert this constrained maximization to an unconstrained one by the Lagrange multiplier method. We are concerned with only the necessary, but not the sufficient condition for the maximization.

2.2.2 Solution

The solution procedure consists of three steps. It should be pointed out before going ahead that the solution procedure presented in the following is by no means the simplest, but very readable.

Step 1. From (2.2.2, 2.2.3), we form the objective functional to be maximized:

$$
G = \int_{-1/2}^{1/2} \log S(f)\mathrm{d}f - \sum_{k=-M}^{M} \lambda_k \left[\int_{-1/2}^{1/2} S(f)\exp(2\pi ifk)\mathrm{d}f - R(k)\right],
\tag{2.2.4}
$$

where λ_k's are the Lagrange multipliers, $\lambda_{-k} = \lambda_k^*$ since $R(-k) = R^*(k)$. The standard variational method is used to solve this problem.

Take the variation of G with respect to $S(f)$ and let it be zero, i.e.,

$$0 = \delta G = \int \delta S(f) \left[\frac{1}{S(f)} - \sum_{k=-M}^{M} \lambda_k \exp(2\pi i f k) \right] \mathrm{d}f \, .$$

Since $\delta S(f)$ is arbitrary, the quantity in the square brackets must be zero. Therefore, we obtain

$$S(f) = \frac{1}{\displaystyle\sum_{k=-M}^{M} \lambda_k \exp(2\pi i f k)} \, .$$

Introducing parameters C_k which are related to λ_k by

$$C_k = \lambda_k^* \, ,$$

and considering the property that $S(f)$ is real, $S(f)$ can be changed to another form:

$$S(f) = \frac{1}{\displaystyle\sum_{k=-M}^{M} C_k \exp(-2\pi i f k)} \, . \tag{2.2.5}$$

This is done because λ_k and C_k are quantities in the time domain; the convention that the forward FT is used in calculating the spectrum from λ_k and C_k is followed. This also facilitates the following mathematical manipulation.

Thus, the spectrum $S(f)$ has been parametrized by C_k. Provided that C_k's are known, $S(f)$ can be calculated easily using (2.2.5). However, determining C_k by the constraints (2.2.3) is not an easy job. Other parameters must be introduced in addition.

Step 2. C_k is a sequence of length $(2M + 1)$ possessing the Hermitian property: $C_{-k} = C_k^*$. C_k can be decomposed into two sequences α_k and β_k such that

$$C_k = \alpha_k * \beta_k \, , \tag{2.2.6}$$

where $*$ denotes convolution (Fig. 2.2).

Generally speaking, the decomposition of C_k is not unique. But the situation may be changed by imposing a set of appropriate conditions on the decomposition. They are:

(1) $\alpha_k = 0$ for $k < 0$, $k > M$, i.e., α_k is a causal sequence of length $(M + 1)$. $\beta_k = 0$ for $k > 0$, $k < -M$, i.e., β_k is an anticausal sequence of length $(M + 1)$.

(2) These two sequences are related by

$$\beta_{-k} = \alpha_k^* \, , \quad (\beta_0 = \alpha_0 \text{ is real}) \, . \tag{2.2.7}$$

(3) α_k is a minimum-phase sequence while β_k is a maximum-phase sequence.

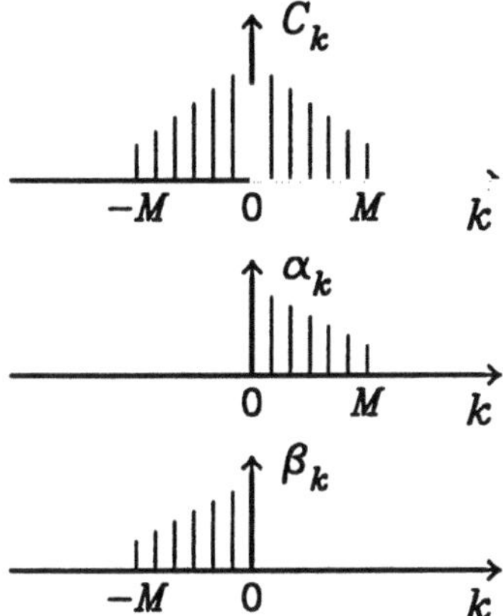

Fig. 2.2. The decomposition of the sequence C_k

Condition (1) matches the two sides of (2.2.6) in length. It can be seen from (2.2.5) that C_k is in fact the ACF of the inverse spectrum $1/S(f)$. By conditions (2) and (3), the zeros of the sequence C_k inside and outside the unit circle on the z-plane are assigned to α_k and β_k, respectively.

Using (2.2.5) and noticing that the summation in its denominator is in fact an FT, we have

$$
S(f) \;=\; \frac{1}{\displaystyle\sum_{k=-M}^{M} (\alpha_k * \beta_k)\exp(-2\pi i f k)} \qquad (\text{by } (2.2.6))
$$

$$
=\; \frac{1}{\displaystyle\sum_{k=-M}^{M} \alpha_k \exp(-2\pi i f k)\cdot \sum_{k=-M}^{M} \beta_k \exp(-2\pi i f k)}
$$

$$
=\; \frac{1}{\displaystyle\sum_{k=-M}^{M} \alpha_k \exp(-2\pi i f k)\cdot \sum_{l=M}^{-M} \beta_{-l} \exp(+2\pi i f l)} \qquad (l=-k)
$$

$$
=\; \frac{1}{\displaystyle\sum_{k=-M}^{M} \alpha_k \exp(-2\pi i f k)\cdot \sum_{l=-M}^{M} \alpha_l^* \exp(+2\pi i f l)}
$$

$$
\hspace{6cm} (\text{by } (2.2.7)) \qquad (2.2.8)
$$

$$
=\; \frac{1}{\displaystyle\sum_{k=-M}^{M} \alpha_k \exp(-2\pi i f k)\cdot \left[\sum_{l=-M}^{M} \alpha_l \exp(-2\pi i f l)\right]^*}
$$

$$= \frac{1}{\left| \displaystyle\sum_{k=-M}^{M} \alpha_k \exp(-2\pi i f k) \right|^2} \cdot \tag{2.2.9}$$

Thus, the spectrum $S(f)$ is parametrized by α_k. In the following, (2.2.8) and the constraints (2.2.3) are used to determine α_k. This is the most difficult step in our solution procedure.

Step 3. From (2.2.8) we get

$$\sum_{k=-M}^{M} \alpha_k \exp(-2\pi i f k) \cdot S(f) = \frac{1}{\displaystyle\sum_{l=-M}^{M} \alpha_l^* \exp(2\pi i f l)} \cdot$$

Taking the IFT of the left-hand side (LHS) and right-hand side (RHS) with respect to f, we obtain

$$\begin{aligned} \text{LHS} \;&=\; \sum_{k=-M}^{M} (\alpha_k \delta_{n-k}) * R(n) \\ &=\; \sum_{k=-M}^{M} \alpha_k R(n-k) \,, \end{aligned} \tag{2.2.10}$$

where $R(n)$ refers to the ACF of the MEM1 spectrum. In the above the following properties of FT and convolution were utilized:

$$\text{IFT}[\exp(-2\pi i f k)] = \delta_{n-k} \,,$$
$$\delta_{n-k} * R(n) = R(n-k) \,.$$

The right-hand side

$$\begin{aligned} \text{RHS} \;&=\; \int_{-1/2}^{1/2} \frac{\exp(2\pi i f n)}{\displaystyle\sum_{l=0}^{M} \alpha_l^* \exp(2\pi i f l)} \mathrm{d}f \\ &\overset{\triangle}{=}\; \int_{-1/2}^{1/2} g(x)\mathrm{d}x \,. \end{aligned} \tag{2.2.11}$$

Note that the real variable f has been changed to x, and the integrand is denoted by $g(x)$. Note also that the lower limit in the summation has been changed from $-M$ to 0 since $\alpha_l = 0$ for $l < 0$.

It is difficult to calculate the above integral with respect to the real variable x. Our approach is to convert it to an integral of a function of a complex variable [2.3]. Consider a contour integral on the w-plane, $\oint_{ABCDA} g(w)\mathrm{d}w$, $w = x + iy$ (Fig. 2.3). We will show first of all that $g(w)$ is analytic on the upper half plane (including the x-axis) so that the contour integral vanishes.

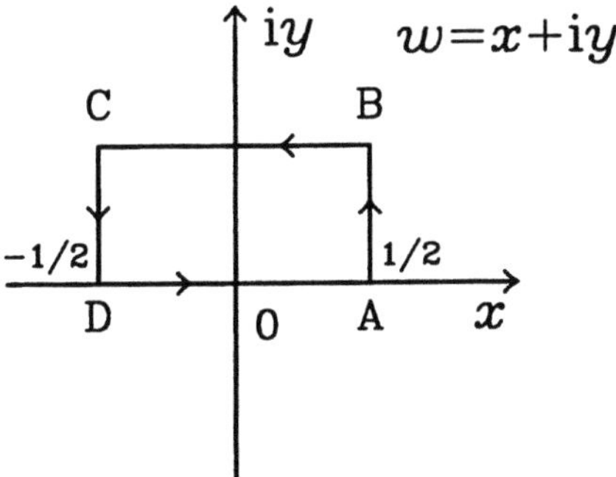

Fig. 2.3. The contour integral on the w-plane

In order to show that $g(w)$ is analytic on the upper half plane, we have only to show that the denominator $\sum_{l=0}^{M} \alpha_l^* \exp(2\pi iwl)$ has no zero when $y \geq 0$, or equivalently, its complex conjugate $\sum_{l=0}^{M} \alpha_l \exp(-2\pi iwl)$ has no zero when $y \leq 0$. As a matter of fact, $\sum_{l=0}^{M} \alpha_l z^{-l}$ has no zero on and outside the unit circle, i.e., when $|z| \geq 1$, since α_l is a minimum-phase sequence. But since $z = \mathrm{e}^{2\pi iw} = \mathrm{e}^{2\pi i(x+iy)} = \mathrm{e}^{2\pi ix - 2\pi y}$, $|z| = \mathrm{e}^{-2\pi y} \geq 1$ is equivalent to $y \leq 0$. Therefore, $\sum_{l=0}^{M} \alpha_l \exp(-2\pi iwl)$ has no zero when $y \leq 0$.

We have shown that

$$\oint_{ABCDA} g(w)\mathrm{d}w = 0 .$$

Since $\exp(2\pi iwn)$ is a constant for $n = 0$ while it is a function with the period 1 (or more precisely speaking, $1/n$) for $n \neq 0$, the line integrals of $g(w)$ along AB and DC are equal; this is in a simplified notation:

$$\int_{AB} = \int_{DC} , \quad \text{i.e.,} \quad \int_{AB} + \int_{CD} = 0 .$$

Consequently,

$$\int_{DA} + \int_{BC} = 0 , \quad \text{i.e.,} \quad \int_{DA} = \int_{CB} .$$

From (2.2.11) it follows that

$$
\begin{aligned}
\mathrm{RHS} &= \int_{DA} g(w)\mathrm{d}w \qquad (w = x,\ \mathrm{d}w = \mathrm{d}x) \\[2ex]
&= \int_{CB} \frac{\exp[2\pi in(x + iy)]}{\displaystyle\sum_{l=0}^{M} \alpha_l^* \exp[2\pi il(x + iy)]} \mathrm{d}x \quad (\mathrm{d}w = \mathrm{d}x) \\[2ex]
&= \int_{CB} \frac{\exp(2\pi inx) \cdot \exp(-2\pi ny)}{\displaystyle\sum_{l=0}^{M} \alpha_l^* \exp(2\pi ilx) \cdot \exp(-2\pi ly)} \mathrm{d}x .
\end{aligned}
$$

The value of the above integral is independent of y. Letting $y \to \infty$, we get

$$\mathrm{RHS} \;=\; \int_{-1/2}^{1/2} \frac{1}{\alpha_0^* + 0}\,\mathrm{d}x = \frac{1}{\alpha_0} \quad \text{for } n = 0 \, ,$$

$$\mathrm{RHS} \;=\; \int_{-1/2}^{1/2} \frac{0}{\alpha_0^* + 0}\,\mathrm{d}x = 0 \quad \text{for } n \neq 0 \, .$$

They can be combined into a single expression:

$$\mathrm{RHS} = \frac{1}{\alpha_0}\delta_n \, , \quad n = 0, 1, \dots \, .$$

From this expression and (2.2.10), we get

$$\sum_{k=0}^{M} \alpha_k R(n - k) = \frac{1}{\alpha_0}\delta_n \, , \quad n = 0, 1, \dots \, . \tag{2.2.12}$$

The factor before the ACF in the first term on the left-hand side can be set to unity. Specifically, the two sides are divided by α_0, then

$$\sum_{k=0}^{M} (\alpha_k/\alpha_0) R(n - k) = \frac{1}{\alpha_0^2}\delta_n \, , \quad n = 0, 1, \dots \, ,$$

and the new factor in the first term becomes $\alpha_0/\alpha_0 = 1$. Denoting α_k/α_0 by a_k, and $1/\alpha_0^2$ by p_M, the above expression is changed to

$$\sum_{k=0}^{M} a_k R(n - k) = p_M \delta_n \, , \quad (a_0 = 1), \quad n = 0, 1, \dots \, . \tag{2.2.13}$$

The formula for calculating the spectrum, (2.2.9), is changed accordingly to

$$S(f) = \frac{p_M}{\left| \displaystyle\sum_{k=0}^{M} a_k \exp(-2\pi i f k) \right|^2} \, . \tag{2.2.14}$$

From (2.2.14) we see that $S(f)$ is parametrized by a_k and p_M. They can be obtained by solving (2.2.13), whose expanded form is

$$\begin{bmatrix} R(0) & R(-1) & R(-2) & \cdots & R(-M) \\ R(1) & R(0) & R(-1) & \cdots & R(1-M) \\ \cdots & \cdots & \cdots & \cdots & \cdots \\ \cdots & \cdots & \cdots & \cdots & \cdots \\ R(M) & R(M-1) & R(M-2) & \cdots & R(0) \\ \hdashline R(M+1) & R(M) & R(M-1) & \cdots & R(1) \\ \cdots & \cdots & \cdots & \cdots & \cdots \\ \cdots & \cdots & \cdots & \cdots & \cdots \end{bmatrix} \begin{bmatrix} 1 \\ a_1 \\ a_2 \\ \cdot \\ \cdot \\ \cdot \\ a_M \end{bmatrix}$$

$$= p_M \begin{bmatrix} 1 \\ 0 \\ \cdot \\ \cdot \\ \cdot \\ 0 \\ --- \\ 0 \\ \cdot \\ \cdot \end{bmatrix} .$$

Obviously, the upper part (above the dashed line) of the matrix is the (given) ACF matrix R. The following equations are called the *normal equations*, which can be used in solving for a_k and p_M:

$$\begin{bmatrix} R(0) & R(-1) & R(-2) & \cdots & R(-M) \\ R(1) & R(0) & R(-1) & \cdots & R(1-M) \\ \cdots & \cdots & \cdots & \cdots & \cdots \\ \cdots & \cdots & \cdots & \cdots & \cdots \\ R(M) & R(M-1) & R(M-2) & \cdots & R(0) \end{bmatrix} \begin{bmatrix} 1 \\ a_1 \\ a_2 \\ \cdot \\ \cdot \\ \cdot \\ a_M \end{bmatrix}$$

$$= p_M \begin{bmatrix} 1 \\ 0 \\ \cdot \\ \cdot \\ \cdot \\ 0 \end{bmatrix} . \tag{2.2.15}$$

Its compact form is

$$\sum_{k=0}^{M} a_k R(n-k) = p_M \delta_n, \quad (a_0 = 1), \quad n = 0, 1, \ldots, M . \tag{2.2.16}$$

The solution can be written formally as

$$(1, a_1, \ldots, a_M)^{\mathrm{T}} = R^{-1} p_M (1, 0, \ldots, 0)^{\mathrm{T}} . \tag{2.2.17}$$

The lower part (below the dashed line) of the equations in the matrix form can be used to extrapolate the ACF. The first formula is

$$R(M+1) + a_1 R(M) + a_2 R(M-1) + \cdots + a_M R(1) = 0 ,$$

or equivalently, after moving the terms,

$$R(M+1) = -\sum_{k=1}^{M} a_k R(M+1-k) .$$

Generally, we have

$$R(M+n) = -\sum_{k=1}^{M} a_k R(M+n-k), \quad n \geq 1. \tag{2.2.18}$$

Summary. We have so far determined the explicit solution and data extension of 1-D MEM1. The results are summarized as follows:

1) Solve the normal equations (2.2.16) for a_k, $(a_0 = 1)$, $k = 0, 1, \ldots, M$ and p_M.
2) Calculate the spectrum $S(f)$ by (2.2.14).
3) Extrapolate the ACF using (2.2.18).
4) Use the given and extrapolated ACF altogether to calculate $S(f)$ by FT. The result is, of course, the same as that in Step 2) above.

2.2.3 Discussion

(1) In practical computation, discrete FT (DFT) is used:

$$S(j) = \sum_{k=0}^{N-1} R(k) \exp(-2\pi i j k/N),$$

$$R(k) = \frac{1}{N} \sum_{j=0}^{N-1} S(j) \exp(+2\pi i j k/N),$$

$j, k = 0, 1, \ldots, N-1$, N being the size of the transform. Conventionally, the scale factor $1/N$ is put in the inverse transform.

Accordingly, the discrete form of (2.2.14) is

$$S(j) = \frac{p_M}{\left| \sum_{k=0}^{M} a_k \exp(-2\pi i j k/N) \right|^2}. \tag{2.2.19}$$

(2) The normal equations (2.2.16) or (2.2.15) contain $(M+1)$ equations and $(M+1)$ unknowns $a_1, \ldots, a_M$ and p_M. The ACF matrix is nonnegative definite. Therefore, the solution exists and is unique. (The case of degeneracy is not under consideration.) The solution takes the form of (2.2.17).

It may be beneficial to point out that $(1, a_1, \ldots, a_M)^{\mathrm{T}}$ is really a minimum-phase sequence [2.1]; its converse is also true, that is, if the solution to a system of linear equations in a similar form as (2.2.15), $(1, a_1, \ldots, a_M)^{\mathrm{T}}$, is a minimum-phase sequence, then the coefficient matrix must be the ACF matrix of some time series.

Often a_M and p_M are called *AR* (*autoregression*) *parameters*, or (*inverse*) *filter coefficients* of order M and the *power of innovation* (Sect. 2.3.3). They are also called the *coefficients* and *output power of an optimal prediction-error filter* (Sect. 2.3.5).

(3) Introducing the parameters $c_k = C_k/\alpha_0^2$ based on (2.2.6), then the discrete form of (2.2.5) is

$$S(j) = \frac{p_M}{\displaystyle\sum_{k=-M}^{M} c_k \exp(-2\pi ijk/N)} .$$

(2.2.20)

It can be shown, after some mathematical manipulation, that

$$c_k = \sum_{k=0}^{M-k} a_n^* a_{k+n}, \quad k \geq 0 ; \quad c_k = c_{-k}^* , \quad k < 0 .$$

Therefore, in addition to (2.2.19), the above two formulae can also be used to calculate $S(j)$ from a_k. A fast convolution algorithm may be used to compute c_k. The Hermitian property of c_k may be used in computing $S(j)$ by means of FT.

(4) The extrapolation formula of the ACF, (2.2.18), provides an alternative way of calculating the spectrum. It is also of theoretical significance. We will discuss this point in considerable detail in the next section.

(5) If the "entropy" expression (2.1.25), i.e.,

$$H1 = \int_{-1/2}^{1/2} m(f) \log[S(f)/m(f)] df$$

is used, the formula corresponding to (2.2.5) is

$$S(f) = \frac{m(f)}{\displaystyle\sum_{k=-M}^{M} C_k \exp(-2\pi ifk)} .$$

Proceeding in the same way as before in the solution procedure and noticing the new factor $m(f)$, it is not hard to work out the following results:

(i) The "normal equations"

$$\sum_{k=0}^{M} a_k R(n-k) = p_M M(n) , \quad (a_0 = 1) , \quad n = 0, 1, \ldots, M ,$$

(2.2.21)

where $M(n)$ is the IFT of $m(f)$.

(ii) The spectrum

$$S(f) = \frac{m(f)p_M}{\left| \displaystyle\sum_{k=0}^{M} a_k \exp(-2\pi ifk) \right|^2} .$$

(2.2.22)

(iii) The ACF extrapolation

$$R(M+n) = M(M+n) - \sum_{k=1}^{M} a_k R(M+n-k) , \quad n \geq 1 .$$

(2.2.23)

When $m(f) \equiv 1$, $M(n) = \delta_n$ and (2.2.21–2.2.23) degenerate to (2.2.16, 2.2.14, 2.2.18), respectively.

If there is only the total power $(R(0))$ constraint, then it is easy to find that

$$p_M = R(0)/M(0) \,, \quad S(f) = m(f)p_M \,, \quad R(n) = p_M M(n) \,, \quad n \geq 1 \,.$$

If $M(0)$ is rescaled so that $M(0) = R(0)$, then $p_M = 1$. On the other hand, if only $m(f)$ $(M(n))$ is given without any measured data, it is natural that we should have $R(0) = M(0)$ and hence $p_M = 1$. Consequently, we have

$$p_M = 1 \,, \quad S(f) = m(f) \,, \quad R(n) = M(n) \,, \quad n = 0, 1, \dots \,.$$

Now we can see from the above why $m(f)$ is called the prior estimate of the spectrum: Without any measured data, the MEM1 spectrum is equal (or proportional) to $m(f)$. This property is attractive, but the solution does not meet the requirement of consistency (Sect. 4.3).

(6) The constraints (2.2.3) are used in the case of incomplete but noiseless data. In the case in which data are incomplete and noisy, the statistical properties of the MEM1 solution can be derived (Sect. 4.3). In practical application, what we do is to relax the constraints (2.2.3). Specifically, we no longer require the ACF fit be exact at each individual point on $[-M, M]$, but require the error of data fit be on average within a certain limit. In this case no explicit solution is available and iterative algorithms have to be used. A 2-D MEM1 algorithm for spectral estimation given partial noisy ACF will be described in Sect. 2.4.

2.3 Equivalents and Signal Model

In this section we present six equivalents to MEM1 and a signal model. The first four will be emphasized. They are: the ACF extension subject to the nonnegativity constraint, the Principle of Minimum Cross-entropy (MCE), Autoregressive (AR) process (signal model), and the Bayesian method. These methods themselves are important in signal processing. They are related to MEM1 in basic ideas, even have the same formulae as MEM1's, and are relevant to our later discussion. Two other equivalents to MEM1 are the Wiener filter and the approximation theoretic approach.

In the following we briefly introduce each method first, then describe the equivalence between each of them and MEM1. The signal model of MEM1 will be obtained in due course.

2.3.1 ACF Extension Subject to the Nonnegativity Constraint

This equivalent relationship was pointed out first by *Burg* [2.4] in 1967. It was this simple idea that the ACF extension must meet the constraint of nonnegativity that resulted in the birth of MEM1.

Suppose that the partial ACF $R(n)$, $|n| \leq M$ are given. The ACF matrix

$$R_M = \begin{bmatrix} R(0) & R(-1) & \cdots & R(-M) \\ R(1) & R(0) & \cdots & R(-M+1) \\ \cdots & \cdots & \cdots & \cdots \\ \cdots & \cdots & \cdots & \cdots \\ R(M) & R(M-1) & \cdots & R(0) \end{bmatrix} \qquad (2.3.1)$$

must be nonnegative definite, i.e.,

$$\det[R_M] \geq 0 .$$

$R(M+1)$ is unknown. But we can assert that the value of $R(M+1)$ must ensure

$$\det[R_{M+1}]$$
$$= \det \begin{bmatrix} R(0) & R(-1) & \cdots & R(-M) & R(-M-1) \\ R(1) & R(0) & \cdots & R(-M+1) & R(-M) \\ \cdots & \cdots & \cdots & \cdots & \cdots \\ \cdots & \cdots & \cdots & \cdots & \cdots \\ R(M) & R(M-1) & \cdots & R(0) & R(-1) \\ R(M+1) & R(M) & \cdots & R(1) & R(0) \end{bmatrix}$$
$$\geq 0 . \qquad (2.3.2)$$

The matrix R_{M+1} is formed by adding one row under R_M and one column to the right of R_M. R_{M+1} contains only one unknown $R(M+1)$ since $R(-M-1) = R^*(M+1)$. With the notation $x = R(M+1)$, the determinant in (2.3.2) is a function in x, $f(x) \geq 0$.

1. For simplicity and intuitiveness, we assume the ACF to be real for the moment. Then $R(-n) = R(n)$, $R(-M-1) = R(M+1) = x$, and $f(x) = ax^2 + bx + c \geq 0$ is a quadratic function with real coefficients.

By the rule of differentiation of the determinant, we get

$$f'(x) = \det \begin{bmatrix} 0 & 0 & \cdots & 0 & 1 \\ R(1) & R(0) & \cdots & R(-M+1) & R(-M) \\ \cdots & \cdots & \cdots & \cdots & \cdots \\ \cdots & \cdots & \cdots & \cdots & \cdots \\ R(M) & R(M-1) & \cdots & R(0) & R(-1) \\ x & R(M) & \cdots & R(1) & R(0) \end{bmatrix}$$
$$+ \det \begin{bmatrix} R(0) & R(-1) & \cdots & R(-M) & x \\ R(1) & R(0) & \cdots & R(-M+1) & R(-M) \\ \cdots & \cdots & \cdots & \cdots & \cdots \\ \cdots & \cdots & \cdots & \cdots & \cdots \\ R(M) & R(M-1) & \cdots & R(0) & R(-1) \\ 1 & 0 & \cdots & 0 & 0 \end{bmatrix}$$

$$= 2(-1)^{M+3} \det \begin{bmatrix} R(1) & R(0) & \cdots & R(-M+1) \\ R(2) & R(1) & \cdots & R(-M+2) \\ \cdots & \cdots & \cdots & \cdots \\ \cdots & \cdots & \cdots & \cdots \\ R(M) & R(M-1) & \cdots & R(0) \\ x & R(M) & \cdots & R(1) \end{bmatrix} ,$$

$$(2.3.3)$$

and

$$f''(x) = 2(-1)^{2M+5} \det[R_{M-1}] \leq 0 . \tag{2.3.4}$$

(Normally, the degeneracy case where $f''(x) = 0$ is not considered.) $f(x)$ is plotted in Fig. 2.4a.

Solving the quadratic equation $f(x) = 0$ we get two roots α and β. That is to say, two values of $R(M+1)$ exist for which $\det[R_{M+1}] = 0$. These two and all the values in between can meet the nonnegativity condition and hence are permissible values in theory. Without any more knowledge about the ACF, we would naturally like to choose the midpoint of the permissible interval, i.e., $(\alpha + \beta)/2$ as $R(M+1)$. Thus, we have extrapolated the ACF by one step. $R(M+2)$, $R(M+3),\ldots$ can be obtained in the same way. Evidently, the extension of this kind is much more reasonable than the zero extension (assigning zero values to the unmeasured $R(M+1),\ldots\ldots$). The wonderful thing is that this extension of choosing the midpoint subject to the nonnegativity constraint is equivalent to the ACF extension in MEM1. In the latter the normal equations

$$\sum_{k=0}^{M} a_k R(n-k) = p_M \delta_n , \quad (a_0 = 1) , \quad (n = 0, 1, \ldots, M) \tag{2.3.5}$$

are solved for a_k, and then ACF can be extrapolated:

$$R(M+n) = -\sum_{k=1}^{M} a_k R(M+n-k) , \quad n \geq 1 . \tag{2.3.6}$$

Another somewhat different interpretation of the data extension in MEM1 was given by *Van den Bos* [2.5] in 1971. In the ACF extension, the entropy of the stochastic process is to be maximized. This is equivalent to maximizing $\det[R_{M+1}]$ because the entropy of the $(M+2)$-dimensional Gaussian process is proportional to $\log\{\det[R_{M+1}]\}$ as shown in (2.1.6). From (2.3.3) we must have

$$A(x) = \det \begin{bmatrix} R(1) & R(0) & \cdots & R(-M+1) \\ R(2) & R(1) & \cdots & R(-M+2) \\ \cdots & \cdots & \cdots & \cdots \\ \cdots & \cdots & \cdots & \cdots \\ R(M) & R(M-1) & \cdots & R(0) \\ x & R(M) & \cdots & R(1) \end{bmatrix} = 0 . \tag{2.3.7}$$

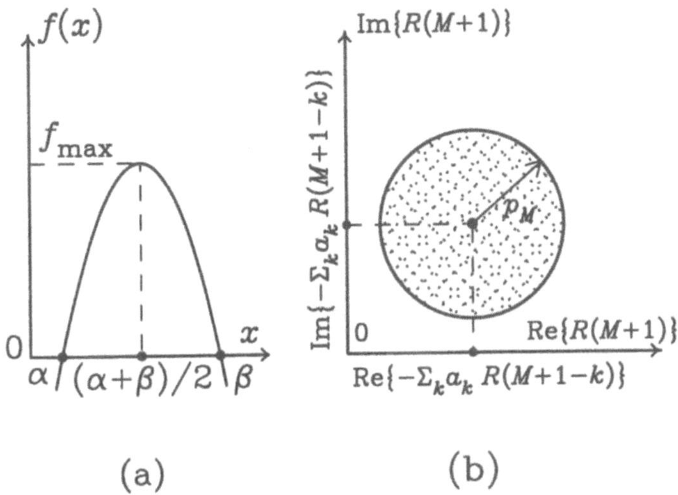

Fig. 2.4. The ACF extension subject to the nonnegativity constraint. (a) Real ACF. (b) Complex ACF

That $R(M+1)$ determined by the above equation is really a maximum point is ensured by (2.3.4). Equation (2.3.7) is the one that the extrapolated $x = R(M+1)$ in MEM1 must satisfy. By direct calculation we can show that this equation is just formula (2.3.6) ($n = 1$) with a_k being determined by (2.3.5).

After determining $R(M+1)$, $R(M+2)$ can be obtained by

$$\partial \det[R_{M+2}]/\partial R(M+2) = 0 \, .$$

$R(M+3), \ldots\ldots$ can be obtained in the same way.

The ideas of maximizing $\det[R_{M+1}] = f(x)$ and taking the midpoint between the two roots of $f(x) = 0$ as the extrapolated $x = R(M+1)$ are consistent with each other. The sum of the two roots of the quadratic form with real coefficients $f(x) = ax^2 + bx + c$ is $(\alpha + \beta) = -b/a$; $f(x)$ attains its maximum at $x = -b/(2a) = (\alpha+\beta)/2$ when $a < 0$ (ensured by (2.3.4)). The wonderful result is by no means an accident.

2. When the ACF is complex, $f(x) \geq 0$ is a function of complex variable. Its terms contain factors $|x|^2$, x, x^* and constant, respectively. Now the problem of maximizing $f(x)$ needs to be discussed in a different way than before. The conclusion is that $x = R(M+1)$ at which $f(x)$ attains its maximum also satisfies (2.3.7), and the formulae for the ACF extension are still (2.3.5, 2.3.6). This can be proved as follows:

Let Δx be a small change of x. Utilizing the determinant in (2.3.2) and $R(-n) = R^*(n)$, it is not hard to get

$$\begin{aligned}
\Delta f(x) &= f(x + \Delta x) - f(x) \\
&= 2(-1)^{M+3}\text{Re}\{\Delta x^* A(x)\} - |\Delta x|^2 \det[R_{M-1}] \, .
\end{aligned}$$

Since Δx is arbitrary and $|\Delta x|^2 \det[R_{M-1}] \geq 0$, it follows that the necessary and sufficient condition for the $f(x)$ maximization is $A(x) = 0$ as shown in (2.3.7).

In the case of complex ACF, it is difficult to determine the permissible range of $x = R(M + 1)$ under the nonnegativity constraint by the $f(x)$ expression. From the theory of linear prediction [2.1] or direct calculation taking the nonnegativity into consideration [2.6], it can be shown that given $R(n)$ for $|n| \leq M$, the permissible values of $R(M+1)$ lie within and on a circle on the complex plane (Fig. 2.4b). The circle is centered at $-\sum_{k=1}^{M} a_k R(M + 1-k)$ with radius p_M. Here a_k and p_M are the solution of the normal equations (2.3.5). Specifically,

$$R(M + 1) = -\sum_{k=1}^{M} a_k R(M + 1 - k) - \rho_{M+1} p_M \ , \tag{2.3.8}$$

where ρ_{M+1} is called the *reflection coefficient*, $|\rho_{M+1}| \leq 1$. (For the proof, see Sect. 2.4.1.)

After choosing $R(M+1)$, the permissible values of $R(M+2)$ also lie within and on a circle. It can be proved that if we choose as $R(M + 1)$ the value at the center of the circle, $-\sum_{k=1}^{M} a_k R(M+1-k)$, i.e., the extrapolated value in MEM1, then the radius of the next circle determining the permissible range of $R(M + 2)$ will be maximum, being $p_{M+1} = p_M (1 - |\rho_{M+1}|^2) \leq p_M$. In contrast, if we choose as $R(M + 1)$ a value on the circle, then the permissible range of $R(M + 2)$ will shrink to a single point. A similar thing will happen to the further extensions. For this reason we say that MEM1 predicts the unknown ACF with maximal flexibility.

2.3.2 Principle of MCE

Shore and Johnson showed that maximizing entropy was a special case of minimizing cross-entropy. (See discussion (5) later for a more precise statement.) They showed that this was true for MEM2 [2.7] in 1980, and also true for MEM1 [2.8] in 1984. We are concerned with the latter in this subsection while the former will be dealt with in Sect. 3.4 concerning MEM2.

In this subsection we introduce the Principle of MCE first, then show the equivalence between MEM1 and the Principle of MCE. Finally, we discuss some issues.

Suppose we have a series (not necessarily a time series) $x = \{x_k\}^{\mathrm{T}}$. Its prior joint p.d.f. is known to be $q(x)$. There are some constraints after measuring some data. Now we want to find its posterior (joint) p.d.f. $p(x)$. The *Principle of MCE* states that *of all the p.d.f's satisfying the constraints, we should choose the one that has the minimum cross-entropy*. The *cross-entropy* is defined as

$$H_c(p, q) \overset{\triangle}{=} \int p(x) \log[p(x)/q(x)]\mathrm{d}x \ . \tag{2.3.9}$$

$H_c(p, q)$ is a measure of the "information dissimilarity" between the prior p.d.f. $q(x)$ and posterior p.d.f. $p(x)$. It can be shown that $H_c(p, q) \geq 0$, the equality holds if and only if $p(x) = q(x)$ (completely similar). Minimizing $H_c(p, q)$ is equivalent to making $p(x)$ be as similar to $q(x)$ as possible under the constraints.

Now we want to show that under the partial ACF constraints, if the signal in the time domain itself is treated as a stochastic process while the spectrum is treated as the expectation, and if prior estimate of the spectrum is uniform, then the Principle of MCE gives results that have been obtained using MEM1 in the previous two sections. At present (2.3.9) is the expression of cross-entropy in the time domain, in which x is viewed as a time series, and $q(x)$ depends on the prior estimate.

Following Approach 2 of deriving the $H1$ expression in the frequency domain in Sect. 2.1, first of all we consider one period of the zero-mean time series, $x = \{x_0, \ldots, x_{N-1}\}^{\mathrm{T}}$. Its FT is $X = \{X_0, \ldots, X_{N-1}\}^{\mathrm{T}}$ (see the definitions in (2.1.14a,b)). By variable substitution, cf. (2.1.15), we get

$$H_c = \int P(X) \log[P(X)/Q(X)] \mathrm{d}X , \tag{2.3.10}$$

where $P(X)$ and $Q(X)$ are the joint p.d.f.'s of X corresponding to $p(x)$ and $q(x)$, respectively. Note that there is no term of $\frac{1}{2} \log N$ because H_c is invariant with regard to the coordinate transformation due to $q(x)$.

Now we can state the problem directly in the frequency domain:

$$\text{Minimization}: \quad H_c \quad \text{in (2.3.10)} ,$$

$$\text{Constraints}: \quad \int P(X) \mathrm{d}X = \int Q(X) \mathrm{d}X = 1 , \tag{2.3.11}$$

$$\sum_{k=0}^{N-1} \int P(X)|X_k|^2 \mathrm{d}X \cdot e^{-2\pi i j k/N} = R^*(j) , \quad j \in D . \tag{2.3.12}$$

We use the notations

$$S_p(k) = S_p(f_k) = \int P(X)|X_k|^2 \mathrm{d}X = \sigma_{Pk}^2 ,$$

$$S_q(k) = S_q(f_k) = \int Q(X)|X_k|^2 \mathrm{d}X = \sigma_{Qk}^2 ,$$

i.e., $S_p(f)$ and $S_q(f)$ are the spectra of x when the p.d.f.'s are $p(x)$ and $q(x)$, respectively. The spectra are expressed in terms of the expectations of $|X_k|^2$ when the p.d.f.'s of X are $P(X)$ and $Q(X)$, respectively.

Introducing the Lagrange multipliers μ and λ_j, $j \in D$ for the normalization constraint (2.3.11) and the ACF constraints (2.3.12), respectively, forming the objective function G, and letting the variation $\delta G = 0$, we can get with little effort the following equation similar to (2.1.18):

$$P(X) = C_1 Q(X) \exp\left(-N \sum_{k=0}^{N-1} \Lambda_k |X_k|^2\right) , \qquad (2.3.13)$$

where $\{\Lambda_k\}$ is the FT of $\{\lambda_j\}$, $(\lambda_j = 0$ for $j \notin D)$, i.e.,

$$\Lambda_k = N^{-1} \sum_{j \in D} \lambda_j \exp(-2\pi i j k / N) .$$

$S_q(k)$ is the prior estimate of the spectrum, and its corresponding p.d.f. is

$$Q(X) = C_2 \exp\left(-\frac{1}{2} \sum_{k=0}^{N-1} \frac{1}{\sigma_{Qk}^2} |X_k|^2\right) . \qquad (2.3.14)$$

$q(x)$ is determined by the above $Q(X)$. From the normalization constraint, it follows that

$$C_2 = \prod_{k=0}^{N-1} \frac{1}{\sqrt{2\pi}\sigma_{Qk}} = \prod_{k=0}^{N-1} [2\pi S_q(k)]^{-1/2} . \qquad (2.3.15)$$

Substituting (2.3.14) in (2.3.13), we get

$$P(X) = C \exp\left[-\frac{1}{2} \sum_{k=0}^{N-1} \left(\frac{1}{S_q(k)} + 2N\Lambda_k\right) |X_k|^2\right] ,$$

$(C = C_1 C_2)$. Therefore,

$$S_p(k) = \sigma_{Pk}^2 = [1/S_q(k) + 2N\Lambda_k]^{-1} , \qquad (2.3.16)$$

and

$$C = \prod_{k=0}^{N-1} \frac{1}{\sqrt{2\pi}\sigma_{Pk}} = \prod_{k=0}^{N-1} [2\pi S_p(k)]^{1/2} .$$

From the above expression and (2.3.15), we have

$$\log C_1 = \log(C/C_2) = -\frac{1}{2} \sum_{k=0}^{N-1} \log \frac{S_p(k)}{S_q(k)} . \qquad (2.3.17)$$

Substituting (2.3.13) in (2.3.10) and utilizing expressions (2.3.17, 2.3.16), we finally obtain the expression of cross-entropy:

$$\begin{aligned}
H_c &= \int P(X) \log\left[C_1 \exp\left(-N \sum_{k=0}^{N-1} \Lambda_k |X_k|^2\right)\right] \mathrm{d}X \\
&= \log C_1 - \sum_{k=0}^{N-1} N\Lambda_k \int P(X) |X_k|^2 \mathrm{d}X \\
&= \log C_1 - \sum_{k=0}^{N-1} N\Lambda_k S_p(k) \\
&= -\frac{1}{2} \sum_{k=0}^{N-1} \log \frac{S_p(k)}{S_q(k)} + \frac{1}{2} \sum_{k=0}^{N-1} \left(\frac{S_p(k)}{S_q(k)} - 1\right) . \qquad (2.3.18)
\end{aligned}$$

Equation (2.3.18) is the discrete expression of the cross-entropy H_c in the frequency domain, in which the factor $1/2$ is usually omitted. Let us come back to the case of continuous frequency. Letting $N \to \infty$, $1/N \to \mathrm{d}f$, $\sum \to \int$, we get the cross-entropy rate

$$h_c \overset{\triangle}{=} \lim_{N \to \infty} \frac{H_c}{N} = \frac{1}{2} \int_{-1/2}^{1/2} \left(\frac{S_p(f)}{S_q(f)} - \log \frac{S_p(f)}{S_q(f)} - 1 \right) \mathrm{d}f \; . \tag{2.3.19}$$

Usually, cross-entropy rate is simply called cross-entropy, the notation H_c is still used and the factor $1/2$ is omitted. Then, (2.3.19) becomes the commonly used cross-entropy expression of continuous form

$$H_c = \int_{-1/2}^{1/2} \left(\frac{S_p(f)}{S_q(f)} - \log \frac{S_p(f)}{S_q(f)} - 1 \right) \mathrm{d}f \; . \tag{2.3.20}$$

The MCE spectrum of discrete form is, by (2.3.16) and the definition of Λ_k,

$$S_p(k) = \frac{1}{\dfrac{1}{S_q(k)} + 2 \sum_{j \in D} \lambda_j \exp(-2\pi \mathrm{i} jk/N)} \; . \tag{2.3.21}$$

The corresponding continuous form is

$$S_p(f) = \frac{1}{\dfrac{1}{S_q(f)} + 2 \sum_{j \in D} \lambda_j \exp(-2\pi \mathrm{i} jf)} \; . \tag{2.3.22}$$

The Lagrange multipliers λ_j are determined by the constraints (2.3.12). (The factor 2 may be merged into λ_j.) Note that factors 2 in the second terms of the denominators of (2.3.21, 2.3.22) stem from the factors $1/2$ in (2.3.18, 2.3.19), respectively. These factors can be omitted altogether.

Now we come to the point of discussing the equivalence between (the Principle of) MCE and MEM1. If the prior estimate $S_q(f)$ is uniform, it can be assumed to be 1 without the loss of generality. Then the cross-entropy expression (2.3.20) becomes

$$H_c = -\int_{-1/2}^{1/2} \log S_p(f) \mathrm{d}f + \int_{-1/2}^{1/2} S_p(f) \mathrm{d}f - 1$$

$$= -H1 + [R(0) - 1] \; ,$$

where the last term is a constant. It turns out, therefore, that minimizing H_c is equivalent to maximizing $H1$ that was defined in Sect. 2.1. In addition, the MCE spectrum (2.3.22) becomes (with the factor 2 being merged into λ_j and then the constant 1 into λ_0)

$$S_p(f) = \frac{1}{\sum_{j \in D} \lambda_j \exp(-2\pi \mathrm{i} jf)} \; .$$

This is just the MEM1 spectrum expression obtained in Sect. 2.2.

We discuss some issues before concluding this subsection.

(1) Without any measured data, D is an empty set. Consequently, all λ_j's are zero and it follows immediately from (2.3.22) that

$$S_p(f) = S_q(f) \, .$$

The meaning of the so-called prior estimate $S_q(f)$ of the spectrum becomes evident. Adding the measured data as constraints means "modifying" $S_q(f)$ to obtain $S_p(f)$. Now we no longer have simple formulae like (2.2.16, 2.2.14, 2.2.18) for calculating $S_p(f)$ and extrapolating the ACF. An iterative algorithm is inevitable.

(2) We can also derive the H_c expression in the frequency domain following Approach 1 in Sect. 2.1. The procedure is briefly described in the following:

Rewrite the definition of $H_c(p,q)$ in (2.3.9) as

$$H_c = \int p(\boldsymbol{x}) \log p(\boldsymbol{x}) \mathrm{d}\boldsymbol{x} - \int p(\boldsymbol{x}) \log q(\boldsymbol{x}) \mathrm{d}\boldsymbol{x} \, .$$

With the assumption of a Gaussian process, for a series of finite length $\boldsymbol{x}_{M+1} = \{x_0, \ldots, x_M\}^{\mathrm{T}}$ we have

$$\int p(\boldsymbol{x}) \log p(\boldsymbol{x}) \mathrm{d}\boldsymbol{x}$$

$$= -\frac{1}{2} \log[\det(R_p)] - \frac{1}{2}(M+1) \log(2\pi) - \frac{1}{2}(M+1) \, ,$$

cf. (2.1.5). By similar calculation we get

$$-\int p(\boldsymbol{x}) \log q(\boldsymbol{x}) \mathrm{d}\boldsymbol{x}$$

$$= \frac{1}{2} \log[\det(R_q)] + \frac{1}{2}(M+1) \log(2\pi) + \frac{1}{2} \sum_{k=0}^{M} \frac{S_p(k)}{S_q(k)} \, .$$

Combining the three expressions above, it follows that

$$H_c = -\frac{1}{2} \log[\det(R_p)] + \frac{1}{2} \log[\det(R_q)] + \frac{1}{2} \sum_{k=0}^{M} \frac{S_p(k)}{S_q(k)} - \frac{1}{2}(M+1) \, .$$

Therefore, the cross-entropy rate is

$$h_c \stackrel{\triangle}{=} \lim_{M \to \infty} \frac{H_c}{M+1}$$

$$= -\frac{1}{2} \int_{-1/2}^{1/2} \log S_p(f) \mathrm{d}f + \frac{1}{2} \int_{-1/2}^{1/2} \log S_q(f) \mathrm{d}f + \frac{1}{2} \int_{-1/2}^{1/2} \frac{S_p(f)}{S_q(f)} \mathrm{d}f$$

$$-\frac{1}{2} \, ,$$

which is equal to h_c in (2.3.19).

(3) The cross-entropy expression in the frequency domain, (2.3.20), is fairly simple. We are not surprised to see that the procedure is also simple if

(2.3.20) is used to find the MCE solution. Specifically, the statement of the problem is

$$\text{Minimization}: \qquad H_c \quad \text{in } (2.3.20),$$

$$\text{Constraints}: \qquad \int_{-1/2}^{1/2} S_p(f)e^{-2\pi ijf}\,df = R^*(j), \quad j \in D.$$

Introducing the Lagrange multipliers λ_j, $j \in D$ for the constraints, forming the objective functional

$$G = H_c + \sum_{j \in D} \lambda_j \left[\int S_p(f)e^{-2\pi ijf}\,df - R^*(j) \right],$$

and letting $\delta G = 0$, it is easily shown that the MCE solution is

$$S_p(f) = \frac{1}{\dfrac{1}{S_q(f)} + \sum_{j \in D} \lambda_j \exp(-2\pi ijf)}.$$

It's done so easily! But we should not forget the hard work in deriving the H_c expression in the frequency domain while appreciating the simplicity of the procedure above.

(4) Since the expression of H_c explicitly includes the prior estimate $S_q(f)$, one may believe that the MCE solution $S_p(f)$ should be superior to the MEM1 solution in Sect. 2.2. This is true if and only if the prior estimate $S_q(f)$ catches important shape features of the *unknown* true spectrum. In this case, these features will appear in the MCE spectrum $S_p(f)$ even if no measured data are available, and will be enhanced by the data collected. In contrast, all features in the MEM1 spectrum comes from the data: no data, no feature.

With more and more data, the features in the MCE spectrum $S_p(f)$ from the data become more and more prominent, but the features introduced to the MCE spectrum through the prior estimate $S_q(f)$ will never be swept out completely. Consequently, once any wrong feature has been included in the prior estimate $S_q(f)$, it will not be eliminated completely even if a large number of ACFs are available.

Experiments [2.9] support the above argument, and lead to the following suggestion in spectral analysis: It is safer to use MEM than MCE unless it is sure that the shape information of the *unknown* true spectrum is correctly caught by the prior spectrum.

(5) Finally, we would like to remark on the statement: MEM (including MEM1 and MEM2) is a special case of (the Principle of) MCE (the prior $S_q(f)$ being uniform). As shown in Sect. 1.2, in the continuous case the entropy expression $H(p(x))$ must contain the measure $m(x)$ to ensure the invariance of H with regard to coordinate transformation:

$$H(p(x)) = - \int p(x) \log[p(x)/m(x)]\,dx,$$

cf. (1.3). Compared with the cross-entropy H_c in (2.3.9), the entropy H here has the same form but different sign. Provided that $m(x)$ is interpreted as the prior p.d.f., maximizing H and minimizing H_c are one and the same, and MEM and MCE are equivalent to each other. Quite often in the MEM (especially MEM1) literature, $m(x)$ is ignored and hence the procedure of solution is simplified. But this results in the popularity of the statement that MEM is a special case of MCE, which is not quite correct. If we call MEM without $m(x)$ *plain* MEM, and call MEM with $m(x)$ *proper* MEM, then the correct statements should be: Proper MEM and MCE are equivalent, and plain MEM is a special case of MCE. The latter is trivial because it is equivalent to the statement that plain MEM is a special case of proper MEM. In fact, with the prior in the entropy definition in mind, researchers of the MEM2 school have never been concerned with MCE. We first described plain MEM in Sects. 2.1, 2.2 in consideration of the common understanding of MEM (without the prior), the possibility of finding the explicit solution and this solution is still in wide use. Hereafter when we talk about MEM, whether plain MEM or proper MEM is referred to depends on the context.

2.3.3 AR Process (Signal Model)

A comprehensive description of AR process is beyond the scope of this monograph. We introduce only some relevant results [2.5, 2.1].

An AR process of order M is represented by

$$x(k) = \alpha_1 x(k-1) + \alpha_2 x(k-2) + \cdots + \alpha_M x(k-M) + \sigma w(k) \, . \quad (2.3.23)$$

From this expression we see that the signal value at time k is a linear combination of the M previous values plus an error term (with variance σ^2). Hence the name *linear prediction* (LP). $w(k)$ is a zero-mean Gaussian stochastic process, and

$$\mathrm{E}[w(i)w^*(j)] = \begin{cases} 1 \, , & i = j \, , \\ 0 \, , & i \neq j \, , \end{cases}$$

i.e., $w(k)$ is white noise with unit power. $w(k)$ and $x(k-n)$ are orthogonal for $n > 0$:

$$\mathrm{E}[w(k)x^*(k-n)] = 0 \, , \quad n > 0 \, .$$

Sometimes $\sigma w(k)$ is called the *innovation* of $x(k)$.

Multiplying the two sides of (2.3.23) by $x^*(k-n)$ and taking the expectation (average), utilizing the two expressions above and the definition of ACF, $R(n) = \mathrm{E}[x(k)x^*(k-n)]$, we get

$$\begin{aligned} R(n) \;=\; & \alpha_1 R(n-1) + \alpha_2 R(n-2) + \cdots + \alpha_M R(n-M) \\ & + \sigma^2 \delta_n \, , \quad n \geq 0 \, . \end{aligned} \quad (2.3.24)$$

Letting $n = 1, 2, \ldots, M+1$, we get $M+1$ equations:

$$
\begin{bmatrix}
R(1) & R(0) & R(-1) & \cdots & R(-M+1) \\
R(2) & R(1) & R(0) & \cdots & R(-M+2) \\
\cdots & \cdots & \cdots & \cdots & \cdots \\
\cdots & \cdots & \cdots & \cdots & \cdots \\
R(M) & R(M-1) & R(M-2) & \cdots & R(0) \\
R(M+1) & R(M) & R(M-1) & \cdots & R(1)
\end{bmatrix}
\begin{bmatrix}
1 \\ -\alpha_1 \\ -\alpha_2 \\ \cdot \\ \cdot \\ \cdot \\ -\alpha_M
\end{bmatrix}
$$

$$
=
\begin{bmatrix}
0 \\ \cdot \\ \cdot \\ \cdot \\ 0
\end{bmatrix} .
\tag{2.3.25}
$$

These are called the *Yule-Walker equations*. These equations can be written in another form:

$$
\begin{bmatrix}
R(0) & R(-1) & \cdots & R(-M+1) \\
R(1) & R(0) & \cdots & R(-M+2) \\
\cdots & \cdots & \cdots & \cdots \\
\cdots & \cdots & \cdots & \cdots \\
R(M) & R(M-1) & \cdots & R(1)
\end{bmatrix}
\begin{bmatrix}
\alpha_1 \\ \alpha_2 \\ \cdot \\ \cdot \\ \cdot \\ \alpha_M
\end{bmatrix}
$$

$$
=
\begin{bmatrix}
R(1) \\ \cdot \\ \cdot \\ \cdot \\ R(M+1)
\end{bmatrix} .
$$

If $\alpha_1, \ldots, \alpha_M$ are viewed as M unknowns, then (2.3.25) is an overdetermined system of linear equations. Hence, the coefficient determinant must be zero:

$$
\det
\begin{bmatrix}
R(1) & R(0) & \cdots & R(-M+1) \\
R(2) & R(1) & \cdots & R(-M+2) \\
\cdots & \cdots & \cdots & \cdots \\
\cdots & \cdots & \cdots & \cdots \\
R(M) & R(M-1) & \cdots & R(0) \\
R(M+1) & R(M) & \cdots & R(1)
\end{bmatrix}
= 0 .
\tag{2.3.26}
$$

This equation can be used as a formula to extrapolate $R(M+1)$, i.e., to calculate $R(M+1)$ from $R(0), \ldots, R(M)$. Equation (2.3.26) is the same as (2.3.7), an equation the extrapolated $R(M+1)$ must satisfy in MEM1. Therefore, we can conclude immediately that the ACF extension for an AR process is the same as that in MEM1; the spectra are the same, of course. In such a way we have established the equivalence between MEM1 and AR process.

We can also show that the formulae for extrapolating $R(M+1)$ derived from (2.3.25) are indeed the same as (2.3.6, 2.3.5) ($n=1$) in MEM1. One way for doing this is to solve the first M equations in (2.3.25) for the coefficients

$\alpha_1, \ldots, \alpha_M$ from the given $R(0), \ldots, R(M)$. The last equation is the formula for extrapolating $R(M+1)$.

Another easier way is by manipulating the coefficient matrix in (2.3.25). Comparing (2.3.24) with (2.3.5, 2.3.6), it is easy to see that $-\alpha_k = a_k$, $k = 1, \ldots, M$; $\sigma^2 = p_M$. In (2.3.25) eliminating the last row of the matrix and adding one row $[R(0) \quad R(-1) \quad \ldots \quad R(-M)]$ at the top, we get the equations

$$
\begin{bmatrix}
R(0) & R(-1) & \cdots & R(-M) \\
R(1) & R(0) & \cdots & R(-M+1) \\
\cdots & \cdots & \cdots & \cdots \\
\cdots & \cdots & \cdots & \cdots \\
R(M) & R(M-1) & \cdots & R(0)
\end{bmatrix}
\begin{bmatrix}
1 \\
a_1 \\
\cdot \\
\cdot \\
\cdot \\
a_M
\end{bmatrix}
=
\begin{bmatrix}
\sigma^2 \\
0 \\
\cdot \\
\cdot \\
\cdot \\
0
\end{bmatrix}, \quad (2.3.27)
$$

where, as mentioned before,

$$
\sigma^2 = \sum_{n=0}^{M} a_n R(-n) = p_M . \tag{2.3.28}
$$

Thus, (2.3.27) and (2.3.5) in MEM1 are one and the same. The formula for the extension is, from the last equation in (2.3.25),

$$
R(M+1) = \sum_{k=1}^{M} \alpha_k R(M+1-k) .
$$

This is the same as (2.3.6) $(n = 1)$.   Q.E.D.

We may look into the equivalence between MEM1 and the AR process in another way. Taking the Z-transform (ZT) of (2.3.23) representing an AR process, we get

$$
X(z) = \alpha_1 X(z)z^{-1} + \alpha_2 X(z)z^{-2} + \cdots + \alpha_M X(z)Z^{-M} + \sigma W(z) ,
$$

where $W(z)$ and $X(z)$ are the ZTs of $w(k)$ and $x(k)$, respectively. From the above equation it follows that

$$
X(z) = \frac{\sigma W(z)}{1 - \sum_{n=1}^{M} \alpha_k z^{-k}} . \tag{2.3.29}
$$

Therefore, the spectrum of the AR process is

$$
S(f) = |X(z)|^2_{z=e^{2\pi i f}} = \frac{\sigma^2}{\left| \sum_{k=0}^{M} (-\alpha_k) \exp(-2\pi i f k) \right|^2} ,
$$

$(\alpha_0 = -1)$, which is the same as the MEM1 spectrum (2.2.14) with $-\alpha_k = a_k$, $k = 0, 1, \ldots, M$ and $\sigma^2 = p_M$. Hence, a_k and p_M are often called the *AR parameters*.

By *signal model* we mean a filter whose output spectrum is the same as that of the signal when the filter is driven by a write Gaussian noise of unit power, as shown in Fig. 2.5.

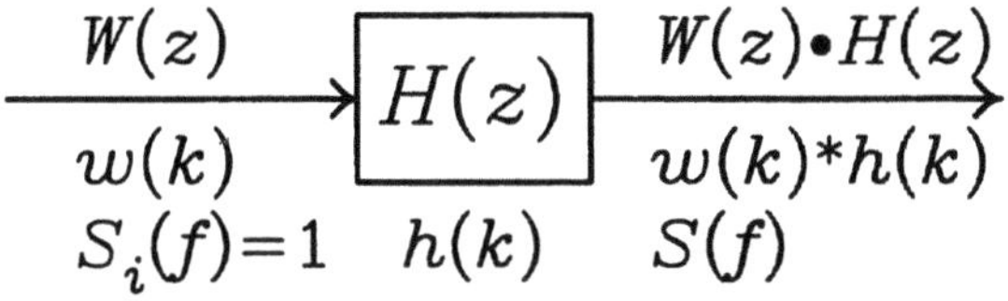

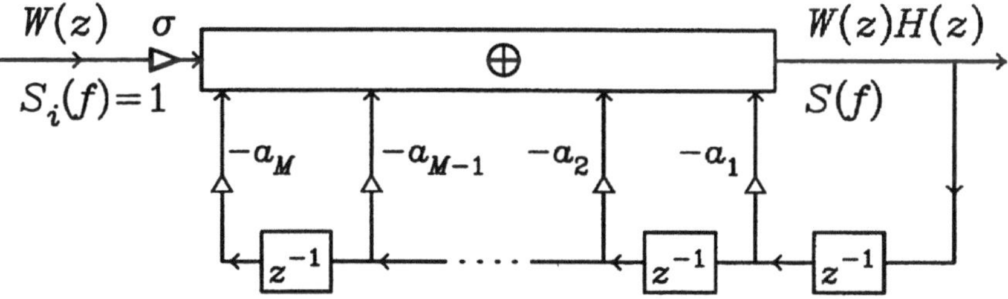

Fig. 2.5. The model of a signal having the spectrum $S(f)$

A filter is characterized by its transfer function $H(z)$ or impulse response $h(k)$. With this understanding we simply call $H(z)$ the signal model. Given the ZT of the signal, $X(z)$, then the output is

$$W(z) \cdot H(z) = X(z) \ .$$

Since the spectrum of the input $S_i(f) = 1$, the spectrum of the output is

$$S(f) = |X(z)|^2_{z=e^{2\pi i f}} = |H(z)|^2_{z=e^{2\pi i f}} \ .$$

We see that we can simply take $X(z)$ as $H(z)$ with the understanding that the factor $W(z)$ in $X(z)$, if there is one, is left out.

To sum up, finding the model of a signal means calculating the ZT of the signal, $X(z)$. Hereafter we simply call $X(z)$ the signal model.

It follows immediately from (2.3.29) that the signal model of order M of MEM1 or an AR process is

$$X(z) = \frac{\sigma}{\displaystyle\sum_{k=0}^{M} a_k z^{-k}} \ , \quad (a_0 = 1) \ . \tag{2.3.30}$$

Here $\sigma^2 = p_M$. The filter is shown in Fig. 2.6. This is a recursive filter of order M. With the above understanding, a_k is also called the (*inverse*) *filter coefficients*, and p_M the *innovation* of $x(k)$. (The AR filter becomes a

Fig. 2.6. AR filter of order M. z^{-1} denotes unit delay in time

whitening filter when operating in the opposite direction. Hence the name inverse filter.)

Expression (2.3.30) shows that the signal model of an AR process is an all-pole one. If $x(k)$ is corrupted by observational noise and becomes $y(k)$, then $Y(z)$ has zeros as well as poles. In this case the model of the noisy signal is an ARMA (AR moving-average) one. It is not discussed here.

2.3.4 Bayesian Method

We have so far dealt with MEM1 with noiseless data, assuming that the measured ACF was incomplete but exact. Now we turn our attention to the case of noisy data. Suppose the measured ACF is incomplete and inexact, i.e., the given data are

$$D(k) = R(k) + N_k , \quad k \in K . \tag{2.3.31}$$

Note that either $R(k)$ or N_k is unobservable.

It has been pointed out in Sect. 1.1 that in this case our strategy is to relax the constraints in maximizing entropy. Specifically, MEM1 with noisy data is formulated as follows:

$$\text{Maximization}: \quad H1 = \sum_{j=0}^{N-1} \log S(j) ,$$

$$\text{Constraint}: \quad \chi^2 = \sum_{k \in K} |D(k) - R(k)|^2 / \sigma_k^2 = C_e .$$

Here the discrete form of $H1$ is used, and the noise

$$N_k = D(k) - R(k) , \quad k \in K \tag{2.3.32}$$

is assumed to be zero-mean Gaussian with variance σ_k^2. Consequently, the statistic in the constraint is χ^2-distributed. Its value is required to be some critical value, say the expectation (the number of datum points).

Now we can form the objective function to be maximized:

$$G = \sum_{j=0}^{N-1} \log S(j) - \lambda \sum_{k \in K} |D(k) - R(k)|^2 / \sigma_k^2 , \tag{2.3.33}$$

where the relative weight of the maximization of the entropy to the data fit depends on the Lagrange multiplier λ.

Before showing the equivalence between MEM1 and the Bayesian method in spectral estimation, we should introduce the Bayesian method first. Spectral estimation can be viewed as a problem of statistical inference. With the incomplete and noisy data $D = \{D(k)\}^{\mathrm{T}}$, $k \in K$, our task is to estimate the "true" spectrum $S = \{S(j)\}^{\mathrm{T}}$. It is natural that we should choose the estimate S that has the maximum posterior (joint) probability density

$p(S|D)$. This is the so-called *MAP* (Maximum A Posteriori probability density) method.

By Bayes' theorem, the posterior p.d.f. is

$$p(\hat{S}|D) = \frac{p(D|\hat{S})P(\hat{S})}{\sum_{(S)} p(D|S)p(S)} = \frac{p(D|\hat{S})p(\hat{S})}{p(D)} \; ,$$

where $\hat{S}$ is some estimate, the summation in the denominator is taken over all the possible S, resulting in a constant $p(D)$ independent of $\hat{S}$. Ignoring $p(D)$ that is of no significance in the maximization of $p(\hat{S}|D)$, the above expression may be simply written as

$$p(\hat{S}|D) \propto p(D|\hat{S})p(\hat{S}) \; . \tag{2.3.34}$$

For simplicity the symbol S is used instead of $\hat{S}$ in the following.

1. Determine the Conditional p.d.f. $p(D|S)$

From (2.3.32) and the assumption of zero-mean white Gaussian noise, we get

$$p(D|S) = K_1 \exp\left[-\frac{1}{2} \sum_{k \in K} |D(k) - R(k)|^2 / \sigma_k^2\right] \; , \tag{2.3.35}$$

where K_1 is a constant.

2. Choose the Prior p.d.f. $p(S)$

Generally speaking, the choice is quite arbitrary [2.10]. This is a dilemma facing everyone and no universal solution has been found. Our aim is quite modest. We just want to show that if we choose

$$p(S) \propto \prod_{j=0}^{N-1} S^\beta(j) = \exp\left[\beta \sum_{j=0}^{N-1} \log S(j)\right] = e^{\beta H1} \; , \quad (S(j) \geq 0) \tag{2.3.36}$$

(the parameter $\beta > 0$), then we have the following equivalence relationships:

Maximizing $p(S|D)$

$$\iff \text{Max} \quad \log p(S|D)$$

$$\iff \text{Max} \quad \log p(D|S) + \log p(S) \; , \quad (\,\text{by } (2.3.34)\,)$$

$$\iff \text{Max} \quad -\frac{1}{2} \sum_{k \in K} |D(k) - R(k)|^2 / \sigma_k^2 + \beta \sum_{j=0}^{N-1} \log S(j) \; ,$$
$$(\,\text{by } (2.3.35, \, 2.3.36)\,)$$

$$\iff \text{Max} \quad \sum_{j=0}^{N-1} \log S(j) - \alpha \sum_{k \in K} |D(k) - R(k)|^2 / \sigma_k^2 \; .$$

Here $R(k) = \text{IFT}[S(j)]$, $\alpha = 1/(2\beta)$. The relative weight of the prior p.d.f. to the conditional p.d.f. depends on α.

The last expression in the above is just the same as (2.3.33) with α in place of λ. Provided that the data constraints are the same, the estimated $\hat{S}$ must also be the same by maximizing the last expression and by (2.3.33). That is to say, MEM1 with noisy data (ACF) is equivalent to the Bayesian method in spectral estimation if the prior p.d.f. $p(S)$ is chosen according to (2.3.36). In Sect. 3.4, we will see that MEM2 with noisy data can also be fitted into the Bayesian framework by appropriately choosing $p(S)$.

2.3.5 Wiener Filter and Approximation Theoretic Approach

Now we revert to MEM1 with noiseless data. Two more equivalents are briefly introduced.

1. Wiener Filter

Consider that a time series x_n is input to a linear digital filter of order M. We want to predict x_{n+1} using this filter and require that the mean square of the prediction error be minimum, i.e.,

$$p_M = \mathrm{E}[|e_n|^2] = \min , \tag{2.3.37}$$

where the prediction error, in turn, is defined as

$$e_n = x_{n+1} + \sum_{k=1}^{M} a_k x_{n-k+1} .$$

The prediction filter coefficients a_k may be determined by the criterion of least mean square, i.e., (2.3.37). It turns out that a_k and p_M satisfy the normal equations in MEM1, and the estimated spectrum and ACF extension are the same as those in MEM1. That is to say, MEM1 is equivalent to the filter defined above. For this reason, a_k and p_M are also called the *coefficients and output power of the optimal prediction-error filter.*

As a matter of fact, the filter is a simplified version of the Wiener filter. For the theory of the Wiener filter the reader is referred to, for example, [2.1].

2. Approximation Theoretic Approach

Consider the reciprocal of the spectrum, $S^{-1}(f)$. Suppose it is approximated by a polynomial of $z = \mathrm{e}^{2\pi i f}$, and this polynomial can be factorized so that the estimated $S^{-1}(f)$ is of the form

$$S^{-1}(f) = p_M^{-1} \left| \sum_{k=0}^{M} a_k z^{-k} \right|^2 = p_M^{-1} \left| \sum_{k=0}^{M} a_k \exp(-2\pi i f k) \right|^2 , \tag{2.3.38}$$

where the scale factor p_M is used to ensure $a_0 = 1$. If we require the following weighted error be minimum, i.e.,

$$\int_{-1/2}^{1/2} \left| S^{-1}(f) - p_M^{-1} \sum_{k=0}^{M} a_k \exp(-2\pi i f k) \right|^2 S(f) \mathrm{d}f = \min , \tag{2.3.39}$$

then it can be shown that a_k and p_M must satisfy the normal equations (2.3.5) in MEM1. Noticing that (2.3.38) is just the formula for calculating the MEM1 spectrum, we conclude that this particular reciprocal-spectrum approximation is equivalent to MEM1 [2.11]. One of the merits of this approach is that a prior estimate $S_0(f)$ of the spectrum can be incorporated into the formulation by introducing an appropriate weight, so that the estimate can be improved. This is done by replacing $S^{-1}(f)$ with $S_0(f)S^{-1}(f)$ in (2.3.39).

2.4 Algorithms and Numerical Example (Given ACF)

In Sect. 2.2, we have obtained the formulae of the explicit solution, (2.2.16, 2.2.14, 2.2.18), for 1-D MEM1 with noiseless data, the partial ACF $R(k)$, $|k| \leq M$. When MEM1 is used for spectral analysis, in most cases the measured data are a segment of the time series, but not its ACF. The formulae mentioned above cannot be used in a straightforward way even in the case of 1-D noiseless data. The situation must be much more difficult in the case of multidimensional and/or noisy data.

Many algorithms have been developed in order to cope with a wide variety of situations in practice. Priority is given to certain performances of algorithms, such as computational speed, sensitivity to noise, easiness of order selection, and even the required space of core memory on a computer and convenience for hardware implementation. An exhaustive investigation into the algorithms is impossible and unnecessary. Our strategy is to describe some basic algorithms in considerable detail and give references. In this way the reader may study material of interest with little difficulty after understanding the basic algorithms.

The MEM1 algorithms are divided into two categories. The first one is applicable to the case where the input data are the ACF, while the second one to the case where the input data are the time series itself. They constitute the subjects of this and the next sections, respectively. Section 2.6 is devoted particularly to the problem of order selection in MEM1.

As mentioned at the beginning of this section, the measured data are usually a segment of the time series. In this case to use the MEM1 algorithms with the ACF as input data to be described in this section, the ACF must be estimated first, normally by time average. Moreover, because the length of the measured segment of the time series is always finite, the unmeasured part of the time series has to be set to zero in the ACF estimation. This violates the stipulation of nonzero extension in MEM. Nevertheless, provided that the segment of the time series is not too short, the scheme in which the ACF is estimated prior to employing MEM1 is usable. It can be expected that in the resolution of spectra this scheme is superior to the correlogram (Fourier transforming the estimated ACF), although inferior to the scheme in which the segment of the time series is used as input to MEM1.

Four common methods are available for estimating the elements $r(i,j)$ of the ACF matrix from the segment of the time series, $\boldsymbol{x} = \{x_1,\ldots,x_N\}^{\mathrm{T}}$ [2.1]. In the following, M denotes the filter order, i.e., the maximum lag of the ACF. How to determine M will be discussed in Sect. 2.6. For the time being we have only to know $M < N$, and normally $M = 0.05 \sim 0.2N$.

1.

$$\begin{cases} R(k) &= \dfrac{1}{N-k} \displaystyle\sum_{n=1}^{N-k} x_{n+k}x_n^* \,, \quad k = 0,1,\ldots,M \,, \\ R(-k) &= R^*(k) \,, \\ r(i,j) &= R(i-j) \,. \end{cases} \tag{2.4.1}$$

This is an unbiased estimation. The resultant ACF matrix possesses the Hermitian property (or Hermitian symmetry as sometimes called) and Toeplitz structure, but is not necessarily nonnegative definite.

2.

$$\begin{cases} R(k) &= \dfrac{1}{N} \displaystyle\sum_{n=1}^{N-k} x_{n+k}x_n^* \,, \quad k = 0,1,\ldots,M \,, \\ R(-k) &= R^*(k) \,, \\ r(i,j) &= R(i-j) \,. \end{cases} \tag{2.4.2}$$

This is a biased estimation. But the variance is smaller compared with that of (2.4.1). The resultant ACF matrix possesses the Hermitian property and Toeplitz structure, and is nonnegative definite.

3.

$$\begin{cases} R(k) &= \dfrac{1}{2(N-M)} \displaystyle\sum_{n=1}^{N-M} \left(x_{n+k}x_n^* + x_{n+M}x_{n+M-k}^* \right) \,, \\ & \qquad\qquad\qquad k = 0,1,\ldots,M \,, \\ R(-k) &= R^*(k) \,, \\ r(i,j) &= R(i-j) \,. \end{cases} \tag{2.4.3}$$

This is an unbiased estimation, and is the same as (2.4.2) otherwise.

4.

$$r(i,j) = \frac{1}{2(N-M)} \sum_{n=1}^{N-M} \left(x_{n+i}x_{n+j}^* + x_{n+M-i}^*x_{n+M-j} \right) \,,$$
$$i,j = 0,1,\ldots,M \,. \tag{2.4.4}$$

This is an unbiased estimation. The resultant ACF matrix possesses the Hermitian property and Hermitian persymmetry ($r(i,j) = r^*(M-i, M-j)$), but no Toeplitz structure (ACF matrix of a nonstationary process), and is nonnegative definite. This is derived on the basis of the criterion of the least sum of the forward and backward prediction error energies (Sect. 2.5.2).

Normally (2.4.2), (2.4.3) or (2.4.4), but not (2.4.1), is used to estimate the ACF. All these formulae can be easily adapted for use in the 2-D case.

Now the input data may be considered as the ACF, which may be obtained by measurement or estimated by the above methods. In this section we describe three MEM1 algorithms, then illustrate a numerical example.

2.4.1 Levinson's Recursion for 1-D Noiseless Data

From Sect. 2.2 we know that the key to the spectral estimation in MEM1 lies in the solution of the normal equations

$$\sum_{k=0}^{M} a_k R(n-k) = p_M \delta_n \,, \quad (a_0 = 1) \,, \quad n = 0, 1, \ldots, M \,. \tag{2.4.5}$$

Once the parameters a_k and p_M have been obtained, calculating $S(f)$ is an easy job. Here by algorithm we mean the one for solving (2.4.5).

Equations (2.4.5) consist of $M+1$ linear equations in $M+1$ unknowns. Its coefficient determinant is positive (the degeneracy case not being under consideration), so the solution exists and is unique. Of course, general methods, e.g., the Gaussian elimination, can be used. However, we can take advantage of the properties of the ACF matrix to improve computational efficiency.

Algorithms and computer programs for solving general and special systems of linear equations can be found in many books. But the simplest thing to do may be to call subroutines on a computer. For example, the NAG and IMSL libraries have a wide variety of such subroutines. They are handy for use. But one should be careful in selecting particular subroutines according to the particular patterns of the coefficient matrix. For instance, if the ACF is estimated by (2.4.4), then Levinson's recursive algorithm [2.1] to be derived in the following is inapplicable.

If we need to solve (2.4.5) only once and not for real time application, the computational speed is not important. On the other hand, if the computation is for real time application or needs to be carried out for the order M from small to large in test, then it is better to use recursive algorithms. Levinson's recursion is a common one, which is efficient and is fundamental to some other recursive algorithms. The central task in this subsection is to derive Levinson's recursion for solving the normal equations (2.4.5).

For convenience (2.4.5) is written in matrix notation:

$$R_M A_M = E_M \,, \tag{2.4.6}$$

where

$$R_M = \begin{bmatrix} R(0) & R(-1) & \cdots & R(-M) \\ \cdots & \cdots & \cdots & \cdots \\ \cdots & \cdots & \cdots & \cdots \\ R(M) & R(M-1) & \cdots & R(0) \end{bmatrix} \,,$$

$$A_M = (1, a_{M,1}, \ldots, a_{M,M})^{\mathrm{T}} \,,$$
$$E_M = (p_M, 0, \ldots, 0)^{\mathrm{T}} \,,$$

M being the filter order. Note that a_k has been replaced by $a_{M,k}$.

The operator I is defined as reversing and then conjugating a vector when I operates on it:

$$A_M^{\mathrm{I}} = (a_{M,M}^*, \ldots, a_{M,1}^*, 1)^{\mathrm{T}},$$
$$E_M^{\mathrm{I}} = (0, \ldots, 0, p_M)^{\mathrm{T}}.$$

From (2.4.6) and the Hermitian property of R_M, it follows that

$$R_M A_M^{\mathrm{I}} = E_M^{\mathrm{I}}. \tag{2.4.7}$$

The procedure to obtain (2.4.7) is as follows:

(1) Exchange the jth and $(M+2-j)$th columns, $j \le \frac{M+1}{2}$ in the coefficient matrix of (2.4.6). This operation can be visualized as rotating the matrix about its vertical axis of symmetry by $180°$. Accordingly, the vector A_M is reversed. The resultant equations may be written in a simplified form:

$$\begin{bmatrix} R(-M) & \cdots & R(0) \\ & & \\ & \ddots & \\ & & \\ R(0) & \cdots & R(M) \end{bmatrix} \begin{bmatrix} a_{M,M} \\ \vdots \\ \vdots \\ \vdots \\ 1 \end{bmatrix} = \begin{bmatrix} p_M \\ 0 \\ \vdots \\ \\ 0 \end{bmatrix}.$$

(2) Then exchange the ith and $(M+2-i)$th rows, $i \le \frac{M+1}{2}$ in the coefficient matrix above. This operation can be visualized as rotating the matrix about its horizontal axis of symmetry by $180°$. Accordingly, the vector E_M is reversed. The resultant equations are

$$\begin{bmatrix} R(0) & \cdots & R(M) \\ & & \\ & \ddots & \\ & & \\ R(-M) & \cdots & R(0) \end{bmatrix} \begin{bmatrix} a_{M,M} \\ \vdots \\ \vdots \\ \vdots \\ 1 \end{bmatrix} = \begin{bmatrix} 0 \\ \vdots \\ \vdots \\ 0 \\ p_M \end{bmatrix}.$$

(3) Finally, take complex conjugation of the two sides of the above equations. From the fact that p_M is real and $R^*(k) = R(-k)$, (2.4.7) results.

Utilizing the Toeplitz structure of the ACF matrix, we get

$$R_{M+1} = \left[\begin{array}{ccc|c} & & & R(-M-1) \\ & & & \vdots \\ & R_M & & \vdots \\ & & & \\ \hline R(M+1) & \cdots & \cdots & R(0) \end{array} \right].$$

Utilizing (2.4.6) we get

$$R_{M+1} \begin{bmatrix} A_M \\ 0 \end{bmatrix} = \begin{bmatrix} E_M \\ \Delta_{M+1} \end{bmatrix}, \tag{2.4.8}$$

where

$$
\begin{aligned}
\Delta_{M+1} &= \sum_{k=0}^{M} a_{M,k} R(M+1-k) \\
&= R(M+1) - \left[-\sum_{k=1}^{M} a_{M,k} R(M+1-k) \right] ,
\end{aligned}
\qquad (2.4.9)
$$

which is just the prediction error of $R(M+1)$ in MEM1, cf. (2.3.6). Similarly, utilizing (2.4.7) we get

$$
R_{M+1} \begin{bmatrix} 0 \\ A_M^{\mathrm{I}} \end{bmatrix} = \begin{bmatrix} \Delta_{M+1}^* \\ E_M^{\mathrm{I}} \end{bmatrix} .
\qquad (2.4.10)
$$

From (2.4.8, 2.4.10) we have

$$
R_{M+1} \left\{ \begin{bmatrix} A_M \\ 0 \end{bmatrix} + \rho_{M+1} \begin{bmatrix} 0 \\ A_M^{\mathrm{I}} \end{bmatrix} \right\} = \begin{bmatrix} p_M \\ 0 \\ \vdots \\ 0 \\ \Delta_{M+1} \end{bmatrix} + \rho_{M+1} \begin{bmatrix} \Delta_{M+1}^* \\ 0 \\ \vdots \\ 0 \\ p_M \end{bmatrix} ,
$$

where ρ_{M+1} is a new parameter. This system of equations is compared with the normal equations of order $M+1$ in the original form:

$$
R_{M+1} A_{M+1} = E_{M+1} = \begin{bmatrix} p_{M+1} \\ 0 \\ \vdots \\ 0 \end{bmatrix} .
$$

Then we get the recursive formula for A_{M+1}:

$$
A_{M+1} = \begin{bmatrix} A_M \\ 0 \end{bmatrix} + \rho_{M+1} \begin{bmatrix} 0 \\ A_M^{\mathrm{I}} \end{bmatrix} ,
$$

i.e.,

$$
\begin{cases}
a_{M+1,k} = a_{M,k} + \rho_{M+1} a_{M,M+1-k}^* , & 1 \le k \le M , \\
a_{M+1,0} = a_{M,0} = 1 , \quad a_{M+1,M+1} = \rho_{M+1} ;
\end{cases}
\qquad (2.4.11)
$$

and the recursive formulae for p_{M+1}:

$$
\begin{bmatrix} p_{M+1} \\ 0 \\ \vdots \\ 0 \end{bmatrix} = \begin{bmatrix} p_M \\ 0 \\ \vdots \\ 0 \\ \Delta_{M+1} \end{bmatrix} + \rho_{M+1} \begin{bmatrix} \Delta_{M+1}^* \\ 0 \\ \vdots \\ 0 \\ p_M \end{bmatrix} ,
$$

i.e.,

$$\begin{cases} p_{M+1} &= p_M + \rho_{M+1}\Delta^*_{M+1} , \\ 0 &= \Delta_{M+1} + \rho_{M+1}p_M . \end{cases}$$

Solving the above equations, we get

$$\rho_{M+1} = -\frac{\Delta_{M+1}}{p_M} , \tag{2.4.12}$$

$$p_{M+1} = p_M(1 - |\rho_{M+1}|^2) . \tag{2.4.13}$$

$|\rho_{M+1}| \leq 1$ since $p_M \geq 0$ for all M.

Substituting $\Delta_{M+1} = -\rho_{M+1}p_M$ in (2.4.9), we have

$$R(M + 1) = -\sum_{k=1}^{M} a_{M,k}R(M + 1 - k) - \rho_{M+1}p_M .$$

Here $R(M + 1)$ as well as $R(n)$, $|n| \leq M$ is known and ρ_{M+1} is to be calculated. On the other hand, if only $R(n)$, $|n| \leq M$ are known, then the above expression can be viewed as a formula for extrapolating $R(M + 1)$. ρ_{M+1} is unknown, but $|\rho_{M+1}| \leq 1$ must hold to ensure the nonnegativity of the ACF matrix. Therefore, above expression and (2.3.8) are one and the same. The parameter ρ_{M+1} here is just the reflection coefficient mentioned in Sect. 2.3.1.

Levinson's recursion with the given ACF is summarized as follows:

1) The recursion begins with $M = 0$. $p_0 = R(0)$ by (2.4.6).

2) Use successively (2.4.9, 2.4.12, 2.4.13) to calculate Δ_{M+1}, ρ_{M+1} and p_{M+1}.

3) Use (2.4.11) to calculate $a_{M+1,k}$, $k = 1, \ldots, M + 1$.

4) Repeat Steps 2, 3 above to calculate $a_{1,k}$, $a_{2,k}, \ldots$ and p_1, $p_2, \ldots$. The recursion is terminated when the required filter order is reached. The maximum order possible depends on the maximum lag of the measured or estimated ACF.

5) Calculate the MEM1 spectrum by (2.2.14).

Usually, (2.4.2) or (2.4.3) is used to estimate the ACF matrix for utilizing Levinson's recursion. The Hermitian property and Toeplitz structure of the matrix ensure that Levinson's recursion holds. The nonnegativity of the matrix ensures that the moduli of all the reflection coefficients are not greater than 1. The computational complexity of Levinson's recursion is of a magnitude of M^2. (It is M^3 for matrix inversion by a general algorithm.)

A FORTRAN source program list is presented in [2.12], which was designed to accomplish two tasks: (1) estimating the ACF from the input, a real-valued time series, by the second method presented at the beginning of this section; (2) solving the Yule-Walker equations for the parameters $a_{M,k}$ by Levinson's recursion. The result is the same as the solution of the normal equations (Sect. 2.3.3). A FORTRAN source program list and a subroutine on a floppy disk for Levinson's recursion with a complex-valued ACF as input are presented in [2.13].

2.4.2 Lim-Malik Algorithm for 2-D Noiseless Data

Suppose the given data are partial 2-D noiseless ACF. In this case no normal equations are available. The AR parameters must be calculated by iterative algorithms. One of them is the Lim-Malik (LM) algorithm [2.14], which is useful in studying the properties of the 2-D MEM1 spectral estimation. Moreover, some of its basic ideas and tricks can be used in 2-D MEM2 algorithms.

In order to describe the LM algorithm, first of all the statement of the MEM1 problem should be changed. The original statement is:

2-D MEM1: Given the ACF $R(n_1, n_2)$, $(n_1, n_2) \in D$, estimate the spectrum $S(f_1, f_2)$ such that

$$H1 = \int \int_{-1/2}^{1/2} \log S(f_1, f_2) \mathrm{d}f_1 \mathrm{d}f_2$$

is maximized, and the constraints

$$\begin{aligned} R(n_1, n_2) &= \mathrm{IFT}[S(f_1, f_2)] \\ &= \int \int_{-1/2}^{1/2} S(f_1, f_2) \exp[2\pi\mathrm{i}(f_1 n_1 + f_2 n_2)] \mathrm{d}f_1 \mathrm{d}f_2 , \\ & \qquad\qquad\qquad\qquad\qquad (n_1, n_2) \in D \end{aligned}$$

are satisfied.

Note that the data support D has replaced $|n_1| \le M_1$, $|n_2| \le M_2$. D must contain the origin. Inspecting the solution structure obtained by the Lagrange multiplier method in (2.2.5), we can make an alternative statement in MEM1:

2-D MEM1: Given the ACF $R(n_1, n_2)$, $(n_1, n_2) \in D$,

(I) Determine the spectrum $S(f_1, f_2)$ of the following form:

$$S(f_1, f_2) = \frac{1}{\sum\limits_{(n_1,n_2)\in D} C_{n_1,n_2} \exp[-2\pi\mathrm{i}(f_1 n_1 + f_2 n_2)]} = \frac{1}{\mathrm{FT}[C_{n_1,n_2}]} .$$

$$(2.4.14)$$

(II) The following constraints are satisfied:

$$R(n_1, n_2) = \mathrm{IFT}[S(f_1, f_2)] = \mathrm{IFT}\left[\frac{1}{\mathrm{FT}[C_{n_1,n_2}]}\right] , \quad (n_1, n_2) \in D ,$$

$$(2.4.15)$$

where

$$C_{n_1,n_2} \begin{cases} \text{to be determined}, & (n_1, n_2) \in D , \\ = 0 , & \text{otherwise} . \end{cases}$$

$$(2.4.16)$$

The parameters C_{n_1,n_2}, $(n_1, n_2,) \in D$ should be adjusted repeatedly until the constraints (2.4.15) are satisfied. This can be done in principle because the number of the given $R(n_1, n_2)$ (the number of the datum points on D) is equal to the number of C_{n_1, n_2} to be determined. This procedure of adjusting repeatedly C_{n_1, n_2} to determine $S(f_1, f_2)$ is the LM iterative algorithm. On the basis of (2.4.14–2.4.16), a computer flowchart in principle is drawn in Fig. 2.7.

In practical computation, the frequency takes discrete values, $f_1 = j_1/N_1$, $f_2 = j_2/N_2$, $j_1 = 0, \ldots, N_1 - 1$, $j_2 = 0, \ldots, N_2 - 1$, N_1 and N_2 being the size of DFT. Some important issues should be discussed.

1. Existence of the Solution and Convergence Criterion

It can be shown that if the given $R(n_1, n_2)$, $(n_1, n_2) \in D$ is part of the ACF of a process, then the solution to the 2-D MEM1 problem exists and is unique. This does not mean, however, that the LM iteration must converge. Moreover, it is usually hard to confirm that the given ACF is really part of the ACF of a process. Therefore, we simply assume that the solution exists and is unique for this algorithm.

In theory, the convergence criterion should be that the given R and calculated R_x are equal at each point $(n_1, n_2) \in D$. However, this cannot be and needs not be done in practice due to computational errors. A commonly used convergence criterion is

$$\mathrm{DIFF}^{(m)} = \left(\frac{\sum_{(D)} |R_x^{(m)}(n_1, n_2) - R(n_1, n_2)|^2}{\sum_{(D)} |R(n_1, n_2)|^2} \right)^{1/2} < \varepsilon \,, \qquad (2.4.17)$$

i.e., the normalized rms error is less than a small positive number, say, $\varepsilon = 0.01 \sim 0.05$.

Practical computation has shown that the LM iteration converges in the general case. That it does not converge or converges very slowly is due to too many points in D or too small an $R(0, 0)$.

2. Initial Values

One of the simplest ways to choose the initial values is setting

$$C_{n_1, n_2}^{(1)} = \frac{1}{R(0, 0)} \delta_{n_1, n_2} \,,$$

and consequently,

$$\begin{aligned} S^{(1)}(j_1, j_2) &= R(0, 0) \,, \\ R_x^{(1)}(n_1, n_2) &= R(0, 0) \delta_{n_1, n_2} \,. \end{aligned}$$

The former equation corresponds to a uniform spectrum, and the latter means that the constraint at the origin, i.e., the total power (TP) constraint, is satisfied.

3. Positive Definiteness (PD) of the ACF

In order that the iteration may proceed, the FT of the calculated ACF must be positive. The PD of the initial ACF can be ensured by choosing $C_{n_1, n_2}^{(1)}$ properly.

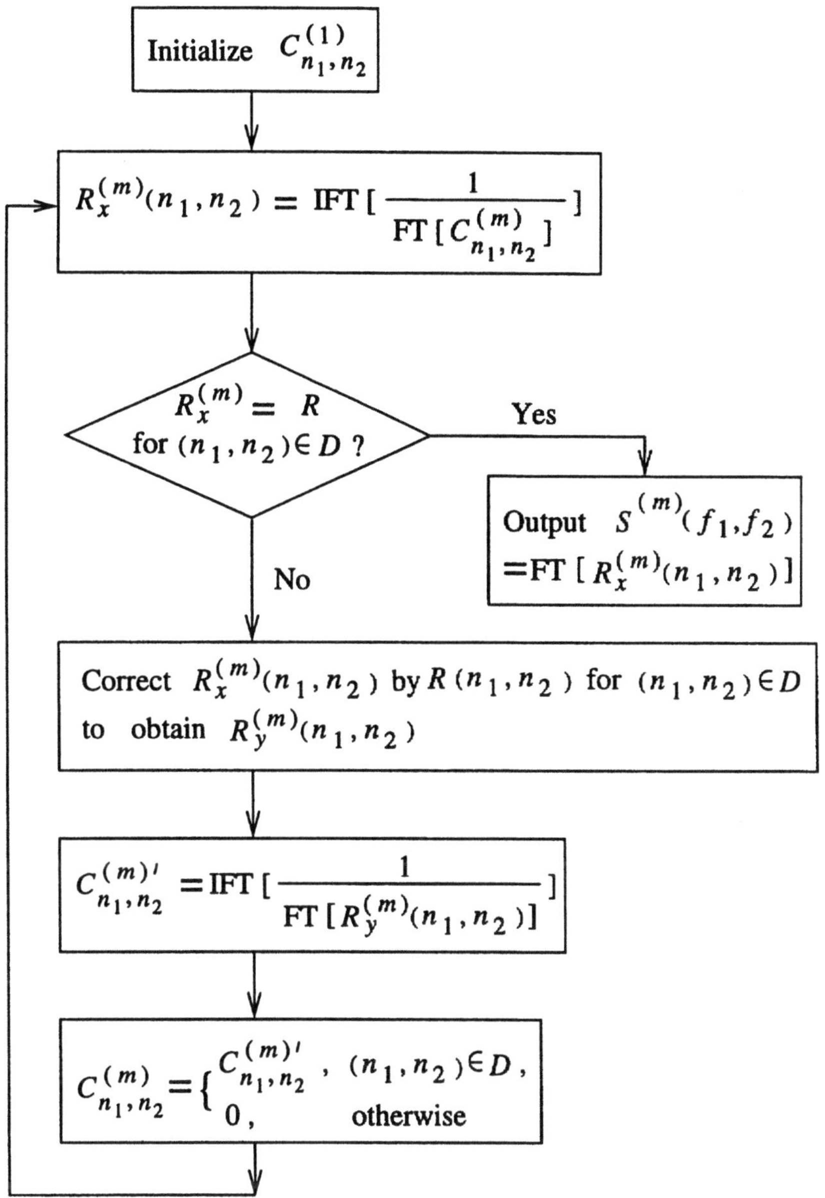

Fig. 2.7. The computer flowchart of the LM algorithm in principle. The superscript m denotes the mth iteration. The iteration begins with $m = 1$

Suppose the calculated $R_x^{(m)}(n_1, n_2)$ in the mth iteration is positive definite, i.e., its FT is positive. After correcting $R_x^{(m)}(n_1, n_2)$ by $R(n_1, n_2)$ for $(n_1, n_2) \in D$, the resulting "ACF" $R_y^{(m)}(n_1, n_2)$ may not be positive definite. Then the calculation of $C_{n_1,n_2}^{(m)\prime}$ may not be done. To resolve this problem, $R_x^{(m)}(n_1, n_2)$ is only "partially corrected" by $R(n_1, n_2)$ for $(n_1, n_2) \in D$.

The "ideal correction" of $R_x^{(m)}$ by R on D is

$$R_y^{(m)}(n_1, n_2) = \begin{cases} R(n_1, n_2) \, , & (n_1, n_2) \in D \, , \\ R_x^{(m)}(n_1, n_2) \, , & \text{otherwise} \, , \end{cases}$$

i.e.,

$$R_y^{(m)}(n_1, n_2) \;\; = \;\; R_x^{(m)}(n_1, n_2) + [R(n_1, n_2) - R_x^{(m)}(n_1, n_2)]W(n_1, n_2) \, , \tag{2.4.18}$$

where the window

$$W(n_1, n_2) = \begin{cases} 1 \, , & (n_1, n_2) \in D \, , \\ 0 \, , & \text{otherwise} \, . \end{cases}$$

The "partial correction" is

$$R_y^{(m)}(n_1, n_2)$$
$$= \begin{cases} (1 - \alpha_m)R(n_1, n_2) + \alpha_m R_x^{(m)}(n_1, n_2) \, , & (n_1, n_2) \in D \, , \\ R_x^{(m)}(n_1, n_2) \, , & \text{otherwise} \, , \end{cases} \tag{2.4.19}$$

i.e.,

$$R_y^{(m)}(n_1, n_2) \;\; = \;\; R_x^{(m)}(n_1, n_2) + (1 - \alpha_m)$$
$$\times [R(n_1, n_2) - R_x^{(m)}(n_1, n_2)]W(n_1, n_2) \, . \tag{2.4.20}$$

When $\alpha_m = 0$, (2.4.20) becomes (2.4.18), that is, the partial correction becomes the ideal correction. When $\alpha_m = 1$, $R_y^{(m)} \equiv R_x^{(m)}$, i.e., there is no correction at all, so that the iteration stagnates. In the light of the above understanding, in order to achieve the most rapid convergence, α should be chosen to be the smallest one permissible by the PD of $R_y^{(m)}$ in (2.4.20).

Now we determine the permissible interval of α_m. Taking the FT of the two sides of (2.4.20), we get

$$\text{FT}[R_y^{(m)}] = \text{FT}[R_x^{(m)}] + (1 - \alpha_m)\text{FT}[(R - R_x^{(m)})W] \, ,$$

where $\text{FT}[R_x^{(m)}] > 0$. If

$$\min_{\forall (n_1,n_2)} \text{FT}[(R - R_x^{(m)})W] \geq 0 \, , \tag{2.4.21}$$

then $\text{FT}[R_y^{(m)}] > 0$ is satisfied. In such a case α_m may take any value on $[0, 1]$. If (2.4.21) is not true, then the PD of $R_y^{(m)}$ requires

$$(1 - \alpha_m)|\text{FT}[(R - R_x^{(m)})W]| < \text{FT}[R_x^{(m)}] , \quad (j_1, j_2) \in \{J_\alpha^-\} ,$$

$$\alpha_m > 1 - \min_{\{J_\alpha^-\}} \frac{\text{FT}[R_x^{(m)}]}{|\text{FT}[(R - R_x^{(m)})W]|} . \tag{2.4.22}$$

Here the subset $\{j_\alpha^-\}$ is defined as

$$\{J_\alpha^-\} = \{(j_1, j_2) : \ \text{FT}[(R - R_x^{(m)})W] < 0\} . \tag{2.4.23}$$

Inequality (2.4.22) is the necessary and sufficient condition concerning α_m for the PD of $R_y^{(m)}$ in (2.4.20).

So far we have determined the inferior limit for α_m. In practice this smallest permissible α_m can never be used. In addition, it was observed that α should not decrease in the course of iteration, otherwise divergence might result. Incorporating these two points into (2.4.22), we obtain the formula for setting α_m in the mth iteration:

$$\alpha_m \ = \ \max \left\{ \alpha_{m-1} \ , \right.$$

$$\left. 1 - \gamma_m \min_{\{J_\alpha^-\}} \frac{\text{FT}[R_x^{(m)}(n_1, n_2)]}{|\text{FT}[\,[R(n_1, n_2) - R_x^{(m)}(n_1, n_2)]\, W(n_1, n_2)\,]|} \right\} ,$$

$$\tag{2.4.24}$$

where γ_m is called *the convergence rate parameter*, $0 < \gamma_m < 1$. A reasonable choice of its initial value is $\gamma_1 = 0.5$. A large γ_m (near 1) may not ensure the PD of $R_y^{(m)}$, while a small γ_m (near 0) will slow down the convergence (insufficiently correcting $R_x^{(m)}$). If the error $\text{DIFF}^{(m)}$ in (2.4.17) increases: $\text{DIFF}^{(m)} > \text{DIFF}^{(m-1)}$, then the value of this parameter is reduced by setting, say, $\gamma_m = \frac{1}{2}\gamma_{m-1}$.

To sum up, determine α_m by (2.4.24), then correct the calculated ACF according to (2.4.20). In this way the PD of $R_y^{(m)}(n_1, n_2)$ is ensured.

4. Zero-Crossing

Another problem which must be resolved in practice is the zero-crossing of $C_{n_1,n_2}^{(m)}$. Although the PD of $R_y^{(m)}$ ensures the PD of $C_{n_1,n_2}^{(m)\prime}$, $\text{FT}[C_{n_1,n_2}^{(m)}]$ may take zero or negative values at some points due to the truncation of $C_{n_1,n_2}^{(m)\prime}$: $C_{n_1,n_2}^{(m)} = 0$ for $(n_1, n_2) \notin D$. In such a case the calculation of $R_x^{(m)}(n_1, n_2)$ at the beginning of the next iteration will be impossible. Therefore, provision must be made to ensure the PD of $C_{n_1,n_2}^{(m)}$. In much the same way as before, the "ideal updating" of $C_{n_1,n_2}^{(m-1)}$ is replaced by the "partial updating", i.e.,

$$C_{n_1,n_2}^{(m)} = \beta_m C_{n_1,n_2}^{(m-1)} + (1 - \beta_m)C_{n_1,n_2}^{(m)\prime} W(n_1, n_2) , \tag{2.4.25}$$

where $\text{FT}[C^{(m-1)}] > 0$, $\beta_m \in [0, 1]$. When $\beta_m = 0$, $C^{(m)} = C^{(m)\prime}W$. This is the ideal updating. When $\beta_m = 1$, $C^{(m)} = C^{(m-1)}$, there is no updating at all and the iteration stands still. In consideration of convergence speed,

β_m should be the smallest permissible. Going in the same way as before, Fourier transforming (2.4.25), utilizing the PD of $C^{(m-1)}$ and considering the requirement of the PD of $C^{(m)}$, it is easily shown that β_m must meet the following condition:

$$\beta_m > \beta_{\min} = \max_{\{J_\beta^-\}} \left\{ \frac{|\mathrm{FT}[\, C_{n_1,n_2}^{(m)\prime} W(n_1,n_2)\,]|}{\mathrm{FT}[C_{n_1,n_2}^{(m-1)}] + |\mathrm{FT}[\, C_{n_1,n_2}^{(m)\prime} W(n_1,n_2)\,]|} \right\} . \tag{2.4.26}$$

Here the subset $\{J_\beta^-\}$ is defined as

$$\{J_\beta^-\} = \{(j_1,j_2): \ \mathrm{FT}[C_{n_1,n_2}^{(m)\prime} W(n_1,n_2)] < 0\} .$$

If $\{J_\beta^-\}$ is an empty set, then $\beta_m = 0$. In consideration of both the PD and convergence speed, we obtain the formula, similar to (2.4.24), for setting β_m in the mth iteration:

$$\beta_m = \max\{\beta_{m-1}, \beta_{\min} + \gamma_m'(1 - \beta_{\min})\} , \tag{2.4.27}$$

where $0 < \gamma_m' < 1$. Its initializing and updating are in the same manner as for γ_m.

To summarize, determine β_m by (2.4.27, 2.4.26), then update the value C according to (2.4.25). In this way the PD of $C_{n_1,n_2}^{(m)}$ is ensured and the problem of zero-crossing is resolved.

Incorporating the four points discussed above into the computer flowchart in Fig. 2.7, a practical flowchart of the LM algorithm is drawn in Fig. 2.8.

In principle the LM algorithm is applicable to any dimensionality. The input ACF may be uniformly or nonuniformly sampled. Its support may have an arbitrary shape but must contain the origin. Because an explicit solution in the 1-D case is available, the higher dimensionality than two is not common, and usually the FFT (fast FT) technique should be used, in practice the LM algorithm is used only for 2-D MEM1 with uniformly sampled ACF as input data. Some FT calculations in an iteration can be combined, so the number of FFT/IFFTs in one iteration is actually four. Usually, the FT length is as large as several times of the given ACF size in each dimension. Taking 64×64, 128×128, etc., as the FT size is a common practice. One way of checking whether the FT size is sufficiently large is to see if the extrapolated ACF is vanishing near the edge (maximum n_1 or n_2). Another way is to see if the result changes insignificantly after doubling the FT length in each dimension. We have learned from experience that the FT-size checking is normally unnecessary.

2.4.3 Wernecke-D'Addario Algorithm for 2-D Noisy Data

Originally the Wernecke-D'Addario (WD) algorithm was designed for image reconstruction for aperture synthesis radio telescopes in astronomy [2.15]. For the uniformity of terminology in this chapter and in order not to involve image restoration here, this algorithm is treated as for 2-D spectral estimation.

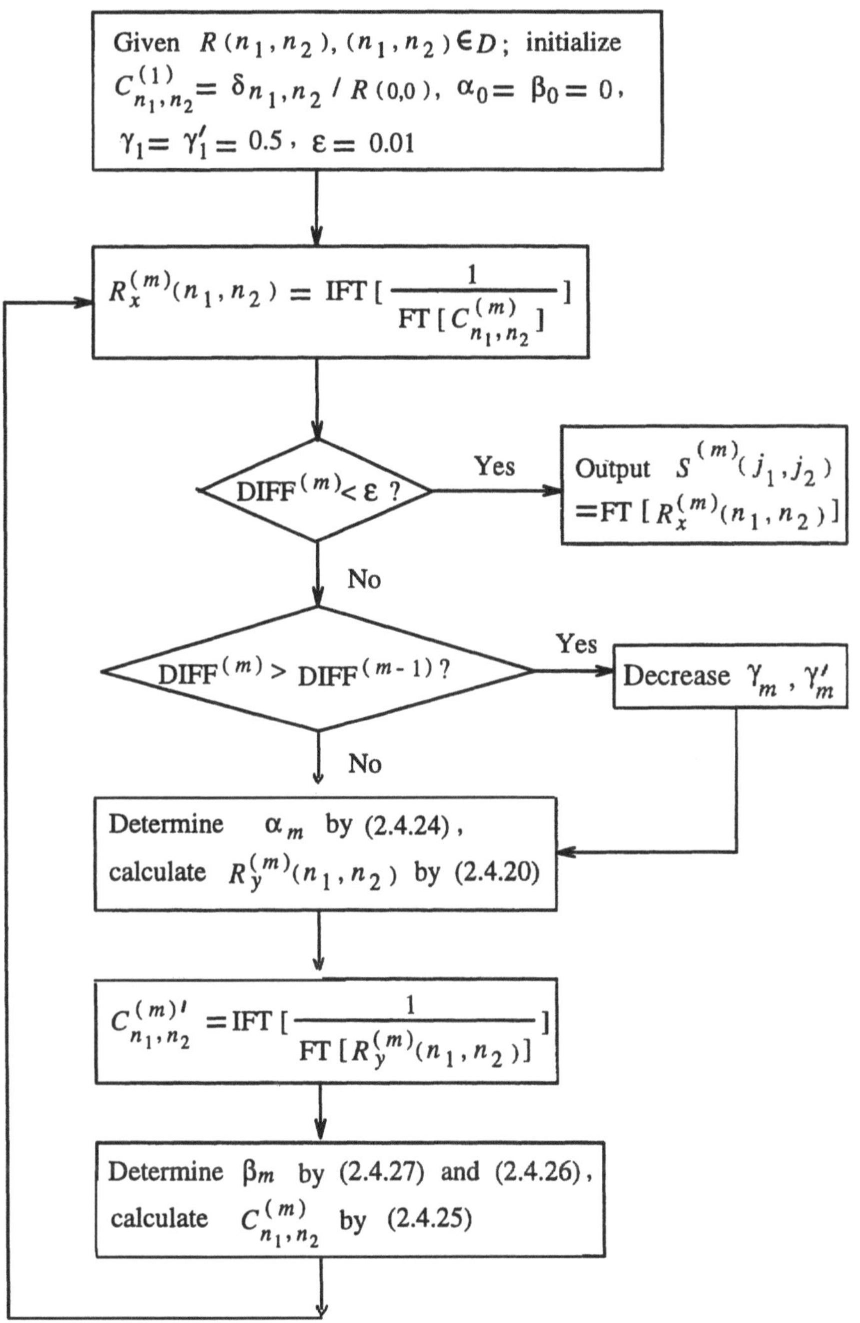

Fig. 2.8. The computer flowchart of the LM algorithm. The iteration begins with $m = 1$

The input is the 2-D noisy ACF:

$$D(n_1, n_2) = R(n_1, n_2) + N(n_1, n_2) , \quad (n_1, n_2) \in D , \tag{2.4.28}$$

cf. (2.3.31). The 2-D spectrum $S(f_1, f_2)$ is to be estimated, $f_1 = j_1/N_1$, $f_2 = j_2/N_2$, $j_1 = 0, \ldots, N_1 - 1$, $j_2 = 0, \ldots, N_2 - 1$. $R(n_1, n_2) = \mathrm{IFT}[S(f_1, f_2)]$. The problem is:

Given $D(n_1, n_2)$, $(n_1, n_2) \in D$, estimate the spectrum $S(f_1, f_2)$ such that

$$H1 = \sum_{(f_1, f_2)} \log S(f_1, f_2) \tag{2.4.29}$$

is maximized, and the constraint

$$\begin{aligned}
\chi^2 &= \sum_{(n_1, n_2) \in D} |D(n_1, n_2) - R(n_1, n_2)|^2 / \sigma^2(n_1, n_2) \\
&= C_e
\end{aligned} \tag{2.4.30}$$

is satisfied. Here σ^2 is the variance of noise, C_e is a constant, say, the expectation of χ^2 (the number of points on D). C_e is called a critical value of the χ^2-statistic.

Utilizing (2.4.29, 2.4.30), we form the objective function

$$G = \sum \log S(f_1, f_2) - \lambda \sum_{(D)} |D(n_1, n_2) - R(n_1, n_2)|^2 / \sigma^2(n_1, n_2) , \tag{2.4.31}$$

(omitting the constant C_e). For a fixed Lagrange multiplier λ, G is an upward convex function of S, so that G has a unique maximum point.

The WD algorithm is actually an optimization algorithm. For a small λ, the maximum of G is found by a 1-D search in the $N (= N_1 \times N_2)$-dimensional space using the gradient or conjugate gradient method. This is an iterative procedure. Then, whether or not the constraint (2.4.30) is satisfied is checked. Normally $\chi^2 > C_e$. Thereafter, λ is increased gradually, the maximum of G is found by iteration and the constraint is checked for each λ. As λ increases, the value of the χ^2-statistic at convergence is decreasing. The iterative procedure continues until the χ^2-statistic falls just below its critical value C_e.

Based on expression (2.4.31), we can carry out the following calculation:

$$\frac{\partial G}{\partial S(f_1, f_2)} = \frac{1}{S(f_1, f_2)} - \lambda \frac{\partial}{\partial S} \sum_{(D)} (D - R)(D - R)^* / \sigma^2 ,$$

where the partial derivative in the last term is equal to

$$\sum (D - R) \left[\frac{\partial}{\partial S} (D - R) \right]^* / \sigma^2 + \sum (D - R)^* \left[\frac{\partial}{\partial S} (D - R) \right] / \sigma^2$$

$$= \sum (D - R) \left[\frac{\partial}{\partial S} (D - R) \right]^* / \sigma^2$$

$$+ \sum \left\{ (D - R) \left[\frac{\partial}{\partial S} (D - R) \right]^* \right\}^* / \sigma^2$$

$$= 2 \sum_{(D)} (D - R) \left[\frac{\partial}{\partial S} (D - R) \right]^* / \sigma^2 \, .$$

In the above the Hermitian property of data (the summation is real) and the fact that S is real were utilized. From the FT relationship between R and S, i.e.,

$$R(n_1, n_2) = \frac{1}{N} \sum_{(f_1, f_2)} S(f_1, f_2) \exp[2\pi i (n_1 f_1 + n_2 f_2)] \, ,$$

the previous expression becomes

$$2 \sum_{(n_1, n_2) \in D} (D - R) \left\{ -\frac{1}{N} \exp[-2\pi i (n_1 f_1 + n_2 f_2)] \right\} / \sigma^2$$

$$= -2 \frac{1}{N} \mathrm{FT} \left[\frac{D}{\sigma^2} W \right] + 2 \frac{1}{N} \mathrm{FT} \left[\frac{R}{\sigma^2} W \right] \, ,$$

where the window $W(n_1, n_2) = 1$, $(n_1, n_2) \in D$; $= 0$, otherwise. Now we obtain

$$\frac{\partial G}{\partial S(f_1, f_2)} = \frac{1}{S(f_1, f_2)} + 2\lambda \frac{1}{N} \mathrm{FT} \left[\frac{D}{\sigma^2} W \right] - 2\lambda \frac{1}{N} \mathrm{FT} \left[\frac{R}{\sigma^2} W \right] \, . \quad (2.4.32)$$

This is a formula for calculating the gradient. If we use the following notation:

$$d_{f_1, f_2} = 2\lambda \frac{1}{N} \mathrm{FT} \left[\frac{D}{\sigma^2} W \right] \, ,$$

$$P_{f_1, f_2} = 2\lambda \frac{1}{N} \mathrm{FT} \left[\frac{1}{\sigma^2} W \right] \, ,$$

(d is a constant independent of S, and P is the impulse response, i.e., point spread function), then (2.4.32) becomes

$$\frac{\partial G}{\partial S(f_1, f_2)} = \frac{1}{S(f_1, f_2)} + d_{f_1, f_2} - P_{f_1, f_2} * S(f_1, f_2) \, ,$$

($*$ denoting 2-D convolution), which corresponds to expression (20) in [2.15]. We use (2.4.32) to calculate the gradient. This was said in [2.15] that convolution was accomplished through FFT. The objective function in (2.4.31) may be expanded to

$$G = \sum \log S(f_1, f_2) + 2\lambda \sum_{(D)} DR^* / \sigma^2 - \lambda \sum_{(D)} |R|^2 / \sigma^2 - \lambda \sum_{(D)} |D|^2 / \sigma^2 \, ,$$

which corresponds to expression (23) in [2.15], but is of not much use in calculation.

So far we have obtained formulae (2.4.31, 2.4.32) for calculating the objective function and gradient, respectively. The remaining problem is how to determine the initial values of S and λ.

For determining the initial value of $S(f_1, f_2)$, the spectrum S_{FT} is calculated from the noisy data $D(n_1, n_2)$, $(n_1, n_2) \in D$ by FT. After its negative and zero values are eliminated by setting them to a small positive value, S_{FT} is used as the initial S. In order to avoid negative and zero values of $S(f_1, f_2)$ in searching for the maximum, the search direction is deflected if necessary. Particularly, the values of spectrum which are below some threshold, say one percent of the peak value, remain unchanged unless they would increase in the search direction.

The determination of the initial λ is somewhat complex. If only one datum $D(0,0)$ and its variance $\sigma^2(0,0)$ were known, the estimated spectrum would be uniform, $S(f_1, f_2) \equiv S_0$. Then the objective function in (2.4.31) would be

$$
\begin{aligned}
G &= N \log S_0 - \lambda[D(0,0) - R(0,0)]^2/\sigma^2(0,0) \\
&= N \log S_0 - \lambda[D(0,0) - S_0]^2/\sigma^2(0,0) \ .
\end{aligned}
$$

To find the maximum of G, let

$$
0 = \partial G/\partial S_0 = N/S_0 + 2\lambda[D(0,0) - S_0]/\sigma^2(0,0) \ .
$$

A reasonable choice for the data fit would be

$$
S_0 - D(0,0) = R(0,0) - D(0,0) = \sigma(0,0) \ ,
$$

(overestimated). Therefore, we should have

$$
\lambda = \frac{N\sigma(0,0)}{2S_0} = \frac{N\sigma(0,0)}{2[D(0,0) + \sigma(0,0)]} = \frac{N}{2[1 + D(0,0)/\sigma(0,0)]} \ . \qquad (2.4.33)
$$

This is the formula for initializing λ.

The main steps of the WD algorithm are summarized as follows:

1) Input the noisy ACF $D(n_1, n_2)$ and its variance $\sigma^2(n_1, n_2)$, $(n_1, n_2) \in D$. Initialize $S(f_1, f_2)$ and λ.

2) Calculate the gradient by (2.4.32). Search for the maximum of G in the direction of gradient or conjugate gradient which is defined as

$$
\mathbf{z} = \nabla G - \mathbf{z}'|\nabla G| / |\nabla G'| \ ,
$$

where primes refer to the values in the previous iteration. Two FT/IFTs are needed in one iteration, one for calculating R from S and the other for calculating the gradient from R. The iterations result in $S(f_1, f_2)$ for which G is maximum for a fixed λ.

3) Check if the constraint (2.4.30) is satisfied. If the answer is *Yes*, then output the MEM1 spectrum $S(f_1, f_2)$; otherwise increase λ and go to Step 2.

One of the drawbacks of the WD algorithm is the slow convergence. For a fixed λ, many iterations are needed in Step 2 because the entropy function is far from quadratic and hence the conjugate gradient method, especially the gradient method, are inefficient. Two FFT/IFFTs are needed in each iteration. Moreover, λ must increase slowly and take a series of values so that χ^2 decreases gradually to just below its critical value. This algorithm of double-iteration type requires a great number of FFT/IFFTs. The WD algorithm has been out of use for quite a long time. One reason, among others, is that MEM1 is no longer used for image restoration. Nevertheless, some of the intermediate results of the WD algorithm are still cited in the literature.

Before concluding this subsection, we point out, by the way, that from $\partial G/\partial S = 0$ another formula for iteration may be obtained:

$$S(f_1, f_2) = \frac{1}{2\lambda\dfrac{1}{N}\mathrm{FT}[(D - R)W]} , \qquad (2.4.34)$$

where $R(n_1, n_2) = \mathrm{IFT}[S(f_1, f_2)]$. A computer program can then be designed.

2.4.4 Numerical Example

Only one numerical example [2.16] concerning MEM1 with the 1-D noiseless ACF as input data is demonstrated. More numerical examples will be shown in Sects. 4.4, 4.6.

The true spectrum shown in Fig. 2.9a consists of three Gaussian peaks. Their magnitude ratio is 10 : 5 : 1. The true spectrum is inverse Fourier transformed to get the ACF sequence, which is then truncated so that only the first 16 lags ($M = 15$) are used as input data. The spectral estimations by FT and MEM1 follow.

Shown in Fig. 2.9b is the FT spectrum with a rectangular window. From the figure it can be seen that the peak of (relative) magnitude 5 degenerates to a suggestive shoulder. Worse still, the peak of magnitude 1 is buried in the strong sidelobes and cannot be identified. Shown in Fig. 2.9c is also the FT spectrum, but with a window of "raised cosine bell" shape:

$$W_n = \frac{1}{2}[1 + \cos(\pi n/M)] , \quad |n| \le M .$$

From the figure it can be seen that there are no more sidelobes and the peak of magnitude 1 can be identified. However, the resolution is further worsened compared with Fig. 2.9b. There is no sign of the peak of magnitude 5.

Shown in Fig. 2.9d is the MEM1 spectrum. From the figure it is clear that compared with the FT spectra, the MEM1 result is much better: The resolution is high and the sidelobes are eliminated. Comparing (d) with (a) we conclude that the difference between the MEM1 and true spectra is not significant.

2.5 Algorithms and Numerical Example (Given Time Series)

In this section, the classical Burg and Marple algorithms are described in considerable detail, then the other fast algorithms are briefly introduced.

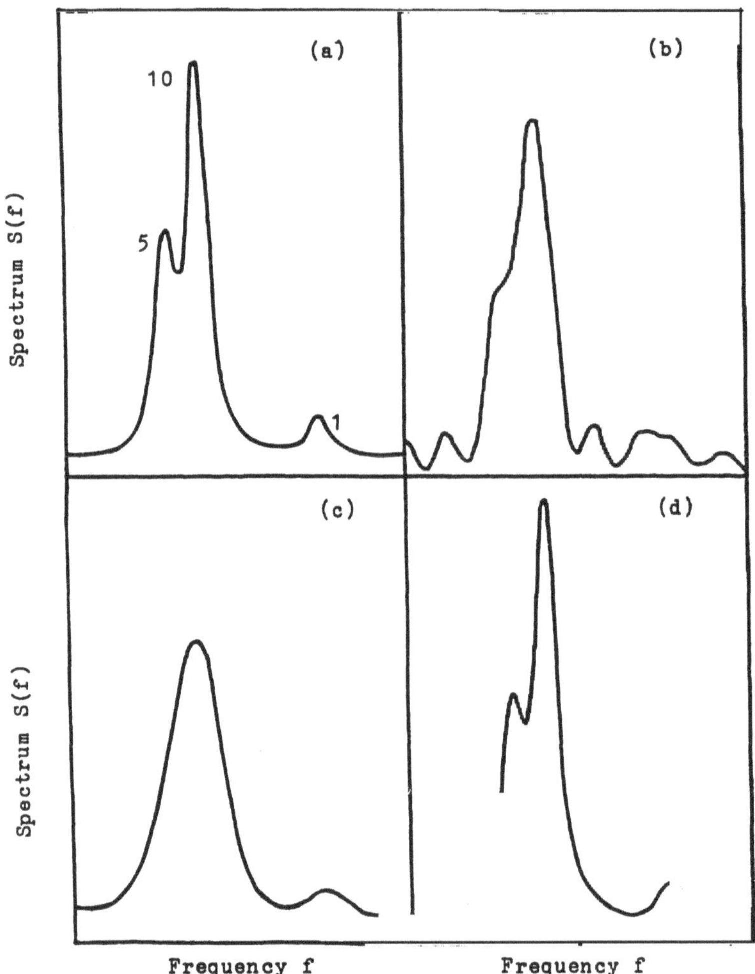

Fig. 2.9. Spectral estimation of Gaussian peaks [2.16]. (a) True spectrum. (b) FT spectrum (rectangular window). (c) FT spectrum (cosine window). (d) MEM1 spectrum

Finally, a numerical example is shown. In the following, the given data are assumed to be a segment of a time series.

2.5.1 Burg Algorithm

The central task in 1-D spectral estimation is to find the parameters $a_{M,k}$ and p_M. In the previous section we see that when these parameters are calculated by Levinson's recursion, the key is to find the reflection coefficient ρ_{M+1}, i.e., $a_{M+1,M+1}$. If a partial ACF sequence is given, Δ_{M+1} can be calculated using (2.4.9) and then $a_{M+1,M+1}$ can be found by (2.4.12). Now we need to determine $a_{M,M}$ directly from the given time series $x = \{x_1, \ldots, x_N\}^{\mathrm{T}}$ without an intermediate step of estimating a partial ACF sequence. To do this, we invoke the LP (linear prediction) theory [2.1, 2.17].

The formula for the forward (linear) prediction of a time series is

$$x_{k+M} = -\sum_{i=1}^{M} a_{M,i} x_{k+M-i} + f_{M,k} \,,$$

where $f_{M,k}$ is the forward prediction error of order M (Sect. 2.3.3). It may be rewritten as

$$f_{M,k} = \sum_{i=0}^{M} a_{M,i} x_{k+M-i} \,, \quad (a_{M,0} = 1) \,, \quad 1 \le k \le N - M \,. \tag{2.5.1}$$

The range of k is determined by the finite length of the series.

Similarly, the formula for the backward prediction is

$$x_k = -\sum_{i=1}^{M} a^*_{M,i} x_{k+i} + b_{M,k} \,,$$

where the complex conjugate of $a_{M,i}$ is used due to the Hermitian property of the ACF. The backward prediction error is

$$b_{M,k} = \sum_{i=0}^{M} a^*_{M,i} x_{k+i} \,, \quad 1 \le k \le N - M \,. \tag{2.5.2}$$

From (2.5.1, 2.5.2), we can calculate the forward and backward prediction error energies, respectively,

$$e_{M,f} = \sum_{k=1}^{N-M} |f_{M,k}|^2 \,, \quad e_{M,b} = \sum_{k=1}^{N-M} |b_{M,k}|^2 \,.$$

For a stationary series, their expectations are equal. But now we have only one observation, i.e., x of finite length, and hence $e_{M,f}$ and $e_{M,b}$ are not equal. Therefore, we take their sum

$$e_M = e_{M,f} + e_{M,b} = \sum_{k=1}^{N-M} [|f_{M,k}|^2 + |b_{M,k}|^2] \,. \tag{2.5.3}$$

Substituting (2.5.1, 2.5.2) in (2.5.3), we see that the expression of e_M contains the parameters of order M, $a_{M,i}$, $i = 1, \ldots, M-1, M$. If we uphold Levinson's recursion (to ensure $a_{M,i}$ to be a minimum-phase sequence)

$$a_{M,i} = a_{M-1,i} + a_{M,M} a^*_{M-1,M-i} \,, \quad 0 \le i \le M - 1 \,, \tag{2.5.4}$$

then e_M contains only one parameter of order M, $a_{M,M}$. The basic idea behind the Burg algorithm is: minimizing e_M with respect to the reflection coefficient $a_{M,M}$ (least sum of the squared errors) under the constraint of Levinson's recursion to determine $a_{M,M}$.

First of all, we derive the recursive formulae for calculating $f_{M,k}$ and $b_{M,k}$. It is easily shown that

$$
\begin{aligned}
f_{M,k} &= f_{M-1,k+1} + a_{M,M} b_{M-1,k} \,, & (2.5.5) \\
b_{M,k} &= b_{M-1,k} + a_{M,M}^* f_{M-1,k+1} \,. & (2.5.6)
\end{aligned}
$$

Taking $f_{M,k}$ as an example, from (2.5.1, 2.5.4) we have

$$
\begin{aligned}
f_{M,k} &= \sum_{i=0}^{M} (a_{M-1,i} + a_{M,M} a_{M-1,M-i}^*) x_{k+M-i} \\
&= \sum_{i=0}^{M} a_{M-1,i} x_{k+M-i} + a_{M,M} \sum_{i=0}^{M} a_{M-1,M-i}^* x_{k+M-i} \\
&= \sum_{i=0}^{M-1} a_{M-1,i} x_{(k+1)+(M-1)-i} + a_{M,M} \sum_{j=0}^{M-1} a_{M-1,j}^* x_{k+j} \,, \\
& \hspace{6cm} (j = M - i) \\
&= f_{M-1,k+1} + a_{M,M} b_{M-1,k} \,.
\end{aligned}
$$

In the above, $a_{M-1,M} = 0$ and the definition of $b_{M,k}$, (2.5.2), were utilized.

Now we derive the expression of e_M. From (2.5.3) and utilizing (2.5.5, 2.5.6), we get

$$
\begin{aligned}
e_M &= \sum_{k=1}^{N-M} \big[\, |f_{M-1,k+1} + a_{M,M} b_{M-1,k}|^2 \\
& \hspace{2cm} + |b_{M-1,k} + a_{M,M}^* f_{M-1,k+1}|^2 \, \big] \\
&= \sum_{k=1}^{N-M} \big[(1 + a_{M,M} a_{M,M}^*)(|f_{M-1,k+1}|^2 + |b_{M-1,k}|^2) \\
& \hspace{2cm} + 2 a_{M,M} f_{M-1,k+1}^* b_{M-1,k} + 2 a_{M,M}^* f_{M-1,k+1} b_{M-1,k}^* \, \big] \,.
\end{aligned}
$$

From $\partial e_M / \partial a_{M,M} = 0$ or $\partial e_M / \partial a_{M,M}^* = 0$, it follows that

$$
a_{M,M} = \frac{-2 \displaystyle\sum_{k=1}^{N-M} f_{M-1,k+1} b_{M-1,k}^*}{\displaystyle\sum_{k=1}^{N-M} \big[|f_{M-1,k+1}|^2 + |b_{M-1,k}|^2 \big]} \,, \quad M = 1, 2, \ldots \,. \quad (2.5.7)
$$

The denominator can be calculated by the following recursive formula:

$$
\begin{aligned}
\mathrm{DEN}_M \\
= [1 - |a_{M-1,M-1}|^2] \, \mathrm{DEN}_{M-1} - |f_{M-1,1}|^2 - |b_{M-1,N-M+1}|^2 \,,
\end{aligned}
$$
$$
(2.5.8)
$$

which can be proved as follows:

$$
\begin{aligned}
\text{DEN}_M &= \sum_{k=2}^{N-(M-1)} |f_{M-1,k}|^2 + \sum_{k=1}^{N-M} |b_{M-1,k}|^2 \\
&= \sum_{k=1}^{N-(M-1)} |f_{M-1,k}|^2 \\
&\quad + \sum_{k=1}^{N-(M-1)} |b_{M-1,k}|^2 - |f_{M-1,1}|^2 - |b_{M-1,N-M+1}|^2 \\
&= \sum_{k=1}^{N-(M-1)} \Big[|f_{M-2,k+1} + a_{M-1,M-1} b_{M-2,k}|^2 \\
&\qquad\qquad + |b_{M-2,k} + a_{M-1,M-1}^* f_{M-2,k+1}|^2 \Big] \\
&\quad -|f_{M-1,1}|^2 - |b_{M-1,N-M+1}|^2 \\
&= [1 + |a_{M-1,M-1}|^2] \sum_{k=1}^{N-(M-1)} \big[|f_{M-2,k+1}|^2 + |b_{M-2,k}|^2 \big] \\
&\quad +2 \sum_{k=1}^{N-(M-1)} \Big[a_{M-1,M-1} f_{M-2,k+1}^* b_{M-2,k} \\
&\qquad\qquad + a_{M-1,M-1}^* f_{M-2,k+1} b_{M-2,k}^* \Big] \\
&\quad -|f_{M-1,1}|^2 - |b_{M-1,N-M+1}|^2 \ .
\end{aligned}
$$

Utilizing (2.5.7) with M being replaced by $M-1$, the above expression is simplified to

$$
\begin{aligned}
\text{DEN}_M &= [1 + |a_{M-1,M-1}|^2]\,\text{DEN}_{M-1} - 2|a_{M-1,M-1}|^2\,\text{DEN}_{M-1} \\
&\quad -|f_{M-1,1}|^2 - |b_{M-1,N-M+1}|^2 \\
&= [1 - |a_{M-1,M-1}|^2]\,\text{DEN}_{M-1} - |f_{M-1,1}|^2 - |b_{M-1,N-M+1}|^2 \ .
\end{aligned}
$$

Q.E.D.

The modulus of the reflection coefficient calculated by (2.5.7) is not greater than 1. This is just what is desirable. In fact, since

$$
\begin{aligned}
[|f_{M-1,k+1}| &- |b_{M-1,k}^*|]^2 \\
&= |f_{M-1,k+1}|^2 + |b_{M-1,k}|^2 - 2|f_{M-1,k+1} b_{M-1,k}^*| \\
&\geq 0 \ ,
\end{aligned}
$$

that the numerator is not greater than the denominator in modulus follows after moving terms and then taking summations.

The Burg algorithm with the given segment of a time series $x = \{x_1,\ldots,x_N\}^{\mathrm{T}}$ is summarized as follows:

1) The recursion begins with $M = 1$. The initial parameters of order zero are

$$f_{0,k} = b_{0,k} = x_k , \quad 1 \le k \le N ,$$

$$p_0 = \hat{R}(0) = \frac{1}{2N} e_0 = \frac{1}{N} \sum_{k=1}^{N} x_k x_k^* .$$

2) Calculate $a_{M,M}$, i.e., ρ_M by (2.5.7).
3) Calculate $a_{M,i}$, $0 \le i \le M - 1$ by Levinson's recursion, i.e., by (2.5.4).
4) Calculate p_M by (2.4.13):

$$p_M = (1 - |a_{M,M}|^2)p_{M-1} .$$

5) Calculate $f_{M,k}$ and $b_{M,k}$ by (2.5.5, 2.5.6) in preparing for the next recursion.
6) Repeat Steps 2–5, obtaining successively $a_{1,k}, a_{2,k}, \ldots \ldots$.
7) Calculate the MEM1 spectrum by (2.2.14).

The filter order M may be determined by the methods to be introduced in Sect. 2.6. The result may not be satisfactory. Experience is essential and a trial-and-error procedure is inevitable. For a short segment of time series, taking $M = N/3 \sim N/2$ is common.

The computational complexity of the Burg algorithm is of a magnitude of M^2 (comparable to Levinson's recursion). The resolution of the spectra estimated is high.

Two of the shortcomings of the Burg algorithm are possible spectral peak shifting and line splitting when it is applied to harmonic processes (the true spectrum consists of isolated spectral lines). The former refers to a significant discrepancy between the locations of the estimated peak and the true line. The latter refers to the appearance of two (or more) peaks in the estimated spectrum around the location of the true line. The amount of peak shift and the seriousness of line splitting depend on the length of the data segment, filter order selected, signal-to-noise ratio and the initial phase of the signal, etc.

These two problems stem mainly from the constraint imposed by Levinson's recursion and from the treatment that the prediction error energy is minimized only with respect to the reflection coefficient $a_{M,M}$. Some modifications to the Burg algorithm aiming to alleviate or eliminate these problems will be introduced in Sect. 4.5.4 where the line splitting of MEM1 is discussed.

In short, the Burg algorithm is simple, fast and enhances resolution. For these reasons, this algorithm is widely used. Computer source programs in FORTRAN can be found in [2.12, 2.13]. Subroutines stored on a floppy disk [2.13] and a ready-to-run software package [2.18] are also available.

2.5.2 Marple Algorithm

In the preceding section we observed that the Burg algorithm is a constrained LS (least square) method. If the constraint of Levinson's recursion is removed and the minimization of the prediction error energy is carried out with respect to all the parameters of order M, $a_{M,i}$, $1 \leq i \leq M$, then a new algorithm can be developed and the performance of spectral estimation is expected to be improved. The reason for this improvement is that algorithms here all result in an approximate maximum entropy solution; the maximum of $H1$ attained in a particular algorithm becomes greater and hence is closer to the "true" maximum if the constraints are fewer [2.19].

The first consequence due to removing Levinson's recursion is that recursive formulae of $f_{M,k}$ and $b_{M,k}$ like (2.5.5, 2.5.6) are no longer available. Substituting (2.5.1, 2.5.2) into (2.5.3), we get

$$e_M = \sum_{k=1}^{N-M} \left[\left(\sum_{j=0}^{M} a_{M,j} x_{k+M-j} \right) \left(\sum_{m=0}^{M} a^*_{M,m} x^*_{k+M-m} \right) \right.$$
$$\left. + \left(\sum_{m=0}^{M} a^*_{M,m} x_{k+m} \right) \left(\sum_{j=0}^{M} a_{M,j} x^*_{k+j} \right) \right] . \qquad (2.5.9)$$

From the necessary condition for minimizing e_M, $\partial e_M / \partial a^*_{M,i} = 0$ (or $\partial e_M / \partial a_{M,i} = 0$), it follows that

$$0 = \partial e_M / \partial a^*_{M,i} ,$$

$$0 = \sum_{k=1}^{N-M} \left(\sum_{j=0}^{M} a_{M,j} x_{k+M-j} x^*_{k+M-i} + \sum_{j=0}^{M} a_{M,j} x_{k+i} x^*_{k+j} \right)$$
$$= \sum_{j=0}^{M} a_{M,j} \sum_{k=1}^{N-M} \left(x_{k+M-j} x^*_{k+M-i} + x^*_{k+j} x_{k+i} \right) , \quad i = 1 , \ldots , M .$$

With the notation

$$r_{M(i,j)} = \sum_{k=1}^{N-M} \left(x_{k+M-j} x^*_{k+M-i} + x^*_{k+j} x_{k+i} \right) , \quad 0 \leq i,j \leq M , \quad (2.5.10)$$

the above equation becomes

$$\sum_{j=0}^{M} a_{M,j} r_M(i,j) = 0 , \quad (a_{M,0} = 1) , \quad i = 1, \ldots , M . \qquad (2.5.11)$$

Replacing m by i, (2.5.9) can be rewritten as

$$e_M = \sum_{k=1}^{N-M} \left(\sum_{j=0}^{M} a_{M,j} \sum_{i=0}^{M} a^*_{M,j} x_{k+M-j} x^*_{k+M-i} \right.$$
$$\left. + \sum_{j=0}^{M} a_{M,j} \sum_{i=0}^{M} a^*_{M,i} x^*_{k+j} x_{k+i} \right)$$

$$= \sum_{i=0}^{M} a_{M,i}^* \sum_{j=0}^{M} a_{M,j} \sum_{k=1}^{N-M} \left(x_{k+M-j} x_{k+M-i}^* + x_{k+j}^* x_{k+i} \right)$$

$$= \sum_{i=0}^{M} a_{M,i}^* \sum_{j=0}^{M} a_{M,j} r_M(i,j) \,, \quad \text{(by (2.5.10))} \,.$$

Referring to (2.5.11), we see that in the above expression, all the terms vanish except that for $i = 0$ in the first summation. Therefore, we have

$$e_M = \sum_{j=0}^{M} a_{M,j} r_M(0,j) \,. \tag{2.5.12}$$

Combining (2.5.11, 2.5.12), we form a system of equations in matrix form:

$$R_M A_M = E_M \,, \tag{2.5.13}$$

where

$$R_M = \begin{bmatrix} r_M(0,0) & r_M(0,1) & \cdots & r_M(0,M) \\ r_M(1,0) & r_M(1,1) & \cdots & r_M(1,M) \\ \cdots & \cdots & \cdots & \cdots \\ \cdots & \cdots & \cdots & \cdots \\ r_M(M,0) & r_M(M,1) & \cdots & r_M(M,M) \end{bmatrix} \,, \tag{2.5.14a}$$

$$A_M = \begin{bmatrix} 1 \\ a_{M,1} \\ \cdot \\ \cdot \\ \cdot \\ a_{M,M} \end{bmatrix} \,, \quad E_M = \begin{bmatrix} e_M \\ 0 \\ \cdot \\ \cdot \\ \cdot \\ 0 \end{bmatrix} \,. \tag{2.5.14b}$$

Thus, the unconstrained LS method results in the "normal equations" (2.5.13) to be solved. Because R_M does not possess the Toeplitz structure, Levinson's recursion is not applicable. If (2.5.13) is solved by inversing the matrix using a general method, the computational complexity will be of a magnitude of M^3 (referring to M^2 for Levinson's recursion and the Burg algorithm).

Quite a number of algorithms have been developed for solving (2.5.13) efficiently. Among them the best-known one is the Marple algorithm of recursive type. Its complexity is of a magnitude of M^2. The rest of this subsection is devoted to the derivation of the Marple algorithm. Other algorithms will be introduced briefly in the next subsection.

The matrix R_M possesses the Hermitian property, i.e., $r_M(i,j) = r^*(j,i)$, and the Hermitian persymmetry, i.e., $r_M(i,j) = r^*(M-i, M-j)$. This can be verified using the definition of $r_M(i,j)$ in (2.5.10). R_M does not possess the Toeplitz structure, i.e., $r_M(i,j)$ are not of the form $c(i-j)$. Nevertheless, it can be decomposed to the sum of two Toeplitz matrices:

$$R_M = (T_M)^{\mathrm{H}} T_M + (T_M^\nu)^{\mathrm{H}} T_M^\nu \ . \tag{2.5.15}$$

Here T_M is an $(N - M) \times (M + 1)$ Toeplitz matrix, whose elements are data samples:

$$T_M = \begin{bmatrix} x_{M+1} & x_M & \cdots & x_1 \\ x_{M+2} & x_{M+1} & \cdots & x_2 \\ \cdots & \cdots & \cdots & \cdots \\ \cdots & \cdots & \cdots & \cdots \\ x_N & x_{N-1} & \cdots & x_{N-M} \end{bmatrix} \ . \tag{2.5.16}$$

T_M^ν is obtained by reversing and then conjugating each row of T_M:

$$T_M^\nu = \begin{bmatrix} x_1^* & x_2^* & \cdots & x_{M+1}^* \\ x_2^* & x_3^* & \cdots & x_{M+2}^* \\ \cdots & \cdots & \cdots & \cdots \\ \cdots & \cdots & \cdots & \cdots \\ x_{N-M}^* & x_{N-M+1}^* & \cdots & x_N^* \end{bmatrix} \ . \tag{2.5.17}$$

Equation (2.5.15) can be verified by writing out the expressions of the matrix elements:

$$\begin{aligned}
[(T_M)^{\mathrm{H}} T_M](i,j) &= \sum_{k=1}^{N-M} x^*_{M-(i-1)+k} x_{M-(j-1)+k} \\
&= \text{first sum in } r_M(i-1, j-1) \ , \ \text{cf. (2.5.10)} \ , \\
[(T_M^\nu)^{\mathrm{H}} T_M^\nu](i,j) &= \sum_{k=1}^{N-M} x_{(i-1)+k} x^*_{(j-1)+k} \\
&= \text{second sum in } r_M(i-1, j-1) \ , \ \text{cf. (2.5.10)} \ , \\
R_M(i,j) &= r_M(i-1, j-1) \ , \ \text{cf. (2.5.14a)} \ .
\end{aligned}$$

The structure of R_M shown in the above plays an essential role in establishing the recursive algorithm. Because R_M itself does not possess the Toeplitz structure, the Marple algorithm is much more complicated and its derivation is much more difficult compared with the Burg algorithm. It is fortunate, however, that the computational complexity is still of a magnitude of M^2. For clarity, the procedure of derivation is divided into some more or less distinctive steps.

Step 1. Introduce the time-index-shifted quantities e_M', e_M'', etc.

We define two additional prediction error energies first:

$$e_M' = \sum_{k=1}^{N-M-1} (|f_{M,k+1}|^2 + |b_{M,k}|^2) \ , \tag{2.5.18}$$

$$e_M'' = \sum_{k=1}^{N-M-1} (|f_{M,k}|^2 + |b_{M,k+1}|^2) \ . \tag{2.5.19}$$

Compared with e_M in (2.5.3), the time index of the forward prediction error in e'_M and that of the backward prediction error in e''_M have been increased by one.

Minimizing e'_M and e''_M with respect to $a_{M,i}$, $1 \leq i \leq M$, in much the same way as deriving (2.5.13), we get two similar equations:

$$R'_M A'_M = E'_M , \tag{2.5.20}$$

$$R''_M A''_M = E''_M . \tag{2.5.21}$$

Here primes and double primes refer to different matrices and vectors. The elements of the $(M + 1) \times (M + 1)$ matrices R'_M and R''_M, $R'_M(i,j)$ and $R''_M(i,j)$, are $r'_M(i - 1, j - 1)$ and $r''_M(i - 1, j - 1)$, respectively, with the definitions

$$r'_M(i,j) \;=\; \sum_{k=1}^{N-M-1} \left(x_{k+M+1-j} x^*_{k+M+1-i} + x^*_{k+j} x_{k+i} \right) ,$$

$$0 \leq i, j \leq M , \tag{2.5.22}$$

$$r''_M(i,j) \;=\; \sum_{k=1}^{N-M-1} \left(x_{k+M-j} x^*_{k+M-i} + x^*_{k+1+j} x_{k+1+i} \right) ,$$

$$0 \leq i, j \leq M . \tag{2.5.23}$$

They were obtained by increasing appropriately the time index by one in the expression of $r_M(i,j)$, (2.5.10). From these definitions it is easy to see the Hermitian persymmetry between R'_M and R''_M:

$$r'_M(i,j) = [r''_M(M - i, M - j)]^* .$$

The new $(M + 1)$-element column vectors are

$$A'_M = \begin{bmatrix} 1 \\ a'_{M,1} \\ \cdot \\ \cdot \\ \cdot \\ a'_{M,M} \end{bmatrix} , \quad E'_M = \begin{bmatrix} e'_M \\ 0 \\ \cdot \\ \cdot \\ \cdot \\ 0 \end{bmatrix} ,$$

$$A''_M = \begin{bmatrix} 1 \\ a''_{M,1} \\ \cdot \\ \cdot \\ \cdot \\ a''_{M,M} \end{bmatrix} , \quad E''_M = \begin{bmatrix} e''_M \\ 0 \\ \cdot \\ \cdot \\ \cdot \\ 0 \end{bmatrix} . \tag{2.5.24}$$

The following relationships among the matrices R_M, R'_M, R''_M and R_{M+1} can be shown by straightforward calculation or inspection:

$$R'_M = R_M - \begin{bmatrix} x^*_{M+1} \\ \cdot \\ \cdot \\ \cdot \\ x^*_1 \end{bmatrix} (x_{M+1},\ldots,x_1)$$

$$- \begin{bmatrix} x_{N-M} \\ \cdot \\ \cdot \\ \cdot \\ x_N \end{bmatrix} (x^*_{N-M},\ldots,x^*_N)\,, \tag{2.5.25}$$

$$R''_M = R_M - \begin{bmatrix} x^*_N \\ \cdot \\ \cdot \\ \cdot \\ x^*_{N-M} \end{bmatrix} (x_N,\ldots,x_{N-M})$$

$$- \begin{bmatrix} x_1 \\ \cdot \\ \cdot \\ \cdot \\ x_{M+1} \end{bmatrix} (x^*_1,\ldots,x^*_{M+1})\,, \tag{2.5.26}$$

$$R_{M+1}$$

$$= \left[\begin{array}{ccc|c} & & & r_{M+1}(0, M+1) \\ & R'_M & & \cdot \\ & & & \cdot \\ & & & \cdot \\ \hline r_{M+1}(M+1,0) & \cdots & \cdots & r_{M+1}(M+1, M+1) \end{array} \right] \tag{2.5.27}$$

$$= \left[\begin{array}{c|c} r_{M+1}(0,0) & \cdots \qquad \cdots \quad r_{M+1}(0, M+1) \\ \hline \cdot & \\ \cdot & R''_M \\ \cdot & \\ r_{M+1}(M+1,0) & \end{array} \right]\,. \tag{2.5.28}$$

Step 2. Introduce the auxiliary $(M + 1)$-element column vectors

$$C_M = \begin{bmatrix} c_{M,0} \\ \cdot \\ \cdot \\ \cdot \\ c_{M,M} \end{bmatrix}, \quad C_M'' = \begin{bmatrix} c_{M,0}'' \\ \cdot \\ \cdot \\ \cdot \\ c_{M,M}'' \end{bmatrix},$$

$$D_M = \begin{bmatrix} d_{M,0} \\ \cdot \\ \cdot \\ \cdot \\ d_{M,M} \end{bmatrix}, \quad D_M'' = \begin{bmatrix} d_{M,0}'' \\ \cdot \\ \cdot \\ \cdot \\ d_{M,M}'' \end{bmatrix}.$$

They are defined by

$$R_M C_M = \begin{bmatrix} x_{M+1}^* \\ \cdot \\ \cdot \\ \cdot \\ x_1^* \end{bmatrix}, \tag{2.5.29}$$

$$R_M'' C_M'' = \begin{bmatrix} x_{M+1}^* \\ \cdot \\ \cdot \\ \cdot \\ x_1^* \end{bmatrix}, \tag{2.5.30}$$

$$R_M D_M = \begin{bmatrix} x_{N-M} \\ \cdot \\ \cdot \\ \cdot \\ x_N \end{bmatrix}, \tag{2.5.31}$$

$$R_M'' D_M'' = \begin{bmatrix} x_{N-M} \\ \cdot \\ \cdot \\ \cdot \\ x_N \end{bmatrix}. \tag{2.5.32}$$

The operator I is defined as reversing and then conjugating a vector when I operates on it, i.e.,

$$A_M^{\mathrm{I}} = \begin{bmatrix} a_{M,M}^* \\ \cdot \\ \cdot \\ \cdot \\ a_{M,1}^* \\ 1 \end{bmatrix}, \quad E_M^{\mathrm{I}} = \begin{bmatrix} 0 \\ \cdot \\ \cdot \\ \cdot \\ 0 \\ e_M \end{bmatrix},$$

$$C_M^{\mathrm{I}} = \begin{bmatrix} c_{M,M}^* \\ \cdot \\ \cdot \\ \cdot \\ c_{M,0}^* \end{bmatrix} , \quad D_M^{\mathrm{I}} = \begin{bmatrix} d_{M,M}^* \\ \cdot \\ \cdot \\ \cdot \\ d_{M,0}^* \end{bmatrix} .$$

The following relationships can be verified using the Hermitian persymmetry of R_M:

$$R_M A_M^{\mathrm{I}} \;=\; E_M^{\mathrm{I}} , \tag{2.5.33}$$

$$R_M C_M^{\mathrm{I}} \;=\; \begin{bmatrix} x_1 \\ \cdot \\ \cdot \\ \cdot \\ x_{M+1} \end{bmatrix} , \tag{2.5.34}$$

$$R_M D_M^{\mathrm{I}} \;=\; \begin{bmatrix} x_N^* \\ \cdot \\ \cdot \\ \cdot \\ x_{N-M}^* \end{bmatrix} . \tag{2.5.35}$$

(Refer to the derivation of (2.4.7), noticing $r_M(i,j) = r_M^*(M - i, M - j)$.)

It can be shown from the definitions of $f_{M,k}$ and $b_{M,k}$, (2.5.1, 2.5.2), respectively, that

$$f_{M,1} = (x_{M+1}, \ldots, x_1) A_M , \tag{2.5.36}$$

$$b_{M,N-M} = (x_{N-M}, \ldots, x_N) A_M^* . \tag{2.5.37}$$

Step 3. Introduce the auxiliary scalars g_M, h_M, etc. They are defined as

$$g_M \;=\; (x_{M+1}, \ldots, x_1) C_M , \tag{2.5.38}$$

$$h_M \;=\; (x_{N-M}^*, \ldots, x_N^*) C_M , \tag{2.5.39}$$

$$s_M \;=\; (x_N, \ldots, x_{N-M}) C_M , \tag{2.5.40}$$

$$u_M \;=\; (x_N, \ldots, x_{N-M}) D_M , \tag{2.5.41}$$

$$v_M \;=\; (x_1^*, \ldots, x_{M+1}^*) C_M , \tag{2.5.42}$$

$$w_M \;=\; (x_{N-M}^*, \ldots, x_N^*) D_M . \tag{2.5.43}$$

Utilizing the identity (a scalar is equal to its transpose)

$$A_M^{\mathrm{H}} R_M C_M = (C_M^{\mathrm{H}} R_M^{\mathrm{H}} A_M)^* = (C_M^{\mathrm{H}} R_M A_M)^* ,$$

we get

$$A_M^{\mathrm{H}} \begin{bmatrix} x_{M+1}^* \\ \cdot \\ \cdot \\ \cdot \\ x_1^* \end{bmatrix} = \left\{ C_M^{\mathrm{H}} \begin{bmatrix} e_M \\ 0 \\ \cdot \\ \cdot \\ \cdot \\ 0 \end{bmatrix} \right\}^* . \tag{2.5.44}$$

Note that (2.5.29, 2.5.13) were utilized in the above. Utilizing (2.5.36), from (2.5.44) we get

$$f^*_{M,1} = c_{M,0} e_M ,$$

i.e.,

$$c_{M,0} = f^*_{M,1} / e_M . \qquad (2.5.45)$$

Similarly, utilizing the identity on the left and the expressions listed on the right in each of the following lines:

$$A^{\mathrm{H}}_M R_M D_M = (D^{\mathrm{H}}_M R_M A_M)^* , \qquad (2.5.31, 2.5.13, 2.5.37);$$

$$D^{\mathrm{H}}_M R_M C_M = (C^{\mathrm{H}}_M R_M D_M)^* , \qquad (2.5.29, 2.5.31, 2.5.39);$$

$$D^{\mathrm{H}}_M R_M C^{\mathrm{I}}_M = (C^{\mathrm{IH}}_M R_M D_M)^* , \qquad (2.5.34, 2.5.31, 2.5.40);$$

$$C^{\mathrm{H}}_M R_M C_M = (C^{\mathrm{H}}_M R_M C_M)^* , \qquad (2.5.29, 2.5.38);$$

$$D^{\mathrm{H}} R_M D_M = (D^{\mathrm{H}}_M R_M D_M)^* , \qquad (2.5.31, 2.5.43);$$

we obtain, respectively,

$$d_{M,0} = b_{M,N-M}/e_M , \qquad (2.5.46)$$
$$h^*_M = (x_{M+1}, \ldots, x_1) D_M , \qquad (2.5.47)$$
$$s_M = (x^*_1, \ldots, x^*_{M+1}) D_M , \qquad (2.5.48)$$
$$g^*_M = g_M , \qquad (2.5.49)$$
$$w^*_M = w_M . \qquad (2.5.50)$$

From the last two expressions, it is clear that g_M and w_M are, in fact, real-valued scalars.

Step 4. Time-shift updates for A'_M, e'_M, C''_M and D''_M .

(1) Utilizing (2.5.29, 2.5.31), then from (2.5.25) we get

$$R'_M = R_M [I - C_M (x_{M+1}, \ldots, x_1) - D_M (x^*_{N-M}, \ldots, x^*_N)] ,$$

where I is the $(M+1) \times (M+1)$ identity matrix. Postmultiplying the above expression by A'_M yields

$$R'_M A'_M = R_M [A'_M - C_M (x_{M+1}, \ldots, x_1) A'_M - D_M (x^*_{N-M}, \ldots, x^*_N) A'_M] .$$

Letting

$$\alpha_M \beta_1 = (x_{M+1}, \ldots, x_1) A'_M ,$$
$$\alpha_M \gamma_1 = (x^*_{N-M}, \ldots, x^*_N) A'_M ,$$
$$e'_M = \alpha_M e_M , \qquad (2.5.51)$$

it follows from the above expression of $R'_M A'_M$ that

$$R_M(A'_M - \alpha_M\beta_1 C_M - \alpha_M\gamma_1 D_M) = \begin{bmatrix} e'_M \\ 0 \\ \cdot \\ \cdot \\ \cdot \\ 0 \end{bmatrix} = \alpha_M \begin{bmatrix} e_M \\ 0 \\ \cdot \\ \cdot \\ \cdot \\ 0 \end{bmatrix}$$

$$= R_M(\alpha_M A_M) .$$

Note that (2.5.20, 2.5.13) were utilized in the above. From the above expression we obtain the time-shift update for A'_M, i.e., the recursive formula from A_M to A'_M:

$$A'_M = \alpha_M(A_M + \beta_1 C_M + \gamma_1 D_M) . \tag{2.5.52}$$

The above expression is premultiplied by R'_M to determine the scalars β_1 and γ_1. Utilizing the relationship (2.5.25), definitions (2.5.29–2.5.32) and (2.5.38–2.5.43), and the relationships (2.5.36, 2.5.37, 2.5.46–2.5.50), after some manipulation we obtain

$$\begin{bmatrix} e'_M \\ 0 \\ \cdot \\ \cdot \\ \cdot \\ 0 \end{bmatrix} = \alpha_M \left\{ \begin{bmatrix} e_M \\ 0 \\ \cdot \\ \cdot \\ \cdot \\ 0 \end{bmatrix} + [-f_{M,1} + (1-g_M)\beta_1 - \gamma_1 h^*_M] \begin{bmatrix} x^*_{M+1} \\ x^*_M \\ \cdot \\ \cdot \\ \cdot \\ x^*_1 \end{bmatrix} \right.$$

$$+ \begin{bmatrix} x_{N-M} \\ x_{N-M+1} \\ \cdot \\ \cdot \\ \cdot \\ x_N \end{bmatrix} [-b^*_{M,N-M} - \beta_1 h_M + (1-w_M)\gamma_1] \left. \right\} .$$

Since $e'_M = \alpha_M e_M$, and $\begin{bmatrix} x^*_{M+1} \\ \vdots \\ x^*_1 \end{bmatrix}$ and $\begin{bmatrix} x_{N-M} \\ \vdots \\ x_N \end{bmatrix}$ are arbitrary,

$$\begin{cases} (1-g_M)\beta_1 - h^*_M \gamma_1 = f_{M,1} , \\ -h_M\beta_1 + (1-w_M)\gamma_1 = b^*_{M,N-M} \end{cases}$$

must hold. Solving these equations we get

$$\begin{cases} \beta_1 = [f_{M,1}(1-w_M) + b^*_{M,N-M} h^*_M]/\text{DEN}_M , & (2.5.53) \\ \gamma_1 = [b^*_{M,N-M}(1-g_M) + f_{M,1} h_M]/\text{DEN}_M , & (2.5.54) \end{cases}$$

where the denominator DEN_M is the coefficient determinant, which is, after being simplified,

$$\text{DEN}_M = (1-g_M)(1-w_M) - |h_M|^2 . \tag{2.5.55}$$

Since $A'_{M,0} = A_{M,0} = 1$, it follows from (2.5.52) that

$$1 = \alpha_M(1 + \beta_1 c_{M,0} + \gamma_1 d_{M,0}) \ .$$

Utilizing (2.5.45, 2.5.46, 2.5.53, 2.5.54), solving the above equation yields

$$\alpha_M =$$
$$\left[1 + \frac{|f_{M,1}|^2(1 - w_M) + |b_{M,N-M}|^2(1 - g_m) + 2\mathrm{Re}\{f_{M,1}h_M b_{M,N-M}\}}{e_M \mathrm{DEN}_M} \right]^{-1} \ . \tag{2.5.56}$$

(2) Utilizing (2.5.35, 2.5.34), then from (2.5.26) we get

$$R''_M = R_M[I - D^{\mathrm{I}}_M(x_N, \ldots, x_{N-M}) - C^{\mathrm{I}}_M(x_1^*, \ldots, x_{M+1}^*)] \ . \tag{2.5.57}$$

Let

$$\beta_2 = (x_1^*, \ldots, x_{M+1}^*)C''_M \ ,$$

$$\gamma_2 = (x_N, \ldots, x_{N-M})C''_M \ .$$

Postmultiplying (2.5.27) by C''_M yields

$$R''_M C''_M = R_M C_M = R_M(C''_M - \gamma_2 D^{\mathrm{I}}_M - \beta_2 C^{\mathrm{I}}_M) \ .$$

Note that the definitions (2.5.29, 2.5.30) were utilized in the above. From the above expression we obtain the time-shift update for C''_M, i.e., the recursive formula from C_M to C''_M:

$$C''_M = C_M + \beta_2 C^{\mathrm{I}}_M + \gamma_2 D^{\mathrm{I}}_M \ . \tag{2.5.58}$$

The above expression is premultiplied by R''_M to determine the scalars β_2 and γ_2. Utilizing the relationship (2.5.26), definitions (2.5.29–2.5.32) and (2.5.38–2.5.43), and the relationships (2.5.34, 2.5.35, 2.5.46–2.5.50), after some manipulation we obtain

$$\begin{bmatrix} x_{M+1}^* \\ \cdot \\ \cdot \\ \cdot \\ x_1^* \end{bmatrix} = \begin{bmatrix} x_{M+1}^* \\ \cdot \\ \cdot \\ \cdot \\ x_1^* \end{bmatrix} + [-s_M - \beta_2 h_M^* + \gamma_2(1 - w_M)] \begin{bmatrix} x_N^* \\ \cdot \\ \cdot \\ \cdot \\ x_{N-M}^* \end{bmatrix}$$

$$+ [v_M + \beta_2(1 - g_M) - \gamma_2 h_M] \begin{bmatrix} x_1 \\ \cdot \\ \cdot \\ \cdot \\ x_{M+1} \end{bmatrix} \ .$$

The quantities in the two sets of square brackets must be zero and a system of equations is formed. The solution is

$$\begin{cases} \beta_2 &= [s_M h_M + v_M(1 - w_M)]/\mathrm{DEN}_M \ , \\ \gamma_2 &= [v_M h_M^* + s_M(1 - g_M)]/\mathrm{DEN}_M \ . \end{cases}$$

$$\text{(2.5.59)}$$
$$\text{(2.5.60)}$$

(3) Postmultiplying (2.5.57) by D_M'' and letting

$$\beta_3 = (x_1^*, \ldots, x_{M+1}^*)D_M'' ,$$
$$\gamma_3 = (x_N, \ldots, x_{N-M})D_M'' ,$$

it follows that

$$R_M'' D_M'' = R_M D_M = R_M(D_M'' - \gamma_3 D_M^{\mathrm{I}} - \beta_3 C_M^{\mathrm{I}}) .$$

Note that the definitions (2.5.31, 2.5.32) were utilized in the above. From the above expression, we obtain the time-shift update for D_M'', i.e., the recursive formula from D_M to D_M'':

$$D_M'' = D_M + \beta_3 C_M^{\mathrm{I}} + \gamma_3 D_M^{\mathrm{I}} . \tag{2.5.61}$$

The scalars β_3 and γ_3 are determined in the same way as for β_2 and γ_2 in (2). Particularly, we have

$$\begin{bmatrix} x_{N-M} \\ \cdot \\ \cdot \\ \cdot \\ x_N \end{bmatrix} = \begin{bmatrix} x_{N-M} \\ \cdot \\ \cdot \\ \cdot \\ x_N \end{bmatrix} + [-u_M - \beta_3 h_M^* + \gamma_3(1 - w_M)] \begin{bmatrix} x_N^* \\ \cdot \\ \cdot \\ \cdot \\ x_{N-M}^* \end{bmatrix}$$

$$+ [-s_M + \beta_3(1 - g_M) - \gamma_3 h_M] \begin{bmatrix} x_1 \\ \cdot \\ \cdot \\ \cdot \\ x_{M+1} \end{bmatrix} .$$

The quantities in the two sets of square brackets must be zero. From this we obtain

$$\begin{cases} \beta_3 = [u_M h_M + s_M(1 - w_M)]/\mathrm{DEN}_M , & \tag{2.5.62} \\ \gamma_3 = [s_M h_M^* + u_M(1 - g_M)]/\mathrm{DEN}_M . & \tag{2.5.63} \end{cases}$$

Step 5. Order updates for A_{M+1}, e_{M+1} and $a_{M+1,i}$.
Utilizing (2.5.20), then from (2.5.27) we get

$$R_{M+1} \begin{bmatrix} A_M' \\ 0 \end{bmatrix} = \begin{bmatrix} E_M' \\ \Delta_{M+1} \end{bmatrix} , \tag{2.5.64}$$

where

$$\Delta_{M+1} = [r_{M+1}(M+1,0), \ldots, r_{M+1}(M+1,M)]A_M' . \tag{2.5.65}$$

On the other hand, from (2.5.28) we have

$$R_{M+1} \begin{bmatrix} 0 \\ (A_M')^{\mathrm{I}} \end{bmatrix} = \begin{bmatrix} \Delta_{M+1}^* \\ (E_M')^{\mathrm{I}} \end{bmatrix} . \tag{2.5.66}$$

In the above the relationship $r_{M+1}(i,j) = r^*_{M+1}(M+1-i, M+1-j)$ was used to obtain the error Δ^*_{M+1}. The relationship $R''_M(A'_M)^{\mathrm{I}} = (E'_M)^{\mathrm{I}}$ derived by $r''_M(i,j) = [r'_M(M-i, M-j)]^*$ was also used.

From (2.5.64, 2.5.66), we have

$$R_{M+1} \left\{ \begin{bmatrix} A'_M \\ 0 \end{bmatrix} + \alpha_2 \begin{bmatrix} 0 \\ (A'_M)^{\mathrm{I}} \end{bmatrix} \right\} = \begin{bmatrix} e'_M \\ 0 \\ \cdot \\ \cdot \\ \cdot \\ 0 \\ \Delta_{M+1} \end{bmatrix} + \alpha_2 \begin{bmatrix} \Delta^*_{M+1} \\ 0 \\ \cdot \\ \cdot \\ \cdot \\ 0 \\ e'_M \end{bmatrix} .$$

Comparing this with the normal equations of order $M+1$, we thus obtain the recursive formula for the order update for A_{M+1}:

$$A_{M+1} = \begin{bmatrix} A'_M \\ 0 \end{bmatrix} + \alpha_2 \begin{bmatrix} 0 \\ (A'_M)^{\mathrm{I}} \end{bmatrix} , \tag{2.5.67}$$

and

$$\begin{bmatrix} e_{M+1} \\ 0 \\ \cdot \\ \cdot \\ \cdot \\ 0 \\ 0 \end{bmatrix} = \begin{bmatrix} e'_M \\ 0 \\ \cdot \\ \cdot \\ \cdot \\ 0 \\ \Delta_{M+1} \end{bmatrix} + \alpha_2 \begin{bmatrix} \Delta^*_{M+1} \\ 0 \\ \cdot \\ \cdot \\ \cdot \\ 0 \\ e'_M \end{bmatrix} .$$

From the last expression, we get a system of equations:

$$\begin{aligned} e_{M+1} &= e'_M + \alpha_2 \Delta^*_{M+1} , \\ 0 &= \Delta_{M+1} + \alpha_2 e'_M . \end{aligned}$$

The solution is

$$\begin{aligned} \alpha_2 &= a_{M+1,M+1} = -\Delta_{M+1}/e'_M , \tag{2.5.68} \\ e_{M+1} &= e'_M[1 - |a_{M+1,M+1}|^2] . \tag{2.5.69} \end{aligned}$$

These are the recursive formulae for the order updates for $a_{M+1,M+1}$ and e_{M+1}, respectively. Note that $|a_{M+1,M+1}| < 1$ since e_{M+1}, $e'_M > 0$.

From (2.5.67) we deduce that the elements satisfy the relationships

$$\left\{ \begin{aligned} a_{M+1,i} &= a'_{M,i} + a_{M+1,M+1}(a_{M,M+1-i})^* , \quad i = 1,\ldots,M , \\ a_{M+1,0} &= a'_{M,0} = 1 . \end{aligned} \right. \tag{2.5.70}$$

These are the recursive formulae for $a_{M+1,i}$. Comparing (2.5.70) with (2.4.11), we see that the recursion for $a_{M+1,i}$ here has the same form as Levinson's recursion with $a'_{M,i}$ replacing $a_{M,i}$.

For the efficiency in calculating Δ_{M+1} by (2.5.65), a recursive formula for calculating $r_{M+1}(M+1, i)$ can be derived from (2.5.10). The result is

$$r_{M+1}(M+1, i) = r_M(M, i-1) - x_{N-i+1}x_{N-M}^* - x_{M+1}x_i^* . \qquad (2.5.71)$$

Step 6. Order updates for C_{M+1}, D_{M+1}, g_{M+1} and w_{M+1} .

(1) From the definition (2.5.29) it follows that

$$R_{M+1}C_{M+1} = \begin{bmatrix} x_{M+2}^* \\ x_{M+1}^* \\ \cdot \\ \cdot \\ \cdot \\ x_1^* \end{bmatrix} . \qquad (2.5.72)$$

From (2.5.13), we have

$$R_{M+1}A_{M+1} = \begin{bmatrix} e_{M+1} \\ 0 \\ \cdot \\ \cdot \\ \cdot \\ 0 \end{bmatrix} . \qquad (2.5.73)$$

Postmultiplying (2.5.28) by $\begin{bmatrix} 0 \\ C_M'' \end{bmatrix}$ and utilizing (2.5.30), we get

$$R_{M+1}\begin{bmatrix} 0 \\ C_M'' \end{bmatrix} = \begin{bmatrix} \varepsilon_M \\ x_{M+1}^* \\ \cdot \\ \cdot \\ \cdot \\ x_1^* \end{bmatrix} , \qquad (2.5.74)$$

where

$$\varepsilon_M = [r_{M+1}(0, 1), \ldots, r_{M+1}(0, M+1)]C_M'' .$$

From (2.5.74, 2.5.73), we obtain

$$R_{M+1}\left\{\begin{bmatrix} 0 \\ C_M'' \end{bmatrix} + \alpha_3 A_{M+1}\right\} = \begin{bmatrix} \varepsilon_M + \alpha_3 e_{M+1} \\ x_{M+1}^* \\ \cdot \\ \cdot \\ \cdot \\ x_1^* \end{bmatrix} .$$

The parameter α_3 is chosen such that $\varepsilon_M + \alpha_3 e_{M+1} = x_{M+2}^*$. Then, comparing the above expression with (2.5.72), we find

$$C_{M+1} = \begin{bmatrix} 0 \\ C''_M \end{bmatrix} + \alpha_3 A_{M+1} \ . \tag{2.5.75}$$

This is the recursive formula for the order update for C_{M+1}. Utilizing (2.5.45), then from the first row in (2.5.75) α_3 can be determined:

$$\alpha_3 = c_{M+1,0} = f^*_{M+1,1}/e_{M+1} \ . \tag{2.5.76}$$

(2) From the definition (2.5.31), it follows that

$$R_{M+1}D_{M+1} = \begin{bmatrix} x_{N-M-1} \\ x_{N-M} \\ \cdot \\ \cdot \\ \cdot \\ x_N \end{bmatrix} \ . \tag{2.5.77}$$

Postmultiplying (2.5.28) by $\begin{bmatrix} 0 \\ D''_M \end{bmatrix}$ and utilizing (2.5.22), we get

$$R_{M+1}\begin{bmatrix} 0 \\ D''_M \end{bmatrix} = \begin{bmatrix} \sigma_M \\ x_{N-M} \\ \cdot \\ \cdot \\ \cdot \\ x_N \end{bmatrix} \ , \tag{2.5.78}$$

where

$$\sigma_M = [r_{M+1}(0,1),\ldots,r_{M+1}(0,M+1)]D''_M \ .$$

From (2.5.78, 2.5.73), we have

$$R_{M+1}\left\{ \begin{bmatrix} 0 \\ D''_M \end{bmatrix} + \alpha_4 A_{M+1} \right\} = \begin{bmatrix} \sigma_M + \alpha_4 e_{M+1} \\ x_{N-M} \\ \cdot \\ \cdot \\ \cdot \\ x_N \end{bmatrix} \ .$$

The parameter α_4 is chosen such that $\sigma_M + \alpha_4 e_{M+1} = x_{N-M-1}$. Comparing the above expression with (2.5.77), we get

$$D_{M+1} = \begin{bmatrix} 0 \\ D''_M \end{bmatrix} + \alpha_4 A_{M+1} \ . \tag{2.5.79}$$

This is the recursive formula for the order update for D_{M+1}. Utilizing (2.5.46), from the first row in (2.5.79) α_4 can be determined:

$$\alpha_4 = d_{M+1,0} = b_{M+1,N-M-1}/e_{M+1} \ . \tag{2.5.80}$$

(3) From the definition (2.5.38) and the relationships (2.5.75, 2.5.58), we get

$$
\begin{aligned}
g_{M+1} &= (x_{M+2}, x_{M+1}, \ldots, x_1) C_{M+1} \\
&= (x_{M+2}, x_{M+1}, \ldots, x_1) \\
&\quad \times \left\{ \begin{bmatrix} 0 \\ C_M + \beta_2 C_M^{\mathrm{I}} + \gamma_2 D_M^{\mathrm{I}} \end{bmatrix} + \alpha_3 A_{M+1} \right\} .
\end{aligned}
$$

Utilizing (2.5.38, 2.5.59, 2.5.42, 2.5.60, 2.5.48, 2.5.76, 2.5.36), then after some manipulation we obtain

$$
\begin{aligned}
g_{M+1} &= g_M + \frac{|f_{M+1,1}|^2}{e_{M+1}} \\
&\quad + \frac{|v_M|^2 (1 - w_M) + |s_M|^2 (1 - g_M) + 2\mathrm{Re}\{s_M h_M v_M^*\}}{\mathrm{DEN}_M} .
\end{aligned}
$$

$$(2.5.81)$$

This is the recursive formula for the order update for the scalar g_{M+1}.

(4) From the definition (2.5.43) and the relationships (2.5.79, 2.5.61), we get

$$
\begin{aligned}
w_{M+1} &= (x_{N-M-1}^*, x_{N-M}^*, \ldots, x_N^*) D_{M+1} \\
&= (x_{N-M-1}^*, x_{N-M}^*, \ldots, x_N^*) \\
&\quad \times \left\{ \begin{bmatrix} 0 \\ D_M + \beta_3 C_M^{\mathrm{I}} + \gamma_3 D_M^{\mathrm{I}} \end{bmatrix} + \alpha_4 A_{M+1} \right\} .
\end{aligned}
$$

Utilizing (2.5.43, 2.5.62, 2.5.40, 2.5.63, 2.5.41, 2.5.80, 2.5.37), then after some manipulation we obtain

$$
\begin{aligned}
w_{M+1} &= w_M + \frac{|b_{M+1,N-M-1}|^2}{e_{M+1}} \\
&\quad + \frac{|s_M|^2 (1 - w_M) + |u_M|^2 (1 - g_M) + 2\mathrm{Re}\{u_M h_M s_M^*\}}{\mathrm{DEN}_M} .
\end{aligned}
$$

$$(2.5.82)$$

This is the recursive formula for the order update for the scalar w_{M+1}. So far we have obtained all the recursive formulae required in the Marple algorithm.

Step 7. Initial values.

The initial values in the Marple algorithm are parameters of order zero or order one. They are listed below. Each line contains on the left the formula(e) for calculating one parameter, and on the right the expression(s) used to derive the formula(e).

$$e_0 = 2 \sum_{k=1}^{N} |x_k|^2 = 2 \times \text{signal energy} , \qquad (2.5.3, \ 2.5.2, \ 2.5.1) \ ;$$

$$r_1(1,0) = 2 \sum_{k=1}^{N-1} (x_{k+1} x_k^*) , \qquad (2.5.10) \ ;$$

$$f_{0,1} = x_1 , \qquad (2.5.36) \ ;$$

$$b_{0,N} = x_N , \qquad (2.5.37) \ ;$$

$$g_0 = |x_1|^2 / e_0 , \qquad (2.5.38, \ 2.5.45) \ ;$$

$$w_0 = |x_N|^2 / e_0 , \qquad (2.5.43, \ 2.5.46) \ ;$$

$$h_0 = (x_1 x_N)^* / e_0 , \qquad (2.5.39, \ 2.5.45) \ ;$$

$$s_0 = x_1^* x_N / e_0 , \qquad (2.5.40, \ 2.5.45) \ ;$$

$$u_0 = x_N^2 / e_0 , \qquad (2.5.41, \ 2.5.46) \ ;$$

$$v_0 = (x_1^*)^2 / e_0 , \qquad (2.5.42, \ 2.5.45) \ ;$$

$$\text{DEN}_0 = 1 - g_0 - w_0 , \qquad (2.5.55) \ ;$$

$$e_0' = e_0 \alpha_0 = e_0 \text{DEN}_0 = e_0 - |x_1|^2 - |x_N|^2 , \qquad (2.5.51, \ 2.5.56) \text{ or}$$
$$(2.5.18, \ 2.5.3) \ ;$$

$$c_{0,0}'' = x_1^* / e_0' , \qquad (2.5.30, \ 2.5.23) \ ;$$

$$d_{0,0}'' = x_N / e_0' , \qquad (2.5.32, \ 2.5.23) \ ;$$

$$a_{1,1} = -r_1(1,0) / e_0' , \qquad (2.5.13, \ 2.5.10) \ ;$$

$$e_1 = e_0' (1 - |a_{1,1}|^2) , \qquad (2.5.69) \ .$$

Now we are ready to describe the iterative procedure of the Marple algorithm. In short, beginning with $M = 1$ one iteration is completed through the following steps:

$$C_{M-1}'' , \ D_{M-1}'' , \ A_M , \ e_M$$

$$\longrightarrow C_M , \ D_M , \ A_M , \ e_M \qquad \text{(using (2.5.75, 2.5.79), etc.)}$$

$$\longrightarrow C_M'' , \ D_M'' , \ A_M' , \ e_M' \qquad \text{(using (2.5.58, 2.5.61,}$$
$$\text{2.5.52, 2.5.51), etc.)}$$

$$\longrightarrow C_M'' , \ D_M'' , \ A_{M+1} , \ e_{M+1} \qquad \text{(using (2.5.70, 2.5.79), etc.) .}$$

The formula for calculating the spectrum is similar to (2.2.14):

$$S(f) = \frac{e_M}{\left| \sum_{k=0}^{M} a_{M,k} \exp(-2\pi i f k) \right|^2} . \qquad (2.5.83)$$

Note that $S(f)$ here is twice the signal energy spectrum. The power spectrum is $S(f)/(2N)$.

A computer flowchart, a FORTRAN source program and numerical examples of the Marple algorithm are presented in [2.19]. It was found that the commonly used order selection criteria FPE and AIC (Sect. 2.6) were not appropriate (resulting in underestimated order). Therefore, the convergence criteria are based on the change of the prediction error energy instead. Particularly, five exits are set in the program:

(1) STATUS=1: $M = M_{\max}$. The number of normal iterations reaches a prescribed maximum.

(2) STATUS=2: $\mathrm{DEN}_M \leq 0$. Numerically ill-conditioned case.

(3) STATUS=3: $|a_{M,M}| > 1$. Also numerically ill-conditioned case.

(4) STATUS=4: $e_M/e_0 < \mathrm{TOL1}$. The prediction error energy is a small fraction of the total signal energy. $\mathrm{TOL1} = 0.001 \sim 0.01$.

(5) STATUS=5: $(e_{M-1} - e_M)/e_{M-1} < \mathrm{TOL2}$. The relative change of the prediction error energy is small. $\mathrm{TOL2} = 0.001 \sim 0.01$.

STATUS=5 is the most frequently encountered case. As M increases, the decay of e_M becomes very slow or in an oscillating manner. Hence it is hard to select a filter order which is considered to be correct.

A possible scheme for improvement is as follows: (a) Consider the contribution of the last term in expression (2.5.12) to e_M for each M. If this contribution is small, the order M is considered to be probably correct. Otherwise M should be increased. In this way the inferior limit of M is determined. (b) Consider the ratio between the forward and backward prediction error energies. If this ratio is near one, the order M is considered to be probably correct. Otherwise M should be decreased. In this way the superior limit of M is determined. For STATUS=5, the inferior and superior limits of M are determined by (a) and (b), respectively. Thus, a reasonable range of M is determined. The above scheme may also be used for the other four cases together with finding the minimum of e_M. But any criterion for order selection does not always work well, and experience and a trial-and-error procedure are necessary.

The Marple and Burg algorithms are more or less the same in the computational complexity and the space of core memory required. But compared with the Burg algorithm, for the Marple algorithm the peak shift is much smaller and not sensitive to the signal's initial phase, the line splitting is much less serious, and the resolution is somewhat improved. It is worthwhile to use the Marple algorithm, although the issue of order selection needs further study.

2.5.3 Other Fast Algorithms

The Burg and Marple algorithms are recursions with respect to the order M. No normal equations or the like are actually formed. On the contrary, in algorithms of another type to be introduced briefly in the following, normal equations are formed and then solved by fast algorithms. They also belong to the category of the unconstrained LS algorithm.

1. Barrodale-Erickson Algorithm

In the Barrodale-Erickson algorithm, the normal equations are generated recursively with respect to the order M. They are then solved nonrecursively with respect to M using the Cholesky-factorization method. Only real-valued series are treated. A FORTRAN source program is presented in [2.20]. (1) The forward, backward, or forward and backward prediction equation may be selectively used. (2) The range of the order M, $[M_s, M_l]$, in which the normal equations are solved can be chosen. An appropriate order M, $M_s \leq M \leq M_l$, is selected by the FPE criterion. Alternatively, an appropriate order M may be selected by other criteria (this is done not in the program), and then let $M_s = M_l = M$. For a fixed M, the number of real multiplications is $NM + M^2/2 + 3M^2/2$ (using the forward and backward prediction equation; including generating all sets of normal equations up to order M and solving the last set of normal equations of order M).

2. LUD Algorithm

The LUD algorithm (Lower and Upper triangular matrix Decomposition) is the same as the preceding one in methodology [2.21]. But a complex-valued time series can be treated. For a fixed M, the computational complexity is $NM + M^3/6 + M^2/2$ (the number of multiplications; including generating and solving the normal equations of order M). It may be beneficial to compare this figure with $3NM - M^2$ of the Burg algorithm and $NM + 9M^2$ of the Marple algorithm (for the recursion up to order M, though). Comparing the above figures, we can see that in the case where the order M is easy to determine, and N is small and hence M is even smaller, the LUD algorithm is the fastest.

The empirical criterion for order selection is

$$\mathrm{MFPE}_M = \frac{(N + M)N}{(N - M)(N - 2M)} e_M = \min \ .$$

(Cf. the next section.) To find the minimum MFPE_M, it is necessary to calculate e_M nonrecursively for $M_s \leq M \leq M_l$. But e_M is known only after evaluating $a_{M,i}$, $i = 1, \ldots, M$. This results in a great increase in the computational complexity. Therefore, in practice M is normally set to a fixed value depending on N.

For a given M, the solutions by the LUD and Marple algorithms are one and the same since the two algorithms are both of the unconstrained LS type. They are different only in the method of solving the normal equations.

Apart from the fast algorithms described above, many more have been developed. The reader is referred to [2.13, 2.18].

3. Sequential Estimation

The algorithms we have introduced so far all fall into the class of batch estimation, i.e., the estimation is carried out after all the data $\{x_1, \ldots, x_N\}^{\mathrm{T}}$ have been collected. Another class of algorithms is sequential estimation, i.e., the estimation is carried out as data are arriving. The estimated parameters are updated by the new data. For details, the reader is referred to [2.22].

2.5.4 Numerical Example

Only one numerical example concerning MEM1 with a given segment of time series is demonstrated here [2.21]. More numerical examples will be shown in Sects. 4.4, 4.6.

The time series is a real-valued AR process of order 4:

$$x_n = 2.7607x_{n-1} - 3.8106x_{n-2} + 2.6535x_{n-3} - 0.9238x_{n-4} + w_n \, ,$$

where w_n is zero-mean white Gaussian noise of unit power.

Only the result for the correctly selected order ($M = 4$) is shown here. The spectra estimated by the Marple (or LUD) and Burg algorithms are plotted in Fig. 2.10, (a) for $N = 16$ and (b) for $N = 8$. From the figure we see that when $N = 16$, the two algorithms both can resolve the two peaks. However, when $N = 8$, the Marple algorithm is superior to the Burg algorithm in resolvability. The latter can no longer resolve the two peaks.

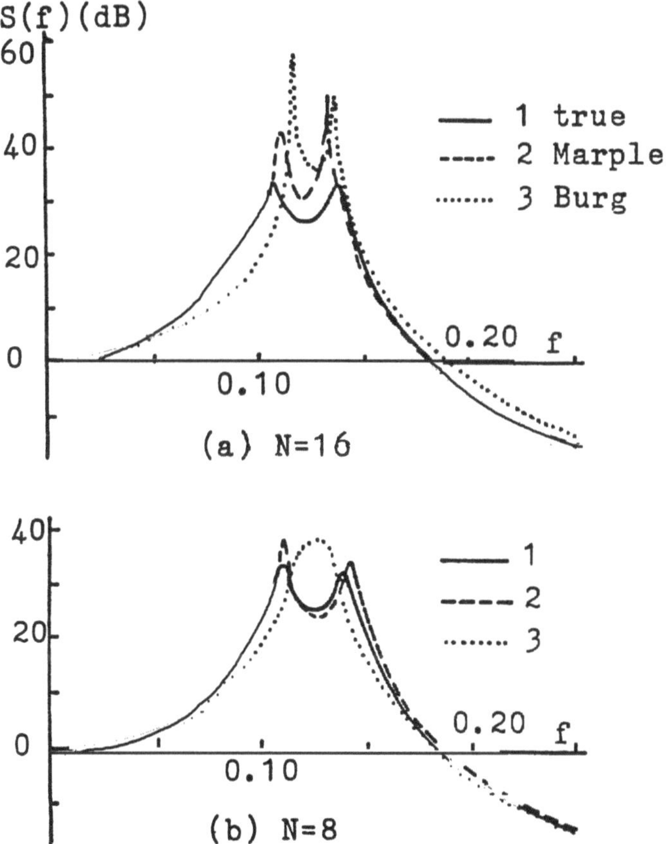

Fig. 2.10. Spectrum estimation of an AR process [2.21]

2.6 Order Selection

In the case where a realization of finite length of the time series is given, the
filter order (AR process order) M must be determined for either estimating
the ACF and then solving the normal equations (or Yule-Walker equations)
in Sect. 2.4 or directly estimating the AR parameters $a_{M,k}$, $k = 1, \ldots, M$
in Sect. 2.5. The order selection especially for the Marple algorithm was
described in Sect. 2.5.2. In this section we are concerned with the order
selection for AR processes in the general case.

Obviously, correctly selecting the AR order is important. If the order is
underestimated (i.e., M is too small), then the resolution of the estimated
MEM1 spectrum is not as high as it should be and the main advantage of
MEM1 may not be observed. On the contrary, if the order is overestimated
(i.e., M is too large), then false peaks and spectral line splitting may result.
Therefore, we must select an "optimal" order, and compromise is usually
necessary.

If the stochastic process, whose spectrum is to be estimated, is really
an AR process of order M (unknown), the problem of order selection is
not so difficult. For example, we may gradually increase the order M in
estimating the AR parameters until the modulus of the reflection coefficient
$|a_{M+1,M+1}| \approx 0$ and the powers $e_{M+1} \approx e_M$, then M is the order of the
process. However, the process under consideration is quite often not a strict
AR process. Generally speaking, e_M decreases rapidly at first and then slowly
with fluctuations as M is increasing. For this reason, the order M, determined
by the conditions concerning $a_{M+1,M+1}$ and e_{M+1} mentioned above, will lie
in a wide range so that the estimated spectra for these M's are significantly
different.

Another kind of way for order selection is concerned with statistical test.
For example, we may test the hypothesis: The residual ε_n, cf. (2.6.1a,b) later,
is white noise. We may also test the hypothesis: $a_{m,k} = 0$, $k \geq M + 1$,
$m \in [M + 1, M_{\max}]$. Methods like the above are usually tedious and may not
be reliable.

In relatively simple and practical methods for order selection, a quantity
("figure of merit") related to M is calculated. The order M is considered to
be correct when this quantity takes the minimum value. The criteria FPE,
AIC and others to be described in this section fall into this category.

2.6.1 FPE Criterion

Suppose the order and parameters of an AR process are unknown. We represent this process by

$$X_n = -\sum_{m=1}^{M} a_{M,m} X_{n-m} + \varepsilon_n \,, \tag{2.6.1a}$$

or

$$\sum_{m=0}^{M} a_{M,m} X_{n-m} = \varepsilon_n \, , \quad (a_{M,0} = 1) \, , \tag{2.6.1b}$$

where the innovation ε_n is a stationary zero-mean Gaussian process, and

$$\mathrm{E}[\varepsilon_m \varepsilon_n^*] = \sigma^2 \delta_{mn} \, , \quad \mathrm{E}[\varepsilon_n X_{n-k}^*] = \sigma^2 \delta_k \, . \tag{2.6.2}$$

X_n is a stationary process with a zero mean. The order M has been selected. ε_n is unobservable, and its variance σ^2 depending on M is also unknown.

The true AR parameters of order M, $a_{M,m}$, satisfy the following Yule-Walker equations:

$$R_M a_M = -r_M \, , \tag{2.6.3a}$$

where the ACF matrix and vectors are

$$R_M = \begin{bmatrix} R(0) & \cdots & R(-M+1) \\ & & \\ & & \\ & & \\ R(M-1) & \cdots & R(0) \end{bmatrix} ,$$

$$a_M = \begin{bmatrix} a_{M,1} \\ \cdot \\ \cdot \\ \cdot \\ a_{M,M} \end{bmatrix} , \quad r_M = \begin{bmatrix} R(1) \\ \cdot \\ \cdot \\ \cdot \\ R(M) \end{bmatrix} . \tag{2.6.3b}$$

These equations may be obtained by reforming (2.3.25).

$$\sigma^2 = \sum_{m=0}^{M} a_{M,m} R(-m) \, , \quad (a_{M,0} = 1) \, , \tag{2.6.3c}$$

which is (2.3.28).

Let $x = \{x_{-M+1}, \ldots, x_0, \ldots, x_N\}^{\mathrm{T}}$ be a sample of X_n. The estimated AR parameters of order M, i.e., the estimates of $a_{M,m}$, $\hat{a}_{M,m}$, satisfy the Yule-Walker equations concerning the estimates:

$$C_M \hat{a}_M = -c_M \, , \tag{2.6.4a}$$

where the elements of the estimated ACF matrix and vectors are

$$c(m,l) = N^{-1} \sum_{n=1}^{N} x_{n-m}^* x_{n-l} \, , \quad c^*(m,l) = c(l,m) \, ,$$

$$\hat{a}_M = \begin{bmatrix} \hat{a}_{M,1} \\ \cdot \\ \cdot \\ \cdot \\ \hat{a}_{M,M} \end{bmatrix} , \quad c_M = \begin{bmatrix} c(1,0) \\ \cdot \\ \cdot \\ \cdot \\ c(M,0) \end{bmatrix} . \tag{2.6.4b}$$

The prediction error power

$$e_M = \sum_{m=0}^{M} \hat{a}_{M,m} c(0,m) \;, \quad (\hat{a}_{M,0} = 1) \;. \tag{2.6.4c}$$

1. The similarity between (2.6.3c) and (2.6.4c) suggests that e_M should be the estimate of σ^2. But this is a biased estimate. In the following we show that

$$\hat{\sigma^2} = \frac{N}{N-M} e_M$$

is an asymptotically ($N \to \infty$) unbiased estimate of σ^2, i.e.,

$$\sigma^2 = \frac{N}{N-M} \mathrm{E}[e_M] \;. \tag{2.6.5}$$

Let $\Delta a_{M,m} = a_{M,m} - \hat{a}_{M,m}$. From (2.6.4c,b) we have

$$
\begin{aligned}
e_M &= \sum_{m=0}^{M} (a_{M,m} - \Delta a_{M,m}) c(0,m) \\
&= \sum_{m=0}^{M} a_{M,m} N^{-1} \sum_{n=1}^{N} x_n^* x_{n-m} - \sum_{m=1}^{M} \Delta a_{M,m} c(0,m) \\
&= H + P \;.
\end{aligned}
\tag{2.6.6}
$$

From (2.6.1b, 2.6.2), the first term above

$$
\begin{aligned}
H &= N^{-1} \sum_{n=1}^{N} x_n^* \sum_{m=0}^{M} a_{M,m} x_{n-m} \\
&= N^{-1} \sum_{n=1}^{N} x_n^* \varepsilon_n \;.
\end{aligned}
$$

The expectation

$$\mathrm{E}[H] = N^{-1} \sum_{n=1}^{N} \mathrm{E}[x_n^* \varepsilon_n] = \sigma^2 \;. \tag{2.6.7}$$

The estimation of the second term P above is rather difficult. From the equation concerning the mth component in (2.6.4a), we get

$$\sum_{l=1}^{M} c(m,l) \hat{a}_{M,l} = -c(m,0) = -c^*(0,m) \;. \tag{2.6.8}$$

Consequently,

$$P = \sum_{m=1}^{M} \Delta a_{M,m} \sum_{l=1}^{M} c^*(m,l) \hat{a}_{M,l}^*$$

$$\begin{aligned}
&= \sum_{l=1}^{M} a_{M,l}^* \sum_{m=1}^{M} \Delta a_{M,m} c(l,m) - \sum_{l=1}^{M}\sum_{m=1}^{M} \Delta a_{M,l}^* c(l,m) \Delta a_{M,m} \\
&= P_1 + P_2 \,,
\end{aligned} \tag{2.6.9}$$

where

$$\Delta a_{M,l}^* = a_{M,l}^* - \hat{a}_{M,l}^* \,.$$

Utilizing (2.6.8) we get

$$\begin{aligned}
\sum_{m=1}^{M} \Delta a_{M,m} c(l,m) &= \sum_{m=1}^{M} a_{M,m} c(l,m) - \sum_{m=1}^{M} \hat{a}_{M,m} c(l,m) \\
&= \sum_{m=1}^{M} a_{M,m} c(l,m) + c(l,0) \,.
\end{aligned}$$

Thus, the first term in (2.6.9)

$$P_1 = \sum_{l=1}^{M}\sum_{m=1}^{M} a_{M,l}^* a_{M,m} c(l,m) + \sum_{l=1}^{M} a_{M,l}^* c(l,0) \,.$$

By the definition of $c(l,m)$ in (2.6.4b), the above expression becomes

$$\begin{aligned}
P_1 &= \sum_{l=1}^{M}\sum_{m=1}^{M} a_{M,l}^* a_{M,m} N^{-1} \sum_{n=1}^{N} x_{n-l}^* x_{n-m} \\
&\quad + \sum_{l=1}^{M} a_{M,l}^* N^{-1} \sum_{n=1}^{N} x_{n-l}^* x_n \\
&= N^{-1} \sum_{n=1}^{N} \left(\sum_{l=1}^{M} a_{M,l}^* x_{n-l}^* \right) \left(\sum_{m=1}^{M} a_{M,m} x_{n-m} \right) \\
&\quad + N^{-1} \sum_{n=1}^{N} x_n \left(\sum_{l=1}^{M} a_{M,l}^* x_{n-l}^* \right) \,.
\end{aligned}$$

The calculation of P_1 can be carried out further by utilizing (2.6.1a):

$$\begin{aligned}
P_1 &= N^{-1} \sum_{n=1}^{N} (\varepsilon_n^* - x_n^*)(\varepsilon_n - x_n) + N^{-1} \sum_{n=1}^{N} x_n(\varepsilon_n^* - x_n^*) \\
&= N^{-1} \sum_{n=1}^{N} (\varepsilon_n^* - x_n^*)\varepsilon_n \\
&= N^{-1} \sum_{n=1}^{N} (\varepsilon_n^* \varepsilon_n - x_n^* \varepsilon_n) \,.
\end{aligned}$$

Therefore, the expectation of P_1 is

$$E[P_1] = N^{-1} \sum_{n=1}^{N} (E[\varepsilon_n^* \varepsilon_n] - E[x_n^* \varepsilon_n]) = 0 .$$

Note that (2.6.2) was utilized in the above. From this result and (2.6.9), we find that the expectation of the second term in (2.6.6) is

$$\begin{aligned} E[P] &= E[P_2] \\ &= -E\left[\sum_{l=1}^{M} \sum_{m=1}^{M} \Delta a_{M,l}^* c(l,m) \Delta a_{M,m} \right] . \end{aligned} \qquad (2.6.10)$$

In order to find the asymptotic value of $E[P]$, we need the following two theorems in probability theory:

Theorem 1. *Suppose that $Z_m, m = 1, \ldots, M$ are independent normal random variables, $Z_m \sim N(0,1)$. Then the random variable*

$$U = \sum_{m=1}^{M} |Z_m|^2$$

has a χ^2-distribution with M degrees of freedom.

In general the random vector $Z = \{Z_1, \ldots, Z_M\}^{\mathrm{T}} \sim N(\mathbf{0}, B)$, B being the covariance (or ACF) matrix, then the quadratic form

$$U = Z^{\mathrm{H}} B^{-1} Z$$

has a χ^2-distribution with M degrees of freedom. The expectation $E[U] = M$.

Theorem 2. *For the AR process represented by X_n in (2.6.1a,b), the limit distribution of $\sqrt{N} \Delta a_{M,m} = \sqrt{N}(a_{M,m} - \hat{a}_{M,m})$ $(m = 1, 2, \ldots, M)$ as $N \to \infty$ is M-dimensional Gaussian: $\sqrt{N} \Delta a_{M,m} \sim N(\mathbf{0}, \sigma^2 R_M^{-1})$, where R_M is the ACF matrix of X_n.*

Now we revert to the estimation in (2.6.10). By Theorems 1, 2, the quadratic form

$$Q = \sum_{l=1}^{M} \sum_{m=1}^{M} \sqrt{N} \Delta a_{M,l}^* (\sigma^2 R_M^{-1})_{l,m}^{-1} \sqrt{N} \Delta a_{M,m}$$

has asymptotically a χ^2-distribution, $E[Q] = M$ $(N \to \infty)$. Q may be written as

$$Q = \sigma^{-2} N \sum_{l=1}^{M} \sum_{m=1}^{M} \Delta a_{M,l}^* (R_M)_{l,m} \Delta a_{M,m} . \qquad (2.6.11)$$

Furthermore, from the assumption of the ergodicity of X_n, $c(l,m) \xrightarrow{\text{a.s.}} (R_M)_{l,m}$ as $N \to \infty$. Comparing (2.6.11) with (2.6.10), we see that, as $N \to \infty$,

$$E[P] = -\frac{\sigma^2}{N} E[Q] = -\frac{M}{N} \sigma^2 . \qquad (2.6.12)$$

Combining (2.6.6, 2.6.7, 2.6.12), it follows that the asymptotic expectation of e_M ($N \to \infty$) is

$$
\begin{aligned}
\mathrm{E}[e_M] &= \mathrm{E}[H] + \mathrm{E}[P] \\
&= \left(1 - \frac{M}{N}\right)\sigma^2 = \frac{N-M}{N}\sigma^2,
\end{aligned}
$$

and (2.6.5) results.

2. When the true AR parameters $a_{M,m}$ are used for the forward prediction of X_n by one step, the prediction error is the innovation ε_n. When we use the estimated AR parameters $\hat{a}_{M,m}$ for the prediction, an additional error will result. The total error depends on $\hat{a}_{M,m}$ and hence on the sample x and the order M selected. When $\hat{a}_{M,m}$ is used to predict X_n, the expectation of the prediction error power is a random variable. The expectation of this random variable with respect to $\hat{a}_{M,m}$ (or equivalently with respect to x) is called the *Final Prediction Error* (FPE). Clearly, the smaller this FPE is, the better it is. So the order M is considered to be correct when the FPE attains its minimum. This is the FPE criterion for order selection [2.23].

Suppose that the segment x (which is used to estimate $\hat{a}_{M,m}$) and X_n (which is to be predicted) are very far apart from each other, so that $\Delta a_{M,m} = a_{M,m} - \hat{a}_{M,m}$ and X_n are completely independent. This assumption is reasonable and necessary for simplifying the calculation. X_n predicted using $\hat{a}_{M,m}$ is

$$
\hat{X}_n = -\sum_{m=1}^{M} \hat{a}_{M,m} X_{n-m} \ .
$$

So we have

$$
\begin{aligned}
\mathrm{FPE} &= \mathrm{E}[|X_n - \hat{X}_n|^2] \\
&= \mathrm{E}\left[\left|\left(X_n + \sum_{m=1}^{M} a_{M,m} X_{n-m}\right) - \sum_{m=1}^{M} \Delta a_{M,m} X_{n-m}\right|^2\right] \\
&= \mathrm{E}\left[\left|\varepsilon_n - \sum_{m=1}^{M} \Delta a_{M,m} X_{n-m}\right|^2\right] \ .
\end{aligned}
$$

Note that (2.6.1a,b) were utilized in the above. Since $\Delta a_{M,m}$ and X_n are independent, we have

$$
\mathrm{E}\left[\varepsilon_n \sum_{m=1}^{M} \Delta a_{M,m}^* X_{n-m}^*\right] = \sum_{m=1}^{M} \mathrm{E}[\Delta a_{M,m}^*]\mathrm{E}[\varepsilon_n X_{n-m}^*] = 0 \ ,
$$

((2.6.2) was used). This means that the crossed product will vanish in expanding the expression of the FPE. Thus, we get

$$\text{FPE} = \text{E}[\varepsilon_n \varepsilon_n^*] + \text{E}\left[\left|\sum_{m=1}^{M} \Delta a_{M,m} X_{n-m}\right|^2\right]$$

$$= \sigma^2 + \sum_{l=1}^{M}\sum_{m=1}^{M} \text{E}[\Delta a_{M,l}^* \Delta a_{M,m}]\text{E}[X_{n-l}^* X_{n-m}]$$

$$= \sigma^2 + \text{E}\left[\sum_{l=1}^{M}\sum_{m=1}^{M} \Delta a_{M,l}^* (R_M)_{l,m} \Delta a_{M,m}\right] .$$

Note that $(R_M)_{l,m} = R(l-m)$ is a deterministic variable, and the expectation is taken only with respect to $\Delta a_{M,m}$. By comparing the second term of the above expression with Q of (2.6.11), it is easy to see that as $N \to \infty$,

$$\text{FPE} = \sigma^2 + \frac{\sigma^2}{N}\text{E}[Q] = \left(1 + \frac{M}{N}\right)\sigma^2 . \tag{2.6.13}$$

Combining (2.6.5, 2.6.13) yields the asymptotic value

$$\text{FPE} = \frac{N+M}{N-M}\text{E}[e_M] . \tag{2.6.14}$$

Generally speaking, given N, the calculated $\text{E}[e_M]$ will decrease but the factor $(N+M)/(N-M)$ will increase as M increases. The optimal order is the M, ranging from zero to a prescribed maximum M_{max}, which minimizes the FPE.

The order selection can also be understood in the following way. It can be seen from the procedure of deriving (2.6.13) that the FPE is composed of two terms. The first, equal to σ^2, is the prediction error power when the true AR parameters of order M are used to predict X_n. σ^2 decreases or remains unchanged as M increases. On the other hand, the second term, equal to $M\sigma^2/N$, is the error power due to the difference between the true and estimated AR parameters of order M. $M\sigma^2/N$ increases as M increases. As a compromise, the order M is selected when the sum of these two terms, i.e., FPE in (2.6.14), is minimum.

Formula (2.6.14) is concerned with the expectation. Usually, only one observation (realization) of the sample of X_n, $x = \{x_{-M+1}, \dots, x_0, \dots, x_N\}^{\text{T}}$, is available. Hence, we can only obtain one value of e_M, but not $\text{E}[e_M]$. $\text{E}[e_M]$ has to be replaced by e_M, and (2.6.14) becomes

$$\text{FPE} = \frac{N+M}{N-M}e_M . \tag{2.6.15}$$

This is the FPE expression used for order selection in practice. Because of the difference between e_M and $\text{E}[e_M]$, the order selected by minimizing the FPE in (2.6.15) may not be correct. Other reasons for the incorrectness of order selection include: N is finite, and the stochastic process is not necessarily a strict AR process.

3. Discussion. (1) In the above the measured data were assumed to be

$$x = \{x_{-M+1}, \ldots, x_0, \ldots, x_N\}^{\mathrm{T}} \, .$$

If the measured data are

$$x' = \{x_1, \ldots, x_N\}^{\mathrm{T}} \, ,$$

then the elements of the ACF matrix can also be estimated by (2.6.4b):

$$c(m, l) = N^{-1} \sum_{n=1}^{N} x_{n-m}^{*} x_{n-l} \, , \quad m, l = 0, \ldots, M \, ,$$

but $x_n = 0$ for $n \leq 0$. Thus, some terms may be missing in the summation. This makes no difference as $N \to \infty$. Expression (2.6.14) is still an asymptotic formula for the FPE, and (2.6.15) can still be used for order selection in practice.

Since the summation is taken actually from $n = M + 1$ to N, we may think that effective data length is $N - M$, and hence modify (2.6.15) to

$$\mathrm{FPE} = \frac{(N - M) + M}{(N - M) - M} e_M = \frac{N}{N - 2M} e_M \, . \tag{2.6.16}$$

We may even go further to combine (2.6.15, 2.6.16) so as to obtain the so-called *modified FPE*:

$$\mathrm{MFPE} = \frac{(N + M)N}{(N - M)(N - 2M)} e_M \, . \tag{2.6.17}$$

It was reported that the MFPE was suitable for a short time series [2.21]. So far we have obtained three formulae (2.6.15–2.6.17) for calculating the FPE. It is hard to say which is generally best for which scheme of AR parameter estimation.

(2) In the above, X_n was assumed to have a zero mean. If this is not true, then there will be an additional term for the mean on the right-hand side of (2.6.1a). As a result, M in (2.6.5, 2.6.13) must be replaced by $M + 1$. (For details, see [2.23].) In such a case (2.6.15) is modified to

$$\mathrm{FPE} = \frac{N + (M + 1)}{N - (M + 1)} e_M \, . \tag{2.6.18}$$

Formulae (2.6.16, 2.6.17) should also be modified accordingly. It should be pointed out that if the mean of X_n is not zero but the average of measured data has been removed before estimating the AR parameters, then the above modification concerning the FPE formula is unnecessary.

2.6.2 AIC Criterion

In the preceding subsection we have derived the asymptotic formula (2.6.15) of the FPE for practical use. From the point of view of information theory, we have only to minimize the logarithm of the FPE in order to minimize the information content of the FPE. Hence we write the following expression for the information content of the prediction error:

$$\text{AIC} \;=\; \log\left(\frac{N+M}{N-M}e_M\right) \approx \log e_M + \frac{2M}{N}\,, \quad (N \gg 1, N \gg M)\,.$$

$$(2.6.19)$$

The order should be selected such that AIC attains its minimum. This is the AIC (An Information theoretic Criterion) [2.24]. Obviously, the FPE criterion and AIC are equivalent as $N \to \infty$. Our task in this subsection is to derive (2.6.19) by the Principle of (generalized) maximum likelihood (ML) and an information theoretic criterion.

1. Maximum Likelihood Estimation of Parameters

For the AR process represented by (2.6.1a,b, 2.6.2), the joint p.d.f. of the sample $X = \{X_1, \ldots, X_N\}^{\mathrm{T}}$ is

$$
\begin{aligned}
f(\boldsymbol{x}; \theta) \\
&= (2\pi)^{-\frac{N}{2}}(\sigma^2)^{-\frac{N}{2}} \\
&\quad \times \exp\left[-\frac{1}{2\sigma^2}\sum_{n=1}^{N}\left(\sum_{l=0}^{M}a_{M,l}x_{n-l}\right)^{*}\left(\sum_{m=0}^{M}a_{M,m}x_{n-m}\right)\right] \\
&= (2\pi)^{-\frac{N}{2}}(\sigma^2)^{-\frac{N}{2}}\prod_{n=1}^{N}\exp\left[-\frac{1}{2\sigma^2}\sum_{l=0}^{M}\sum_{m=0}^{M}a_{M,l}^{*}a_{M,m}x_{n-l}^{*}x_{n-m}\right]\,,
\end{aligned}
$$

$$(2.6.20)$$

where the parameter vector $\theta = (\sigma^2, a_{M,1}, \ldots, a_{M,M})^{\mathrm{T}}$. $f(\boldsymbol{x}; \theta)$ is called the *likelihood function* of θ while $\log f(\boldsymbol{x}; \theta)$ the logarithmic likelihood function L:

$$
\begin{aligned}
L(\theta) &= \log f(\boldsymbol{x}; \theta) \\
&= -\frac{N}{2}\log(2\pi) - \frac{N}{2}\log\sigma^2 \\
&\quad + \sum_{n=1}^{N}\left(-\frac{1}{2\sigma^2}\sum_{l=0}^{M}\sum_{m=0}^{M}a_{M,l}^{*}a_{M,m}x_{n-l}^{*}x_{n-m}\right)\,.
\end{aligned}
$$

$$(2.6.21)$$

The Maximum Likelihood Method (MLM) for parameter estimation is to estimate θ by maximizing f or equivalently $L = \log f$. Specifically, let

$$
\begin{aligned}
0 &= \frac{\partial L}{\partial a_{M,m}} = -\frac{2}{2\sigma^2}\sum_{l=0}^{M}a_{M,l}^{*}\sum_{n=1}^{N}x_{n-l}^{*}x_{n-m} \\
&= -\frac{N}{\sigma^2}\sum_{l=0}^{M}a_{M,l}^{*}c(l,m)\,,
\end{aligned}
$$

where $c(l,m) = \frac{1}{N}\sum_{n=1}^{N} x_{n-l}^* x_{n-m}$ are the elements of the estimated ACF matrix. The estimated $\hat{a}_{M,l}$ satisfy the following equations:

$$\sum_{l=0}^{M} \hat{a}_{M,l} c(m,l) = 0, \quad m = 1, \ldots, M.$$

Note that $c(m,l) = c^*(l,m)$ was utilized in the above. Expanding the above equations, we see immediately that $\hat{a}_{M,l}$ satisfy the Yule-Walker equations (2.6.4a).

Let

$$0 = \frac{\partial L}{\partial \sigma^2} = -\frac{N}{2}\frac{1}{\sigma^2} + \frac{N}{2(\sigma^2)^2} \sum_{l=0}^{M}\sum_{m=0}^{M} a_{M,l}^* a_{M,m} c(l,m).$$

We obtain the ML estimate

$$\begin{aligned}
\hat{\sigma^2}_L &= \sum_{m=0}^{M} \hat{a}_{M,m} \sum_{l=0}^{M} \hat{a}_{M,l}^* c(l,m) \\
&= \sum_{l=0}^{M} \hat{a}_{M,l}^* c(l,0) \\
&= \sum_{l=0}^{M} \hat{a}_{M,l} c(0,l).
\end{aligned}$$

Therefore, $\hat{\sigma^2}_L$ satisfies (2.6.4c). To sum up, when the estimated ACF matrix is known, the parameters estimated by MLM in $\hat{\theta} = (\hat{\sigma^2}_L, \hat{a}_{M,1}, \ldots, \hat{a}_{M,M})^{\mathrm{T}}$ with $\hat{\sigma^2}_L = e_M$ are the solution of the estimated Yule-Walker or normal equations. $c(l,m) \xrightarrow{\text{a.s.}} (R_M)_{l,m}$ as $N \to \infty$; so that if the (true) ACF matrix is known, then the parameters estimated by MLM satisfy the Yule-Walker equations (2.6.3a) with probability one, and hence are equal to the true parameters in $\theta_0 = (\sigma^2, a_{M,1}, \ldots, a_{M,M})^{\mathrm{T}}$. Equation (2.6.21) is the expression of $L(\theta_0)$:

$$\begin{aligned}
L(\theta_0) \\
&= \log f(x; \theta_0) \\
&= -\frac{N}{2}\log(2\pi) - \frac{N}{2}\log\sigma^2 + N\left[-\frac{1}{2\sigma^2}\sum_{l=0}^{M}\sum_{m=0}^{M} a_{M,l}^* a_{M,m} c(l,m)\right].
\end{aligned}$$

$$(2.6.22)$$

On the other hand,

$$\begin{aligned}
L(\hat{\theta}) \\
&= \log f(x; \hat{\theta}) \\
&= -\frac{N}{2}\log(2\pi) - \frac{N}{2}\log e_M + N\left[-\frac{1}{2e_M}\sum_{l=0}^{M}\sum_{m=0}^{M} \hat{a}_{M,l}^* \hat{a}_{M,m} c(l,m)\right]
\end{aligned}$$

$$
\begin{aligned}
&= -\frac{N}{2}\log(2\pi) - \frac{N}{2}\log e_M + N\left(-\frac{1}{2e_M}\cdot e_M\right) \\
&= -\frac{N}{2}\log(2\pi) - \frac{N}{2}\log e_M - \frac{N}{2} \ .
\end{aligned}
\tag{2.6.23}
$$

Combining the above two expressions yields

$$
\begin{aligned}
&L(\hat{\theta}) - L(\theta_0) \\
&\quad = -\frac{N}{2}\log\frac{e_M}{\sigma^2} - \frac{N}{2}\left[1 - \frac{1}{\sigma^2}\sum_{l=0}^{M}\sum_{m=0}^{M}a_{M,l}^{*}a_{M,m}c(l,m)\right] \ .
\end{aligned}
$$

Utilizing (2.6.5),

$$
\begin{aligned}
\text{the first term} &= -\frac{N}{2}\log\left(1 - \frac{M}{N}\right) \approx -\frac{N}{2}\left(-\frac{M}{N}\right) \\
&= \frac{M}{2}\ , \quad (N \gg 1, N \gg M)\ .
\end{aligned}
$$

The estimation of the second term is rather difficult. But the result is simply zero. Specifically,

$$
\begin{aligned}
&\sum_{l=0}^{M}\sum_{m=0}^{M} a_{M,l}^{*}a_{M,m}c(l,m) \\
&= \sum_{l=0}^{M}\sum_{m=0}^{M}(\hat{a}_{M,l}^{*} + \Delta a_{M,l}^{*})(\hat{a}_{M,m} + \Delta a_{M,m})c(l,m) \\
&= \sum_{l=0}^{M}\sum_{m=0}^{M}\hat{a}_{M,l}^{*}\hat{a}_{M,m}c(l,m) + \sum_{l=1}^{M}\sum_{m=1}^{M}\Delta a_{M,l}^{*}\Delta a_{M,m}c(l,m) \\
&\quad + \sum_{l=1}^{M}\Delta a_{M,l}^{*}\sum_{m=0}^{M}\hat{a}_{M,m}c(l,m) + \sum_{m=1}^{M}\Delta a_{M,m}\sum_{l=0}^{M}a_{M,l}^{*}c(l,m)\ ,
\end{aligned}
$$

where

the first term $= e_M$,

the second term $\approx \frac{M}{N}\sigma^2$, (cf. (2.6.10, 2.6.12))),

the third term $=$ the fourth term $= 0$, (calculating by the Yule-Walker equations (2.6.4a)).

From all the results above, we have

$$
\begin{aligned}
L(\hat{\theta}) - L(\theta_0) &\approx \frac{M}{2} - \frac{N}{2}\left[1 - \frac{1}{\sigma^2}\left(e_M + \frac{M}{N}\sigma^2\right)\right] \\
&\approx \frac{M}{2}\ .
\end{aligned}
\tag{2.6.24}
$$

Note that $e_M \approx (1 - M/N)\sigma^2$ was utilized in the above.

2. Information Theoretic Criterion

Denote the true distribution by $g(x) = f(x; \theta_0)$, and define the information content

$$
\begin{aligned}
I(g, f(\cdot\,; \hat{\theta})) &= \int g(x) \log[g(x)/f(x; \hat{\theta})] \mathrm{d}x \\
&= S(g, g) - S(g, f(\cdot\,; \hat{\theta})) \geq 0\,,
\end{aligned}
\tag{2.6.25}
$$

where

$$
S(g, g) = \int g(x) \log g(x) \mathrm{d}x = \mathrm{E}[L(\theta_0)]\,,
$$

$$
S(g, f(\cdot\,; \hat{\theta})) = \int g(x) \log f(x; \hat{\theta}) \mathrm{d}x = \mathrm{E}[L(\hat{\theta})]\,.
$$

I is called the *Kullback information* which is in fact the cross-entropy of $g(x)$ and $f(x; \hat{\theta})$ (Sect. 2.3.2). This is a measure of the distance between g and f. The smaller I is, i.e., the larger $S(g, f(\cdot\,; \hat{\theta}))$ is, the closer the estimated $f(x; \hat{\theta})$ is to the true distribution $g(x)$. But $S(g, f(\cdot\,; \hat{\theta}))$ is the expectation of $L(\hat{\theta})$. Therefore, the information theoretic criterion minimizing I is in line with the *Principle of "generalized" ML* maximizing $\mathrm{E}[L(\hat{\theta})]$.

$S(g, g)$ is the ensemble average of $L(\theta_0)$ and the true parameters in θ_0 are independent of N. So $S(g, g)$ is equal to the time average:

$$
S(g, g) = \mathrm{E}[L(\theta_0)] \approx \frac{1}{N} L(\theta_0)\,.
\tag{2.6.26}
$$

The equality is approximate because N is finite. This approximation cannot be used for $\mathrm{E}[L(\hat{\theta})]$ because the estimated parameters in $\hat{\theta}$ depend on N.

Now we estimate $S(g, f(\cdot\,; \hat{\theta})) = \mathrm{E}[L(\hat{\theta})]$. Expanding $S(g, f(\cdot\,; \hat{\theta}))$ in a Taylor series, we get

$$
\begin{aligned}
S(g, f(\cdot\,; \hat{\theta})) = {}& S(g, g) \\
&+ \frac{1}{2} \sum_{i=0}^{M} \sum_{j=0}^{M} (\hat{\theta} - \theta_0)^*)(\hat{\theta} - \theta_0) \frac{\partial^2}{\partial \theta_i^* \partial \theta_j} \mathrm{E}[\log f(x; \theta)] \bigg|_{\theta = \theta_0}\,.
\end{aligned}
\tag{2.6.27}
$$

The term containing the first partial derivatives vanishes since θ_0 is the ML estimate. At $\theta = \theta_0$, the ensemble average may be replaced by the time average, so the second derivative is equal approximately to

$$
\frac{1}{N} \frac{\partial^2}{\partial \theta_i^* \partial \theta_j} \log f(x; \theta) \bigg|_{\theta = \theta_0}\,.
$$

We use (2.6.21) to calculate:

$$
\frac{\partial^2 L(\theta)}{\partial (\sigma^2)^2} \bigg|_{\theta = \theta_0} = \frac{N}{2} \frac{1}{(\sigma^2)^2} - \frac{N}{(\sigma^2)^3} \sum_{l=0}^{M} \sum_{m=0}^{M} a_{M,l}^* a_{M,m} c(l, m)
$$

$$\approx \; \frac{N}{2}\frac{1}{(\sigma^2)^2} - \frac{N}{(\sigma^2)^3}\left(e_M + \frac{M}{N}\sigma^2\right)$$

$$= \; -\frac{N}{2}\frac{1}{(\sigma^2)^2}\;,$$

cf. the derivation of (2.6.24);

$$\left.\frac{\partial^2 L(\theta)}{\partial a_{M,i}^* \partial a_{M,j}}\right|_{\theta=\theta_0} = \left.\frac{\partial^2 L(\theta)}{\partial a_{M,j}\partial a_{M,i}^*}\right|_{\theta=\theta_0} = -\frac{N}{\sigma^2}c(i,j)\;;$$

$$\left.\frac{\partial^2 L(\theta)}{\partial \sigma^2 \partial a_{M,j}}\right|_{\theta=\theta_0} = \frac{N}{2(\sigma^2)^2}\cdot 2\sum_{l=0}^{M} a_{M,l}^* c(l,j) \approx 0\;,\quad j=1,\dots,M\;;$$

$$\left.\frac{\partial^2 L(\theta)}{\partial a_{M,i}^* \partial \sigma^2}\right|_{\theta=\theta_0} = \frac{N}{(\sigma^2)^2}\sum_{m=0}^{M} a_{M,m}c(l,m) \approx 0\;,\quad i=1,\dots,M\;.$$

Substituting the above expressions of the second derivatives in (2.6.27) yields

$$S(g,f(\cdot\,;\hat{\theta})) \;\approx\; S(g,g) + \frac{1}{2}(e_M - \sigma^2)^2\left[-\frac{1}{N}\cdot\frac{N}{2}\frac{1}{(\sigma^2)^2}\right]$$

$$+\frac{1}{2}\sum_{i=1}^{M}\sum_{j=1}^{M}\Delta a_{M,i}^* \Delta a_{M,j}\left[-\frac{1}{N}\cdot\frac{N}{\sigma^2}c(i,j)\right]$$

$$\approx\; -\frac{1}{4}\left(\frac{M}{N}\right)^2 - \frac{1}{2\sigma^2}\cdot\frac{M}{N}\sigma^2$$

$$\approx\; -\frac{M}{2N}\;,$$

cf. the derivation of (2.6.24). Putting the above expression and (2.6.24) together and noting (2.6.26), we have

$$\begin{cases} S(g,f(\cdot\,;\hat{\theta})) \;=\; \mathrm{E}[L(\hat{\theta})] \approx \dfrac{1}{N}L(\theta_0) - \dfrac{M}{2N}\;, \\[2mm] \dfrac{1}{N}L(\hat{\theta}) \;\approx\; \dfrac{1}{N}L(\theta_0) + \dfrac{M}{2N}\;. \end{cases}$$

The solution is

$$\mathrm{E}[L(\hat{\theta})] = \frac{1}{N}L(\hat{\theta}) - \frac{M}{N}\;. \tag{2.6.28}$$

From this we see that the ensemble average of $L(\hat{\theta})$, $\mathrm{E}[L(\hat{\theta})]$, is less than its time average by M/N. This is because $\mathrm{E}[L(\hat{\theta})]$ is less while $1/N\cdot L(\hat{\theta})$ is greater than $\mathrm{E}[L(\theta_0)] = 1/N\cdot L(\theta_0)$ both by $M/(2N)$. The small quantities of higher order than M/N have been discarded in estimating $\mathrm{E}[L(\hat{\theta})]$, etc.

According to the information theoretic criterion, we require I be minimum or equivalently $\mathrm{E}[L(\hat{\theta})]$ be maximum. Substituting (2.6.23) into (2.6.28) and discarding the constants, we see that this requirement is equivalent to select the order M by minimizing

$$\text{AIC} = \log e_M + \frac{2M}{N} \,. \tag{2.6.29}$$

This is the AIC for order selection. Note that e_M is the prediction error power of the prediction filter of order M. As for the FPE criterion, here e_M has substituted $\mathrm{E}[e_M]$.

2.6.3 Other Criteria

1. AICα (BIC)

In the preceding subsection the quantity to be maximized is

$$2N\mathrm{E}[L(\hat{\theta})] = 2L(\hat{\theta}) - 2 \cdot M \,, \tag{2.6.30}$$

which was obtained by the principle of ML. This principle is a special case of the Bayesian method: The prior p.d.f. is assumed to be constant. In the Bayesian method the prior p.d.f. is assigned (generally not constant) and the posterior p.d.f., instead of $2N\mathrm{E}[L(\hat{\theta})]$ in (2.6.30), is maximized. Then the factor before M in the second term on the right-hand side of (2.6.30) is not necessarily 2, and may be 3, 4, $\log N$, etc. depending on the prior p.d.f.. Denoting this factor by α and we have the criteria AICα for order selection. They are listed below:

$$
\begin{aligned}
\text{AIC2} &= \log e_M + \tfrac{2M}{N} \,, \quad \text{(Akaike AIC)} \,, \\
\text{AIC3} &= \log e_M + \tfrac{3M}{N} \,, \\
\text{AIC4} &= \log e_M + \tfrac{4M}{N} \,, \\
\text{AIC}\ell &= \log e_M + \tfrac{\ell M}{N} \,, \quad (\ell = \log N) \,.
\end{aligned}
\tag{2.6.31}
$$

Normally, the order M selected by the Akaike AIC is the highest. (This requires $N \geq 8$ for AICℓ).

The AICα presented here is also called BIC because the use of the Bayesian method in the derivation [2.25].

2. CAT Criterion

From the formula of the MEM1 spectral estimation

$$S(f) = \sigma^2 / |g_L(\mathrm{e}^{-2\pi \mathrm{i}f})|^2$$

or

$$\sigma^2 = S(f)|g_L(\mathrm{e}^{-2\pi \mathrm{i}f})|^2 \,,$$

we see that $g_L(\mathrm{e}^{-2\pi \mathrm{i}f})$ is the transfer function of a filter which converts the signal with the power spectrum $S(f)$ to white Gaussian noise with power σ^2 (Sect. 2.3.3). Here the polynomial of $\mathrm{e}^{-2\pi \mathrm{i}f}$

$$g_L(\mathrm{e}^{-2\pi \mathrm{i}f}) = \sum_{l=0}^{L} a_{L,l}(\mathrm{e}^{-2\pi \mathrm{i}f})^l \,,$$

where L is the true order of the AR process. ($L = \infty$ for an ARMA process.) The estimated transfer function is

$$\hat{g}_M(e^{-2\pi i f}) = \sum_{l=0}^{M} \hat{a}_{M,l}(e^{-2\pi i f})^l .$$

The distance between g_L and $\hat{g}_M$ is defined by

$$J_g = \int_{-1/2}^{1/2} \left| \frac{g_L(e^{-2\pi i f}) - \hat{g}_M(e^{-2\pi i f})}{g_L(e^{-2\pi i f})} \right|^2 \mathrm{d}f , \tag{2.6.32}$$

whose asymptotic value is, after discarding the constants,

$$\mathrm{CAT} = \frac{1}{N} \sum_{m=1}^{M} \frac{N - m}{N e_m} - \frac{N - M}{N e_M} , \tag{2.6.33}$$

where N is the number of the datum points, e_m is the prediction error power of the prediction filter of order m.

The CAT (Criterion of Autoregressive Transfer function) states that the order M, ranging from 1 to the prescribed maximum $M_{\max}$, should be selected such that distance J_g between the true and estimated transfer functions g_L and $\hat{g}_M$, or equivalently CAT in (2.6.33), is minimum [2.26].

3. PLS Criterion

Given data $\boldsymbol{x}_N = \{x_1, \ldots, x_N\}^{\mathrm{T}}$. For $\boldsymbol{x}_n = \{x_1, \ldots, x_n\}^{\mathrm{T}}$, $n \le N$, the "true" prediction error is defined as the error in linearly predicting x_n from $\boldsymbol{x}_{n-1} = \{x_1, \ldots, x_{n-1}\}^{\mathrm{T}}$:

$$\hat{e}_{M,n} = x_n + \sum_{i=1}^{M} \hat{a}_{M,i,n} x_{n-i} , \tag{2.6.34}$$

where M is the filter order, $\hat{a}_{M,i,n}$ are calculated from $\boldsymbol{x}_{n-1}$ by the LS method. Here "true" refers to the fact that only the "past" data (before n) are used in estimating $\hat{a}_{M,i,n}$.

The PLS (Predictive Least Squares) criterion states that the order M ($1 \le M \le M_{\max}$) should be selected such that ($1/N$ times) the sum of the squares of the true prediction errors, PLS(M), is minimum [2.27]:

$$\mathrm{PLS}(M) = \frac{1}{N} \sum_{n=1}^{N} (\hat{e}_{M,n})^2 = \min . \tag{2.6.35}$$

Here we see that in order to select M, PLS(M) needs to be calculated by (2.6.35) for each M ranging from 1 to $M_{\max}$. In doing this for a fixed M, $\hat{e}_{M,n}$ in turn needs to be calculated by (2.6.34) for each n ($n = 1, \ldots, N$) using the LS method. Hence the computational complexity would be very high. Fortunately, a very fast algorithm (lattice filters) recursive in both M and n is available for calculating $\hat{e}_{M,n}$ and other parameters. This makes the PLS criterion usable in practice.

The calculation of $\hat{e}_{M,n}$ and other parameters may be carried out as data are coming in, since no "future" data are needed in the calculation. After collecting all the data, the optimal order M is selected and the AR parameters $\hat{a}_{M,i}$ and $\hat{e}_M$ are calculated recursively from $\hat{e}_{M,n}$ and other parameters.

4. ALF Method

Assuming for the moment that the filter order M is known, an AR process is represented by

$$y_n = \sum_{i=1}^{M} \alpha_{M,i} y_{n-i} + v_n ,$$

where $\{v_n\}$ is zero-mean white noise. Let $\boldsymbol{x}_n = (\alpha_{M,1}, \ldots, \alpha_{M,M})^{\mathrm{T}}$ (constant) and $\boldsymbol{h}_n = (y_{n-1}, \ldots, y_{n-M})^{\mathrm{T}}$, the above equation can be reformulated in the state-space form:

$$\begin{aligned} \boldsymbol{x}_{n+1} &= \boldsymbol{x}_n , && \text{(state transition equation)} , \\ y_n &= \boldsymbol{h}_n^{\mathrm{T}} \boldsymbol{x}_n + v_n , && \text{(observation equation)} . \end{aligned}$$

Now let us drop out the assumption that the order M is known, and suppose only that $1 \leq M \leq M_{\max}$. Then we can utilize a state-space estimation method, called the Adapt Lainiotis Filter (ALF), to estimate M and $\boldsymbol{x}_n$ simultaneously [2.28]. Given the data up to and including y_n, the conditional posterior probability density $P(M|n)$ can be calculated. The maximization of $P(M|n)$ gives the most likely order M, and the corresponding estimate $\hat{\boldsymbol{x}}(n|n, M)$ is calculated subsequently. Finally, the parameter $\hat{\boldsymbol{x}}(n|n)$ having the estimated $\alpha_{M,i}$ as its components is obtained by the integration (summation) of $\hat{\boldsymbol{x}}(n|n, M)P(M|n)$ over M.

5. PDC Criterion

Consider an AR process represented by

$$y_n = -\sum_{i=1}^{M} a_{M,i} y_{n-i} + w_n ,$$

for which the order is unknown but its range $[0, M_{\max}]$. Denote the probability density of y_n predicted from the past data $y_{1 \cdot n-1} \overset{\triangle}{=} (y_1, \ldots, y_{n-1})^{\mathrm{T}}$ by $f(y_n | y_{1 \cdot n-1}, M)$, where M is the assumed order.

Suppose we have a sample of the process, $y_{1 \cdot N} \overset{\triangle}{=} (y_1, \ldots, y_N)^{\mathrm{T}}$. It is possible to write out the expression of $f(y_n | y_{1 \cdot n-1}, M)$ utilizing the Bayesian method based on some assumptions like noninformative priors. Then the optimal estimate of the order M is determined by the condition

$$\sum_{n=2}^{N} -\log f(y_n | y_{1 \cdot n-1}, M) = \min .$$

This is called the Predictive Density Criterion (PDC) for order selection. The detailed calculations and experimental comparison between PDC and a number of other criteria can be found in [2.29].

2.6.4 Summary

Various criteria for the order selection for AR processes have been described in this section. They are based on some kinds of statistical methods. In practice only one segment of data is available. As a result, the correctness of the selected order is not guaranteed even for true AR processes. Monte Carlo experiments have shown that various criteria for order selection are different in reliability. In brief, the problem of order selection needs further study. At present, experience and the trial-and-error method are good remedies for the existing theoretical results.

3. Maximum Entropy Method MEM2 and Its Application in Image Restoration

This chapter is concerned with MEM2, namely, MEM with the entropy expression

$$H2 = -\int S(f) \log[S(f)/m(f)]\mathrm{d}f \ ,$$

or its plain form

$$H2 = -\int S(f) \log S(f)\mathrm{d}f \ .$$

The organization of this chapter is more or less parallel to that of Chap. 2 but with more emphasis on algorithms. In signal processing, MEM2 is sometimes used for 1-D spectral analysis, but is mainly used for image restoration (virtually 2-D spectral analysis). Technical terms and notations are of a wide variety due to a wide range of applications. We will use terms and notations of spectral analysis and those of image processing alternatively, depending on the problem addressed.

In this chapter, Sect. 3.1 is devoted to the definition of the entropy expression $H2$. Then, in Sect. 3.2 the problem of spectral analysis by MEM2 is formulated and its implicit (indirect, or iterative) solution in the 1-D case is determined utilizing the Cepstral Analysis Method (CAM). An iterative algorithm is presented. In Sect. 3.3 an explicit (direct, or noniterative) solution is worked out for a real, causal and minimum-phase 1-D series. Its applicability and limitation are discussed in detail and demonstrated by numerical examples. Section 3.4 is concerned with five equivalents to MEM2 and a signal model. The remaining four sections are devoted to numerical algorithms and examples; namely, Sect. 3.5 is for the $R - \lambda$ procedure of spectral analysis, and Sects. 3.6–3.8 are for a variety of algorithms of image restoration.

Two appendices for this chapter are included at the end of this monograph: A.) Cepstral Analysis, and B.) Image Restoration. They provide the minimum relevant knowledge for understanding this chapter. For readers who lack the background of cepstral analysis and image restoration, reading these two appendices first is essential. As pointed out at the beginning of Chap. 2, readers who are interested mainly in the MEM application should concentrate on algorithms.

3.1 Definition and Expressions of Entropy *H*2

The entropy $H2$ is directly defined in the frequency domain. Since MEM2 was first introduced and is used mainly in image restoration, technical terms and notations in image processing are used in deriving the expressions of $H2$. Then, the equivalent expressions in spectral analysis are simply written out by analogy. In the following, the Maximum Likelihood (meaning here maximum multiplicity) Method (MLM) and then the direct definition method are presented. This section concludes with a discussion.

3.1.1 MLM

Suppose that a (1-D or 2-D) object or true image is composed of J pixels numbered by $1, \ldots, J$. The intensity of each pixel is quantized, the intensity unit being ΔB. Then, the intensity of pixel j is $B(j)\Delta B$, where $B(j)$ is a nonnegative integer. The intensity of pixel j may be represented simply by $B(j)$ when the unit ΔB has been chosen.

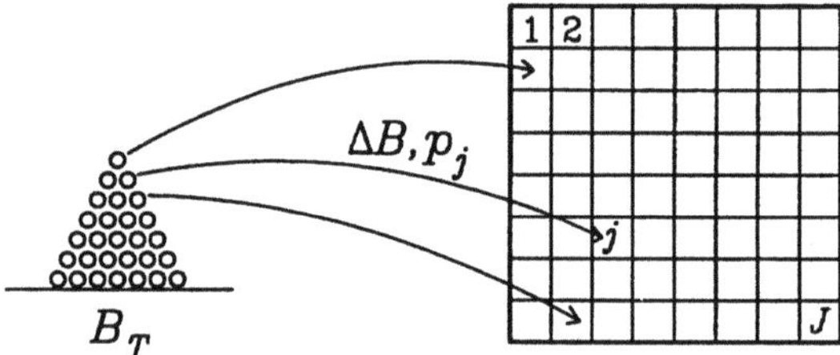

Fig. 3.1. Formation of an image whose intensities are randomly distributed

We imagine that an image is formed as follows (Fig. 3.1): At the beginning, all the pixels are zero in intensity, and the image is "blank". Many silver grains, each with unit intensity, are heaped up nearby. A monkey (or other reasonable being) picks up a grain and throws it to the above of the image. Then, the monkey does the same thing with the second, third, ... grains until the heap disappears. Suppose that the monkey is fair and square, and not biased in favor of any particular pixel. Suppose also that the gravity field is uniform and there is no mutual interaction between grains. Therefore, which pixel j a grain falls into is completely random. That the intensity of pixel j is $B(j)$ means $B(j)$ grains have fallen into this pixel.

Suppose that the total number of the grains, B_T, is very large. B_T is the total intensity, total power (TP) or total flux of the image:

$$B_T = \sum_{j=1}^{J} B(j) \, . \tag{3.1.1}$$

An image is formed, or in other words, a realization of the image with randomly distributed intensities is obtained after the B_T grains have been thrown and have fallen onto the image. In the same way the second, third, ... images are formed. An image ensemble consists of all these images. By the formula of multinomial distribution, the multiplicity of images with a particular intensity distribution $[B(1), \ldots, B(J)]$, i.e., the number of ways in which such images are formed, is

$$W(B(1), \ldots, B(J)) = \frac{B_T!}{B(1)!B(2)! \cdots B(J)!} .\tag{3.1.2}$$

The most likely distribution of $B(j)$ is obtained by maximizing W. This is the MLM.

To facilitate mathematical operations, $\log W$ is used instead of W in the maximization. We also need Stirling's formula:

$$\log N! \sim N(\log N - 1) , \quad N \to \infty .\tag{3.1.3}$$

Assuming that B_T and $B(j)$ are very large, taking the logarithm of (3.1.2) and using (3.1.3), we get

$$
\begin{aligned}
\log W &= \log \left(B_T! \bigg/ \prod_{j=1}^{J} B(j)! \right) \\
&= \log B_T! - \sum_{j=1}^{J} \log B(j)! \\
&\approx B_T(\log B_T - 1) - \sum B(j)[\log B(j) - 1] \\
&= -\sum B(j) \log B(j) + \beta(B_T) ,
\end{aligned}
\tag{3.1.4}
$$

where $\beta(B_T) = B_T \log B_T$. On condition that B_T is constant, β is also constant, and maximizing $\log W$ is equivalent to maximizing $-\sum B(j) \log B(j)$. This quantity is of the same form as $-\sum p_j \log p_j$, the entropy expression in information theory, and is also called entropy. To show the difference from $H1$, this quantity is denoted by $H2$. Hence,

$$H2 = -\sum_{j=1}^{J} B(j) \log B(j) .\tag{3.1.5}$$

Now we no longer require that $B(j)$ should be a nonnegative integer. $B(j)$ has only to be a nonnegative real number.

Thus, we have derived the expression of the entropy $H2$ by MLM [3.1]. MEM with $H2$ as the entropy definition is called MEM2. Note that going from (3.1.4) to (3.1.5) requires B_T to be constant. Therefore, normally the condition $B_T = \sum B(j) = $ constant (TP constraint) should be imposed in maximizing $H2$ in MEM2. This condition may be implicitly included in given data, or ignored in special cases.

3.1.2 Direct Definition Method

Because the events in which "a silver grain falls into particular pixels" are mutually independent, the proportion $B(j)/B_T$ converges (in probability) to the probability that a grain falls into pixel j according to the law of large numbers. Hence, this probability may be approximated by $B(j)/B_T$, and we have

$$H2 = -\sum_{j=1}^{J} p_j \log p_j \, , \tag{3.1.6}$$

where p_j is the proportion,

$$p_j = B(j)/B_T \, . \tag{3.1.7}$$

This is the direct definition of $H2$ [3.2, 3.3]. $H2$ defined as such is equivalent to (3.1.5). Since B_T is assumed to be constant, from (3.1.6, 3.1.7) we have

$$\begin{aligned}
H2 &= -\sum [B(j)/B_T] \log[B(j)/B_T] \\
&= -\frac{1}{B_T} \sum B(j)[\log B(j) - \log B_T] \\
&= -\frac{1}{B_T} \sum B(j) \log B(j) + \log B_T \, ,
\end{aligned}$$

which is different from (3.1.5) only by a constant factor and a constant term, which are of no significance in the maximization.

The distribution $\{p_j\}$ represents the shape, pattern or configuration of an image, but not its absolute intensities. Hence the name *configurational entropy* for $H2$.

It should be noticed that p_j in (3.1.7) is proportion, but not probability. As stated in [3.4], it makes no sense to talk about probability with a single image. To avoid the conceptual difficulty, $H2$ should take the form

$$H2 = -\sum B(j) \log[B(j)/m(j)] \, , \tag{3.1.8}$$

where $m(j)$ is the prior estimate and $B(j)$ is the posterior estimate of the intensity distribution. $H2$ should be called "relative entropy". Maximizing $H2$ under constraints means making the posterior similar to the prior to the greatest extent possible (the point of view of MCE). A reply to the above argument is [3.5]: the concept entropy is applicable not only to probability, but also to a set of proportions. So is the idea of MEM. We may simply say that a set of proportions $\boldsymbol{p} = \{p_1, \ldots, p_n\}$ has its configurational entropy $H = -\sum p_j \log p_j$. A well-known example of this is the entropy of the proportions or frequencies of the occurrence of the English letters (or other characters as well) in a book.

To summarize, putting the argument in concept aside, the entropy expression is just the same regardless of whether p_j is interpreted as probability or proportion, and $H2$ as probabilistic or configurational entropy in image restoration by MEM.

3.1.3 Discussion

(1) *Expressions of H2 in Spectral Analysis and Discrete Forms.* By analogy with (3.1.5), $H2$ in spectral analysis is

$$H2 \;=\; -\sum_{j} S(j) \log S(j) , \quad \text{(discrete form)} , \tag{3.1.9}$$

$$H2 \;=\; -\int S(f) \log S(f) \mathrm{d}f , \quad \text{(continuous form)} . \tag{3.1.10}$$

Changing the arguments to $(j_1,\ j_2)$ or $(f_1,\ f_2)$ yields the 2-D expressions.

It is worthwhile to point out that $H2$ is defined originally in the frequency domain. Its definition has nothing to do with the time domain. In contrast, $H1$ is originally defined in the time domain and its simple expression in the frequency domain is derived. Then a question arises: As $H2$ is defined in the frequency domain, what is its expression in the time domain? This is an interesting and open question, which is related to the data extension in the time domain.

(2) *H2 Expressions with Measure.* In Sect. 1.2 we obtained the entropy expressions (1.3, 1.4) with measure in consideration of the invariance of entropy with regard to coordinate transformation. Now we would like to derive the entropy expression with measure:

$$H2 = -\sum_{j=1}^{J} p_j \log(p_j/m_j) \tag{3.1.11}$$

specifically in image processing. Here p_j is interpreted as the probability that a silver grain falls into pixel j.

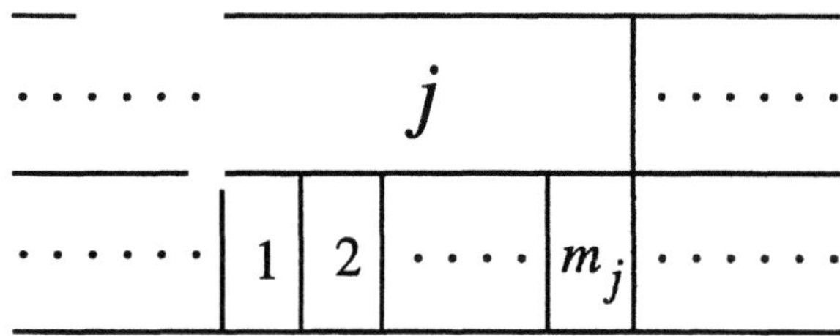

Fig. 3.2. Pixel j and its subpixels $1, 2, \ldots, m_j$

In deriving the $H2$ expression by MLM, it was assumed that all the pixels were equal in the sense that p_j was independent of j. Now we assume that pixel j consists of m_j subpixels (Fig. 3.2). This time it is all the subpixels that are equal, i.e., a grain falls into each subpixel with equal probability. By the composition law of entropy (Sect. 1.2), we have

$$H2 = H(p_1, \ldots, p_J) + \sum_{j=1}^{J} p_j H(p_{j1}, \ldots, p_{jm_j} | j) \,,$$

where $H(\ldots | j)$ in the last term represents the uncertainty as to which sub-pixel of pixel j a grain falls into, given that this grain is known to have fallen into pixel j. The conditional probability is

$$p_{ji}|j = \frac{p_j/m_j}{p_j} = \frac{1}{m_j} \,, \quad i = 1, \ldots, m_j \,.$$

So we have

$$
\begin{aligned}
H2 &= H(p_1, \ldots, p_J) + \sum_{j=1}^{J} p_j H \left(\overbrace{\frac{1}{m_j}, \ldots, \frac{1}{m_j}}^{m_j \text{ items}} \right) \\
&= -\sum_{j=1}^{J} p_j \log p_j + \sum_{j=1}^{J} p_j \log m_j \\
&= -\sum_{j=1}^{J} p_j \log(p_j/m_j) \,.
\end{aligned}
$$

Q.E.D.   In terms of $B(j)$, $H2$ takes the form of (3.1.8).

Maximizing $H2$ in (3.1.11) or (3.1.8), without any measured data, yields

$$p_j \propto m_j$$

or

$$B(j) \propto m(j) \,.$$

Hence m_j or $m(j)$ is the prior knowledge about the intensity of pixel j. If we know nothing about the intensity distribution of an image beforehand, we take $m_j = 1$ or $m(j) = 1$ (or other constants). In this case we say that the measure is uniform, and (3.1.11, 3.1.8) degenerate to (3.1.6, 3.1.5), respectively. Here we discuss the problem of the measure m_j in some detail because the prior knowledge about an image is usually taken into consideration in image restoration by MEM2.

Now the $H2$ expressions in spectral analysis are

$$H2 = -\sum S(j) \log[S(j)/m(j)] \,, \quad \text{(discrete form)} \,, \tag{3.1.12}$$

$$H2 = -\int S(f) \log[S(f)/m(f)]\mathrm{d}f \,, \quad \text{(continuous form)} \,. \tag{3.1.13}$$

In image restoration, other $H2$ expressions of somewhat different forms are also in use. They will be introduced in due course in describing MEM2 algorithms.

3.2 Formulation and Implicit Solution

In the previous section we have obtained the $H2$ expressions in the frequency domain. In this section we derive an implicit MEM2 solution, given 1-D noiseless data, the ACF (autocorrelation function), by CAM (Cepstral Analysis Method). A computer flowchart is presented. We will use technical terms and notation of spectral analysis as used in Chap. 2.

3.2.1 Formulation

First of all, MEM2 is formulated. This is a problem of constrained maximization. Specifically,

$$\text{Maximization}: \quad H2 = -\int_{-1/2}^{1/2} S(f) \log S(f) \mathrm{d}f , \tag{3.2.1}$$

$$\text{Constraints}: \quad R(k) = \int_{-1/2}^{1/2} S(f) \exp(2\pi i f k) \mathrm{d}f , \quad |k| \le M . \tag{3.2.2}$$

(The case in which the measure $m(f)$ is not uniform will be dealt with later.) The TP constraint is included in (3.2.2) since

$$R(0) = \int_{-1/2}^{1/2} S(f) \mathrm{d}f , \quad (k = 0) \tag{3.2.3}$$

is just the TP.

3.2.2 Implicit Solution

It can be shown that the solution to the above problem is unique, i.e., the maximal point must be unique (Sect. 4.3). The statement of the MEM2 problem is much the same as that of MEM1 except the entropy expression. However, this difference is significant so that the MEM1 and MEM2 solutions are different in structure.

As in MEM1, now the Lagrange multiplier method is used. The objective functional is

$$\begin{aligned}
G &= -\int_{-1/2}^{1/2} S(f) \log S(f) \mathrm{d}f \\
&\quad + \sum_{k=-M}^{M} \lambda_k \left[\int_{-1/2}^{1/2} S(f) \exp(2\pi i f k) \mathrm{d}f - R(k) \right] ,
\end{aligned}$$

where λ_k's are the Lagrange multipliers, $\lambda_{-k} = \lambda_k^*$ since $R(-k) = R^*(k)$. λ_0 is the Lagrange multiplier related to the TP ($R(0)$) constraint.

Taking the variation of G with respect to $S(f)$ and letting it be zero, we have

$$0 = \delta G = -\int \left[\log S(f) + S(f) \cdot \frac{1}{S(f)} \right] \delta S(f) \mathrm{d}f$$

$$+ \sum_{k=-M}^{M} \lambda_k \int \exp(2\pi \mathrm{i} f k) \delta S(f) \mathrm{d}f$$

$$= \int \delta S(f) \left\{ -[\log S(f) + 1] + \sum_{k=-M}^{M} \lambda_k \exp(2\pi \mathrm{i} f k) \right\} \mathrm{d}f .$$

Since $\delta S(f)$ is arbitrary, the quantity in the braces must be zero. Hence,

$$\log S(f) = -1 + \sum_{l=-M}^{M} \lambda_l \exp(2\pi \mathrm{i} f l) ,$$

i.e.,

$$S(f) = \exp \left[-1 + \sum_{l=-M}^{M} \lambda_l \exp(2\pi \mathrm{i} f l) \right]$$

$$= \exp \left[-1 + \sum_{l=-M}^{M} \lambda_l^* \exp(-2\pi \mathrm{i} f l) \right] . \tag{3.2.4}$$

Adding $\lambda_l = 0$, $|l| > M$ and merging the constant -1 into λ_0, (3.2.4) becomes

$$S(f) = \exp \left[\sum_{l=-\infty}^{\infty} \lambda_l^* \exp(-2\pi \mathrm{i} f l) \right] , \tag{3.2.5}$$

or, equivalently,

$$\sum_{l=-\infty}^{\infty} \lambda_l^* \exp(-2\pi \mathrm{i} f l) = \log S(f) , \tag{3.2.6}$$

where $\lambda_{-M}, \ldots, \lambda_0, \ldots, \lambda_M$ are to be determined.

Thus, we have parametrized the spectrum $S(f)$ by λ_l, $|l| \leq M$. Having determined λ_l for $|l| \leq M$, we can then easily calculate $S(f)$. But the determination of λ_l from the given ACF is very difficult as we will see below.

Taking the IFT of (3.2.6) with respect to f yields

$$\sum_{l=-\infty}^{\infty} \lambda_l^* \delta_{n-l} = \mathrm{IFT}[\log S(f)] ,$$

i.e.,

$$\lambda_n^* = \lambda_{-n} = \mathrm{IFT}[\log S(f)] = \mathrm{IFT}[\log \mathrm{FT}[R(n)]] . \tag{3.2.7}$$

If the logarithm were not in (3.2.7), the FT and IFT would cancel each other and the right-hand side would be equal to $R(n)$, which is the IFT of the spectrum $S(f)$. This difficult operator *log* makes trouble. However, recalling the definition of cepstrum, we see immediately that the right-hand side is just the (complex) cepstrum of the ACF sequence. Equation (3.2.7) shows an important relationship: In MEM2 Lagrange multipliers and the (complex) cepstrum of the ACF sequence are complex conjugates of each other. An equivalent statement is: In MEM2 the complex conjugates of the Lagrange multipliers are the inverse Fourier components of the logarithmic spectrum. Because of the positivity of the ACF, the operand of logarithm is real-valued. Hence, the complex and real cepstra are identical. They are simply called the cepstrum.

Clearly, the above formulation and the structure of the MEM2 solution are applicable to the 2-D case. The only thing to do is to change the frequency and subscripts of all the quantities from 1-D to 2-D.

Thus, we have built a bridge between MEM2 and CAM. They seemed before to be far apart from each other. Through this bridge we can apply to MEM2 the existing results in CAM. Specifically, we will utilize the relationship between the input $y(n)$ and the output $\hat{y}(n)$ (complex cepstrum) in a cepstrum analysis system:

$$y(n) = \sum_{k=-\infty}^{\infty} \left(\frac{k}{n}\right) \hat{y}(k)y(n-k) \,, \quad n \neq 0 \tag{3.2.8}$$

to obtain an implicit solution in MEM2, that is, to calculate the spectrum $S(f)$ iteratively from the given partial ACF.

Taking $R(n)$ and $\hat{R}(k) = \lambda_k^*$ as $y(n)$ and $\hat{y}(k)$, respectively, and noticing $\lambda_k = 0$ for $|k| > M$, (3.2.8) becomes

$$R(n) = \sum_{k=-M}^{M} \left(\frac{k}{n}\right) \lambda_k^* R(n-k) \,, \quad n \neq 0 \,. \tag{3.2.9}$$

From (3.2.9) a formula for the ACF extension follows. Splitting the summation on the right-hand side into two parts:

$$R(n) = \frac{-M}{n}\lambda_{-M}^* R(M+n) + \sum_{k=-M+1}^{M} \left(\frac{k}{n}\right) \lambda_k^* R(n-k) \,,$$

we get

$$R(M+n) \;=\; \frac{n}{M\lambda_M}\left[\sum_{k=-M+1}^{M} \left(\frac{k}{n}\right) \lambda_k^* R(n-k) - R(n)\right] \,, \quad (n \geq 1) \,.$$
$$\tag{3.2.10a}$$

With n replaced by $-n$ in (3.2.9), we get similarly

$$R(-M-n) \;=\; \frac{n}{M\lambda_M^*}\left[\sum_{k=-M+1}^{M} \left(\frac{k}{n}\right) \lambda_k R(k-n) - R(-n)\right] \,,$$
$$(n \geq 1) \,. \tag{3.2.10b}$$

Equations (3.2.10a, b) are, respectively, the formulae for extrapolating the ACF rightwards from M and leftwards from $-M$.

It is easy to see that the extrapolated ACF is Hermitian: $R(-M - n) = R^*(M + n)$. So only (3.2.10a) needs considering. It should be noticed that the above formulae represent implicit relationships because λ_k^* cannot be determined explicitly from the given $R(n)$, $|n| \leq M$.

$\hat{R}(0) = \lambda_0$ cannot be determined by (3.2.9) since λ_0 is not included in (3.2.9) at all: $k/n = 0$ when $k = 0$. It is interesting to notice that by definition $(k = 0)$,

$$\lambda_0 = \hat{R}(0) = \int_{-1/2}^{1/2} \log S(f) \exp(2\pi i f 0) \mathrm{d}f = \int_{-1/2}^{1/2} \log S(f) \mathrm{d}f \,,$$

which is just the $H1$ expression. It cannot be used to determine $\hat{R}(0)$ since the spectrum $S(f)$ is just what is sought.

From (3.2.5), we get

$$S(f) = \mathrm{e}^{\lambda_0} \exp \left[\sum_{\substack{k=-M \\ k \neq 0}}^{M} \lambda_k^* \exp(-2\pi i f k) \right] .$$

Substituting this $S(f)$ in (3.2.3) yields, after a simple manipulation,

$$\mathrm{e}^{\lambda_0} = R(0) \left\{ \int_{-1/2}^{1/2} \exp \left[\sum_{\substack{k=-M \\ k \neq 0}}^{M} \lambda_k^* \exp(-2\pi i f k) \right] \mathrm{d}f \right\}^{-1} . \tag{3.2.11}$$

Therefore, e^{λ_0}, and hence λ_0, can be determined by the TP constraint after obtaining λ_k, $|k| = 1, \ldots, M$. However, e^{λ_0} is only a scale factor and has nothing to do with the shape of the spectrum. For this reason we are not concerned with the determination of λ_0 at present. We need only to determine λ_k, $|k| = 1, \ldots, M$.

When $N \to \infty$, system (3.2.9) consists of infinitely many equations and unknowns. For practical application, n must be and can be truncated at a certain value, say, $N/2$, i.e., $n = 1, \ldots, N/2$ (N being the FT length). (For the time being the negative n's are of no concern since $R(n)$ is Hermitian.) Expanding (3.2.9) with the truncated n yields (3.2.12), which is formidable in size and therefore needs to be put on a separate page (Fig. 3.3). However, it is worthwhile since the structure of the system of equations is somewhat apparent.

From (3.2.12) in Fig. 3.3 it can be seen that we still have a problem even after truncating n. There are $N/2$ equations but $N/2 + 2M$ unknowns $\lambda_{-M}^*, \ldots, \lambda_{-1}^*, \lambda_1^*, \ldots, \lambda_M^*, R(M+1), R(N/2+M)$. (The number of unknowns is actually $N/2+M$ since $\lambda_{-k}^* = \lambda_k$.) This happens because the system (3.2.9)

$$
\begin{array}{lllllll}
R(1) & = \frac{-M}{1}\lambda^*_{-M}R(1+M) & +\cdots & +\frac{-1}{1}\lambda^*_{-1}R(2) & +\frac{1}{1}\lambda^*_{1}R(0) & +\cdots & +\frac{M}{1}\lambda^*_{M}R(1-M) \\
R(2) & = \frac{-M}{2}\lambda^*_{-M}R(2+M) & +\cdots & +\frac{-1}{2}\lambda^*_{-1}R(3) & +\frac{1}{2}\lambda^*_{1}R(1) & +\cdots & +\frac{M}{2}\lambda^*_{M}R(2-M) \\
\vdots & \quad\vdots & & & & & \quad\vdots \\
R(M-1) & = \frac{-M}{M-1}\lambda^*_{-M}R(-1+2M) & +\cdots & +\frac{-1}{M-1}\lambda^*_{-1}R(M) & +\frac{1}{M-1}\lambda^*_{1}R(M-2) & +\cdots & +\frac{M}{M-1}\lambda^*_{M}R(-1) \\
R(M) & = \frac{-M}{M}\lambda^*_{-M}R(2M) & +\cdots & +\frac{-1}{M}\lambda^*_{-1}R(M+1) & +\frac{1}{M}\lambda^*_{1}R(M-1) & +\cdots & +\frac{M}{M}\lambda^*_{M}R(0) \qquad (a) \\
\hdashline
R(M+1) & = \frac{-M}{M+1}\lambda^*_{-M}R(2M+1) & +\cdots & +\frac{-1}{M+1}\lambda^*_{-1}R(M+2) & +\frac{1}{M+1}\lambda^*_{1}R(M) & +\cdots & +\frac{M}{M+1}\lambda^*_{M}R(1) \qquad (b) \\
R(M+2) & = \frac{-M}{M+2}\lambda^*_{-M}R(2M+2) & +\cdots & +\frac{-1}{M+2}\lambda^*_{-1}R(M+3) & +\frac{1}{M+2}\lambda^*_{1}R(M+1) & +\cdots & +\frac{M}{M+2}\lambda^*_{M}R(2) \\
\vdots & \quad\vdots & & & & & \quad\vdots \\
R(N/2) & = \frac{-M}{N/2}\lambda^*_{-M}R(N/2+M) & +\cdots & +\frac{-1}{N/2}\lambda^*_{-1}R(N/2+1) & +\frac{1}{N/2}\lambda^*_{1}R(N/2-1) & +\cdots & +\frac{M}{N/2}\lambda^*_{M}R(N/2-M)
\end{array}
$$

Fig. 3.3. Equations (3.2.12), the expanded form of (3.2.9), $n = 1, \ldots, N/2$. The upper part (3.2.12a) is above while the lower part (3.2.12b) is below the dashed line

is noncausal, $n-k > n$ for $k < 0$. To avoid this undesirable situation, $R(n)$ for $n > N/2$ must be assumed to be zero; that is, we are satisfied with a finite extension of the ACF. This simplification can be justified by the "modest extension" of data in MEM2. To improve the accuracy, $N/2$ can be taken to be indefinitely large.

After the simplification above, the solution of (3.2.12) is a vector

$$V = [\,\lambda^*_{-M}, \ldots, \lambda^*_{-1}, \lambda^*_1, \ldots, \lambda^*_M, R(M+1), \ldots, R(N/2)\,]^{\mathrm{T}} , \qquad (3.2.13)$$

with constraints

$$\lambda^*_{-k} = \lambda_k = (\lambda^*_k)^* , \quad k = 1, \ldots, M . \qquad (3.2.14)$$

For complex $R(n)$, we have $N/2+M$ equations (3.2.12, 3.2.14) in $N/2+M$ unknowns in (3.2.13). For real $R(n)$, λ_k are also real, $\lambda_{-k} = \lambda_k$. So we have $N/2$ equations (3.2.12) in $N/2$ unknowns in the vector

$$V_R = [\,\lambda_1, \ldots, \lambda_M, R(M+1), \ldots, R(N/2)\,]^{\mathrm{T}} . \qquad (3.2.15)$$

After solving the equations, the MEM2 spectrum can be calculated by the correlogram method (using the given and extrapolated ACF) or by formula (3.2.5) (using $\lambda^*_k, |k| = 1, \ldots, M$; λ_0 may assume an arbitrary value, say zero, or be determined by the TP constraint, cf. (3.2.11)). The results by the two methods are consistent. (In the latter method, the absolute value of $S(f)$ depends on λ_0.) The summation in the exponent on the right-hand side of (3.2.5) is accomplished by FFT.

3.2.3 Iterative Algorithm

The remaining problem now is the solution of (3.2.12). Since they are quadratic forms in the unknowns, an explicit solution is impossible and iteration is necessary. An iterative algorithm is described as follows:

In the Case of Real $R(n)$

Step 1. Initialize $R(M+1), \ldots, R(2M), \ldots, R(N/2)$. Then solve the upper part (3.2.12a) for λ_k using $R(M+1), \ldots, R(N/2)$ and the given $R(0), \ldots, R(M)$ (M linear equations in M unknowns).

Step 2. Calculate the linear combinations of $R(1), \ldots, R(M), R(M+1), \ldots, R(N/2)$ on the right-hand sides of the $N/2 - M$ equations in the lower part (3.2.12b) using λ_k obtained in Step 1. $R(M+1), \ldots, R(N/2)$ on the left-hand sides are updated.

Repeat the two steps above until convergence is achieved, that is, the values of $R(M+1), \ldots, R(N/2)$ do not change with the linear combinations on the right-hand sides of (3.2.12b). A computer flowchart is shown in Fig. 3.4.

In the Case of Complex $R(n)$

Minor modifications to the iterative procedure described above are needed. In Step 1, the solution V is found using (3.2.12a, 3.2.14). In Step 2, the linear

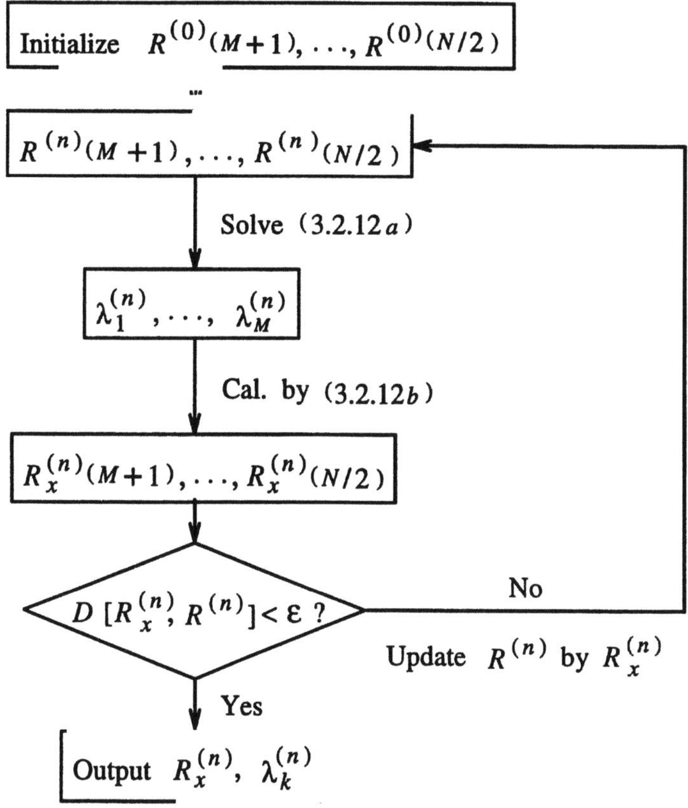

Fig. 3.4. Computer flowchart for solving (3.2.12) (for real ACF)

combinations on the right-hand sides of (3.2.12b) are evaluated using V from Step 1.

Now we discuss some issues in writing a computer program to implement this kind of iterative algorithm.

(1) *The FT Length N.* There is no point in using too large an N due to the modest extension of the ACF. It was observed that $N/2 = 4M$ or even $N/2 = 2M$ gave satisfactory results. Note that the burden in computation depends slightly on the value of $N/2$ because the number of equations in (3.2.12a) is fixed (given M). Calculating a linear combination is much faster than solving a set of simultaneous linear equations.

(2) *Solution of the Linear Equations (3.2.12a).* The linear equations need to be solved once in each iteration. It would be ideal if an efficient algorithm could be found on the basis of the pattern of the coefficient matrix of (3.2.12a). Unfortunately, up to now no positive answer has been found. It is convenient to use general purpose subroutines in, say, the NAG library.

(3) *Convergence Consideration.* (a) The initial values of $R(n)$. It was observed that convergence speed was almost independent of the initial values of $R(n)$ ($M \leq n \leq N/2$). Using the true $R(n)$ or zero as $R^{(0)}(n)$ made no difference. So zero initial values are used for simplicity.

(b) Convergence criterion. We define the difference between $R_x^{(n)}$ and $R^{(n)}$ by

$$D[R_x^{(n)}(n), R^{(n)}(n)] = \left(\frac{\sum_{n=M+1}^{N/2} |R_x^{(n)}(n) - R^{(n)}(n)|^2}{\sum_{n=M+1}^{N/2} |R^{(n)}(n)|^2} \right)^{1/2} . \qquad (3.2.16)$$

Convergence is achieved when $D < \varepsilon$. Usually $\varepsilon = 0.01 \sim 0.05$.

It was observed in many numerical examples that convergence could be achieved as long as the zero-lag, $R(0)$, was sufficiently large. But we are not able to quantitatively answer the question: How large should $R(0)$ be exactly? The exact zero-lag required for convergence depends on the specific $R(n)$ ($|n| \leq M$) given. Furthermore, the larger $R(0)$ was, the faster the convergence was, and the closer the MEM2 results were to those obtained by the simple FT method with zero extension of the ACF.

Numerical examples concerning CAM described in this section will be shown in Sect. 4.4. The results will be compared with those from the $R - \lambda$ procedure in MEM2 and from MEM1.

3.2.4 Discussion

We discuss the case where the measure is nonuniform before concluding this section. The entropy expression is now

$$H2 = - \int_{-1/2}^{1/2} S(f) \log[S(f)/m(f)]\mathrm{d}f . \qquad (3.2.17)$$

The constraints under which $H2$ is maximized are still (3.2.2). Following the variational procedure for deriving (3.2.5), it is easily shown that

$$S(f) = m(f) \exp \left[\sum_{l=-\infty}^{\infty} \lambda_l^* \exp(-2\pi \mathrm{i}fl) \right] , \qquad (3.2.18)$$

where $\lambda_l = 0$ for $|l| > M$. Define a sequence C_l by

$$C_l = \mathrm{IFT}[\log m(f)] = \int_{-1/2}^{1/2} \log m(f) \exp(2\pi \mathrm{i}fl)\mathrm{d}f . \qquad (3.2.19)$$

C_l is the cepstrum of the ACF sequence corresponding to the prior spectrum $m(f)$.

From (3.2.18) it follows that

$$
\begin{aligned}
S(f) &= \exp\left[\log m(f) + \sum_{l=-\infty}^{\infty} \lambda_l^* \exp(-2\pi i f l)\right] \\
&= \exp\left[\sum_{l=-\infty}^{\infty} (\lambda_l^* + C_l)\exp(-2\pi i f l)\right].
\end{aligned}
\tag{3.2.20}
$$

Comparing (3.2.20) with (3.2.5), we see here that $\lambda_l^* + C_l$ plays the same role as λ_l^* does in the case of uniform measure. Now we have

$$
\begin{aligned}
\lambda_n^* &= \mathrm{IFT}[\,\log \mathrm{FT}[R(n)]\,] - C_n \\
&= \hat{R}(n) - C_n\,,
\end{aligned}
\tag{3.2.21}
$$

which is similar to (3.2.7). Equation (3.2.21) shows the relationship between the Lagrange multipliers and the cepstrum of the ACF sequence.

On the analogy of (3.2.9), we now have

$$
R(n) = \sum_{k=-\infty}^{\infty} \left(\frac{k}{n}\right) (\lambda_k^* + C_k) R(n-k)\,, \quad n \neq 0\,.
\tag{3.2.22}
$$

Note that in general, $C_k \not\equiv 0$ for $|k| > M$, and the summation cannot be truncated. Splitting the right-hand side of (3.2.22) into two terms and noticing $\lambda_k = 0$ for $|k| > M$, we have

$$
R(n) = \sum_{k=-M}^{M} \left(\frac{k}{n}\right) \lambda_k^* R(n-k) + \sum_{k=-\infty}^{\infty} \left(\frac{k}{n}\right) C_k R(n-k)\,, \quad n \neq 0\,.
\tag{3.2.23}
$$

As contrasted with (3.2.9), there is an additional term on the right-hand side above. This time we are also satisfied with a finite extension of $R(n)$, i.e., let $R(n) = 0$ for $|n| > N/2$. Moreover, we assume that the prior estimate $m(f)$ is fairly flat, then C_k is decaying fairly rapidly as k increases. Hence, it is reasonable to set $C_k = 0$ for $|k| > N/2$. The error due to truncation may be reduced by increasing N. Consequently, (3.2.23) becomes

$$
R(n) = \sum_{k=-M}^{M} \left(\frac{k}{n}\right) \lambda_k^* R(n-k) + \sum_{k=-N/2}^{N/2} \left(\frac{k}{n}\right) C_k R(n-k)\,, \quad n \neq 0\,.
\tag{3.2.24}
$$

Let $n = 1,\ldots,N/2$ and expand (3.2.24). Then, the left-hand sides of the resultant equations are the same as those of (3.2.12); the first part on the each right-hand side is the same as that of (3.2.12), while the second part is calculable in the iteration. The procedure for solving (3.2.12) can be used here after appropriate modification. Since nonuniform measure is not common in spectral analysis, we will not go further in the discussion.

3.3 Explicit Solution

In the previous section the implicit solution in MEM2 was derived, given the partial ACF as data. In this section we investigate the possibility and conditions of determining an explicit solution [3.6, 3.7]. A signal model will be obtained in due course in deriving the explicit solution. Then some discussion will follow. This section concludes with numerical examples.

3.3.1 Explicit Solution

For convenience, the formula for calculating the MEM2 spectrum, (3.2.5), is repeated below:

$$S(f) = \exp\left[\sum_{k=-\infty}^{\infty} \lambda_k^* \exp(-2\pi i f k) \right] , \tag{3.3.1}$$

where $\lambda_k = 0$ for $|k| > M$; λ_k, $|k| \leq M$ satisfy the relationships (3.2.9), i.e.,

$$R(n) = \sum_{k=-M}^{M} \left(\frac{k}{n}\right) \lambda_k^* R(n-k) , \quad n \neq 0 . \tag{3.3.2}$$

The key to calculating the spectrum is to find λ_k^*, which is the complex cepstrum of the ACF sequence $R(n)$. In the preceding section we saw that (3.3.2) after simplification must be solved for λ_k^* iteratively.

Equations (3.3.2) come from the relationships

$$y(n) = \sum_{k=-\infty}^{\infty} \left(\frac{k}{n}\right) \hat{y}(k) y(n-k) , \quad n \neq 0 , \tag{3.3.3}$$

which represent a noncausal system. In general , $\hat{y}(k)$, $|k| \leq M$ cannot be solved out explicitly, given the ACF $y(n)$, $|n| \leq M$.

The only way out for determining an explicit solution is to use formula (A.12), i.e.,

$$\hat{y}(n) = \begin{cases} 0 , & n < 0 , \\ \log y(0) , & n = 0 , \\ \dfrac{y(n)}{y(0)} - \displaystyle\sum_{k=0}^{n-1} \left(\frac{k}{n}\right) \hat{y}(k) \dfrac{y(n-k)}{y(0)} , & n > 0 . \end{cases} \tag{3.3.4}$$

The conditions are that $y(n)$ is a causal and minimum-phase sequence. (A similar formula for an anticausal and maximum-phase sequence is available. But this kind of sequence is rare in practice, so no consideration is given to it.)

However, the ACF sequence $R(n)$ possesses the Hermitian symmetry, and hence cannot be either causal or of minimum-phase. Therefore, it would be hopeless if one wants to determine an explicit MEM2 solution by CAM from the given partial ACF.

We turn our attention to the time series $x(n)$, whose ACF is $R(n)$. $x(n)$ may satisfy the aforementioned conditions (causal and minimum-phase). The key to deriving an explicit solution is to express the spectrum $S(f)$ in terms of the real cepstrum of $x(n)$, and to utilize the relationship between the real and complex cepstra of $x(n)$.

According to the correlogram method,

$$S(f) = \mathrm{FT}[R(n)] \ . \tag{3.3.5}$$

On the other hand, by the periodogram method,

$$S(f) = |\mathrm{FT}[x(n)]|^2 \ . \tag{3.3.6}$$

Equate the two expressions above, take algorithm and perform IFT; then

$$\mathrm{IFT}[\,\log \mathrm{FT}[R(n)]\,] = 2\,\mathrm{IFT}[\,\log |\mathrm{FT}[x(n)]|\,] \ ,$$

where the left-hand side is the complex cepstrum of $R(n)$, λ_n^*, while the right-hand side contains the real cepstrum of $x(n)$, which is denoted by $C(n)$. The relationship is

$$\lambda_n^* = 2C(n) \ . \tag{3.3.7}$$

Substituting it in (3.3.1) yields

$$S(f) = \exp\left[2\sum_{k=-\infty}^{\infty} C(k)\exp(-2\pi ifk)\right] \ , \tag{3.3.8}$$

where $C(k) = 0$ for $|k| > M$. This is the formula for calculating the spectrum by the real cepstrum of the time series.

The signal model in MEM2 is, from (3.3.8),

$$X(z) = \exp\left[C(0) + 2\sum_{k=1}^{M} C(k)z^{-k}\right] \ . \tag{3.3.9}$$

This can be verified as follows:

$$
\begin{aligned}
|X(z)|^2_{z=e^{2\pi if}} &= \exp\left[2C(0) + 4\,\mathrm{Re}\left\{\sum_{k=1}^{M} C(k)\exp(-2\pi ifk)\right\}\right] \\
&= \exp\left[2\sum_{k=-M}^{M} C(k)\exp(-2\pi ifk)\right] \\
&= S(f) \ .
\end{aligned}
$$

In order to utilize the relationship between the complex and real cepstra, (A.13), the condition that $x(n)$ is real is necessary. Then,

$$C(n) = \begin{cases} 1/2\hat{x}(n) \ , & n > 0 \ , \\ \hat{x}(0) \ , & n = 0 \ , \\ 1/2\hat{x}(-n) \ , & n < 0 \ , \end{cases} \tag{3.3.10}$$

where $\hat{x}(n)$ and $C(n)$ are both real-valued. $C(n) = \hat{x}(n) = 0$ for $|n| \geq M+1$.

Thus, we have sufficient conditions for determining an explicit solution: $x(n)$ is a real, causal and minimum-phase series. Specifically, the procedure for finding the solution includes the following steps:

1) Calculate $\hat{x}(n)$, $n = 0, 1, \ldots, M$ from the given $x(n)$, $n = 0, 1, \ldots, M$ by the recursive formula (3.3.4):

$$\hat{x}(n) = \begin{cases} \log x(0) , & n = 0 , \\ \dfrac{x(n)}{x(0)} - \displaystyle\sum_{k=0}^{n-1} \left(\dfrac{k}{n}\right) \hat{x}(k)\dfrac{x(n-k)}{x(0)} , & n = 1, \ldots, M . \end{cases} \qquad (3.3.11)$$

If $x(0) < 0$, we must reverse the polarity of $x(n)$. This has no effect on the spectrum.

2) Calculate $C(n)$, $|n| = 0, \ldots, M$ by (3.3.10).

3) Calculate the spectrum $S(f)$ by (3.3.8).

In practical computation, some steps may be combined; the frequency is discretized to be j/N, $j = 0, \ldots, N - 1$, N being the FT length.

The procedure is completed if one wants only to calculate the spectrum. Alternatively, one can also extrapolate the time series $x(n)$, and then calculate the spectrum.

4) Extrapolate $x(n)$ by $\hat{x}(n)$ calculated above and the given $x(n), n = 1, \ldots, M$, utilizing the recursive formula (A.11):

$$x(n) = \sum_{k=0}^{M} \left(\frac{k}{n}\right) \hat{x}(k)x(n-k) , \quad n \geq M + 1 . \qquad (3.3.12)$$

Since $\hat{x}(n) = 0$ for $n \geq M + 1$, the first term in the bottom line of the right-hand side of (A.11) is zero and hence discarded. For the same reason, the terms containing the factor $\hat{x}(k)$, $k \geq M + 1$ in the summation are also discarded.

5) Calculate the spectrum $S(f)$ from the given and extrapolated $x(n)$ by the periodogram or correlogram method.

3.3.2 Discussion

Now we discuss some issues before demonstrating numerical examples.

(1) The conditions for the explicit solution are: $x(n)$ is a real, causal and minimum-phase series. The last condition is rather restrictive. It is not usually satisfied. The blind use of the explicit solution may lead to completely unacceptable results, cf. Example 3 later.

(2) As shown in (3.3.9), the signal model in MEM2 is an exponential function. The exponent is a polynomial of finite order in z^{-1}. Series of this model are not common. Usually, the Z-transform of a series is a rational function. In this case, a causal and minimum-phase series $x(n)$ can be modeled by the following exponential function:

$$X(z) = \exp\left[C(0) + 2\sum_{k=1}^{\infty} C(k)z^{-k} \right] , \tag{3.3.13}$$

which is different from (3.3.9) in that here the exponent is a polynomial of infinite degrees since

$$\begin{cases} C(0) &= \log A , \\ C(k) &= \dfrac{1}{2}\left(-\sum_{l=1}^{m_i} \dfrac{a_l^k}{k} + \sum_{l=1}^{p_i} \dfrac{c_l^k}{k} \right) , \quad (|a_l|,\ |c_l| < 1),\quad k = 0, 1, \dots \end{cases} \tag{3.3.14}$$

by (A.8, 3.3.10).

When a signal of a rational model (or equivalently (3.3.13)) is processed as if of an exponential model (3.3.9), error will result due to the truncation of $C(k)$. This error may be reduced in some way [3.8]. The longer the given data segment is, the smaller the error is. On the other hand, when the length of data, M, is fixed, the smaller $|a_l|$ and $|c_l|$ are, the faster $C(k)$ decays and hence the smaller the error is. This conclusion is in line with the experimental result: The closer the zeros and poles of a signal are to the origin, the more precise the MEM2 result is [3.9].

As a matter of fact, this kind of problem that a signal to be processed is not of the due model is also encountered in MEM1. There signals are taken as AR processes whether it is true or not.

(3) As stated before, "minimum-phase" is a rather restrictive condition, which considerably limits the applicability of the explicit solution.

Suppose that $y(n)$ is a real and causal series. Perform the following exponential transformation:

$$x(n) = \alpha^n y(n) , \quad (n \ge 0) , \tag{3.3.15}$$

where α is real, $0 < \alpha \le 1$. The Z-transform of $x(n)$ is

$$X(z) = \sum_{n=0}^{\infty} \alpha^n y(n) z^{-n} = Y(\alpha^{-1}z) .$$

$x(n)$ is a minimum-phase series provided that α is sufficiently small. (If $y(n)$ itself is of minimum-phase, then $\alpha = 1$.) Consequently, $x(n)$ can be extrapolated by (3.3.12). Then $y(n)$ can be recovered by the minus-exponential transformation:

$$y(n) = \alpha^{-n} x(n) , \quad (n \ge 0) . \tag{3.3.16}$$

Finally, the spectrum can be calculated by the given and extrapolated $y(n)$.

The scheme proposed above seems to be reasonable. However, we are faced with the following problems:

From the point of view of mathematical operation, the value of α cannot be determined since $y(n)$ is only partially given. It is safe to use a small α in consideration of "damping". But if α is too small, the errors for the extrapolated $y(n)$, caused by errors in the given $x(n)$ and computation, will be enlarged to a great extent in the minus-exponential transformation.

In principle, in the proposed scheme the entropy of $x(n)$, but not $y(n)$, has been maximized. Maximizing the entropy of $x(n)$ does not mean maximizing the entropy of $y(n)$ if the two series are related by (3.3.15) or (3.3.16). In other words, if we are concerned strictly with the maximization of the entropy of $y(n)$, then $y(n)$ cannot be recovered simply by (3.3.16). In short, MEM2 in this scheme is in the sense that the entropy of $x(n)$ related to $y(n)$ by (3.3.15) has been maximized.

Because formulae (3.3.11, 3.3.12) concerning the extension are homogeneous in $x(n)$ and $\hat{x}(n)$, it is easily shown that even if $y(n)$ is not of minimum-phase, $y(n)$ and the formally defined $\hat{y}(n) = \alpha^{-n}\hat{x}(n)$ for $n \geq 0$ (this $\hat{y}(n)$ is not necessarily the true complex cepstrum of $y(n)$) satisfy those formulae as well.

Proof. Multiplying the two sides of (3.3.12) by α^{-n} yields

$$\alpha^{-n}x(n) = \sum_{k=0}^{M} \left(\frac{k}{n}\right) [\hat{x}(k)\alpha^{-k}][x(n-k)\alpha^{-(n-k)}] \, ,$$

i.e.,

$$y(n) = \sum_{k=0}^{M} \left(\frac{k}{n}\right) \hat{y}(k)y(n-k) \, . \tag{3.3.17}$$

Similarly, multiplying the two sides of (3.3.11) by α^{-n} yields ($x(0) = y(0)$)

$$\hat{y}(n) = \begin{cases} \log y(0) \, , & n = 0 \, , \\ \dfrac{y(n)}{y(0)} - \displaystyle\sum_{k=0}^{n-1} \left(\frac{k}{n}\right) \hat{y}(k)\dfrac{y(n-k)}{y(0)} \, , & n = 1, \ldots, M \, . \end{cases} \tag{3.3.18}$$

Q.E.D.

To summarize, we may extrapolate $y(n)$, $n \geq M+1$ from the given $y(n)$, $n = 0, 1, \ldots, M$ by (3.3.18, 3.3.17) without involving α explicitly to convert $y(n)$ to $x(n)$ first. But we must keep in mind that the resultant MEM2 spectrum is not in the strict sense (maximizing the entropy of $y(n)$) unless $y(n)$ itself is of minimum-phase. The blind use of the two formulae above may lead to completely unacceptable results, cf. Example 3 later. Circumspection is necessary.

(4) In the formulation of MEM2 in the preceding section, we started off with the maximization of $H2$ under the constraints on the ACF $R(n)$, and end up here with formulae based on the given $x(n)$. Following the development of the argument, one question might arise in one's mind: Is it not a double-definition of the problem? You are thoughtful but your worry is unnecessary.

The same thing happened to MEM1. There one started off with the maximization of $H1$ also under the constraints on the ACF $R(n)$. At the end $S(f)$ was parametrized in terms of $a_k, k = 0, \ldots, M$ and p_M, cf. (2.2.14). If we are given the ACF $R(n), n = 0, \ldots, M$, we can then solve the normal equations (2.2.16) for a_k and p_M. However, what can we do if we are given a segment of the series $x(n)$ instead? In this case one way is to first estimate $R(n)$ from the given $x(n)$ and then to solve the normal equations. But in doing this the maximum entropy principle is doomed to be violated since $x(n)$ for $n > M$ must be assumed to be zero in the estimation of $R(n)$. Then, another way has been thought out, that is, to estimate a_k and p_M directly from $x(n)$, e.g., by the Burg and Marple algorithms. This is based on viewing a_k and p_M as the AR parameters of $x(n)$.

In MEM2, if we are given $R(n)$ we have no choice but to use an iterative algorithm (implicit solution) to find λ_k^*, $k = 1, \ldots, M$. If we are given $x(n)$ which does not meet the three conditions (real, causal and especially minimum-phase) for the explicit solution (at least we do not know), we still have no choice but to estimate $R(n)$ first and then find λ_k^* iteratively. However, if we know beforehand that $x(n)$ is real, causal and especially of minimum-phase, or we would like to take a risk, we may use the explicit solution, which is based on CAM.

The possibility of finding an explicit solution is based on the fact that a minimum-phase signal can be uniquely specified by its magnitude spectrum, while the phase information is redundant. To see this specifically, we use the constraints on $x(n)$, $n = 0, \ldots, M$ in maximizing $H2$. For a real, causal and minimum-phase series $x(n)$, its real cepstrum $C(k)$ is calculable from $x(n)$ $(k, n = 0, \ldots, M)$ and vice versa. The constraints, therefore, can be passed on $C(k)$. Considering the relationship

$$\frac{1}{2}\text{IFT}[\,\log S(f)\,] = \text{IFT}[\,\log |\text{FT}[x(n)]|\,] = C(k)\,, \tag{3.3.19}$$

we form the objective functional

$$\begin{aligned}
Q \;=\; & -\int_{-1/2}^{1/2} S(f)\log S(f)\mathrm{d}f \\
& + \sum_{k=-M}^{M} \mu_k \left[\frac{1}{2}\int_{-1/2}^{1/2} \log S(f)\exp(2\pi ifk)\mathrm{d}f - C(k)\right].
\end{aligned}$$

Let $\delta Q = 0$; then it is easily shown that

$$\log S(f) + 1 \;=\; \sum_{k=-M}^{M} \frac{\mu_{-k}}{2S(f)}\exp(-2\pi ifk)\,, \tag{3.3.20}$$

$$S(f) \;=\; \exp\left[-1 + \sum_{k=-M}^{M} \frac{\mu_{-k}}{2S(f)}\exp(-2\pi ifk)\right]. \tag{3.3.21}$$

Performing IFT on the two sides of (3.3.20) yields

$$\frac{\mu_k}{2S(f)} = \frac{\mu_{-k}}{2S(f)} = \text{IFT}[\,\log S(f) + 1\,] = 2C(k) + \delta_k\,. \tag{3.3.22}$$

Substituting (3.3.22) in (3.3.21), we get

$$S(f) = \exp\left[-1 + \sum_{k=-M}^{M} (2C(k) + \delta_k)\exp(-2\pi ifk)\right]$$

$$= \exp\left[2 \sum_{k=-M}^{M} C(k)\exp(-2\pi ifk)\right], \qquad (3.3.23)$$

which is exactly the same as (3.3.8). So we see that using the partially given time series or ACF as the constraints leads to the same result. The conditions are, as we have emphasized again and again, that $x(n)$ is a real, causal and minimum-phase series.

3.3.3 Examples

At the end of this section we demonstrate four numerical examples.

Example 1. The time series is a damping cosine sequence:

$$x(n) = r^n \cos(2\pi fn), \quad (0 < r < 1), \quad n = 0, 1, \dots .$$

The Z-transform of a (nondamping) cosine sequence has one simple zero at the origin, and two poles and one zero on the unit circle. Damping (multiplying by the factor r^n) means shifting the poles and zero on the unit circle radially towards the origin so that they fall inside the unit circle. As a result, $x(n)$ is a (real, causal and) minimum-phase series. Since the Z-transform of $x(n)$ is rational but not exponential, the explicit solution and data extension are approximate.

Let $f = f_1 = 0.25$, and $r = 0.99$. A part of $x(n)$ is shown in Fig. 3.5a. Shown in Fig. 3.5b are the first 16 data ($M = 15$) given and the extrapolated series. Figure 3.5c shows the MEM2 spectra, including that calculated by (3.3.8) (explicit solution) and that by the periodogram method with the given and extrapolated data (extension). For the purpose of comparison, the true spectrum (true) and the FT spectrum (FT) calculated by the periodogram method with the first 16 data given are also plotted. From the figure it can be seen that compared with the FT result, the MEM2 spectrum has a sharper peak (higher resolution) and lower sidelobes. This is due to the nonzero data extension in MEM2.

If $f = f_2 = 0.1$ and the calculation is repeated, then the results will be similar with the peak location shifted in the spectra.

Example 2. The time series is the sum of two damping cosine sequences with $f_1 = 0.25$ and $f_2 = 0.1$. The other parameters are the same as those in Example 1. The results are shown in Fig. 3.6.

From Fig. 3.6c it can be seen that MEM2 is superior over the FT method as far as the resolution and sidelobes are concerned.

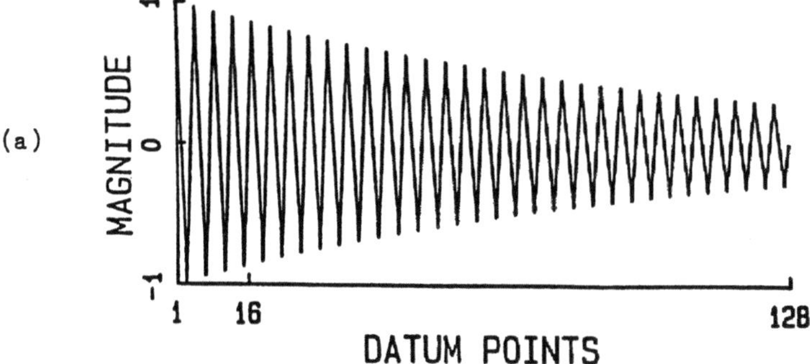

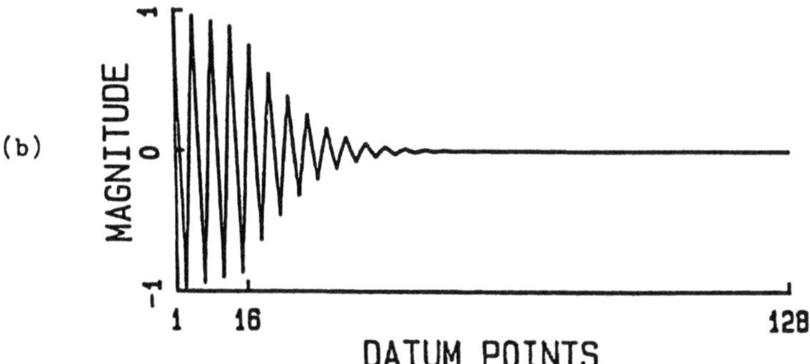

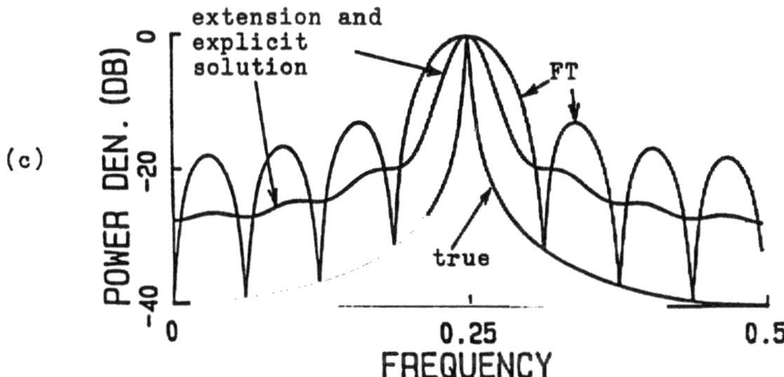

Fig. 3.5. The true series, extrapolated series and spectra in Example 1. (a) The first 128 datum points of the true real and minimum-phase series. $f_1 = 0.25, r = 0.99$. (b) The 16 given data and extrapolated data in MEM2. (c) The spectra computed from the true series (*true*), from the given and extrapolated data (*extension*), by the *explicit solution*, and by *FT* with the 16 data (rectangular window)

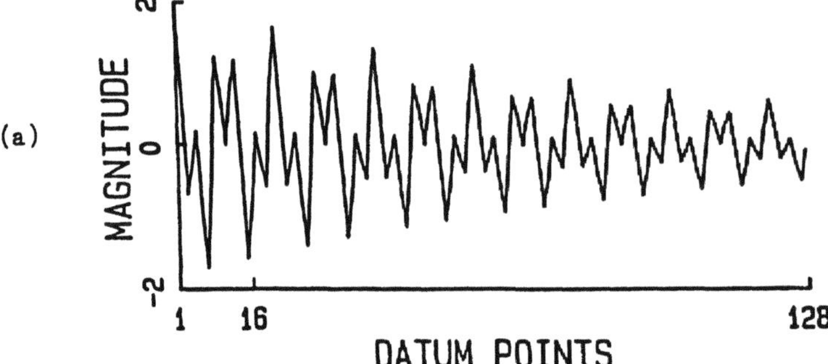

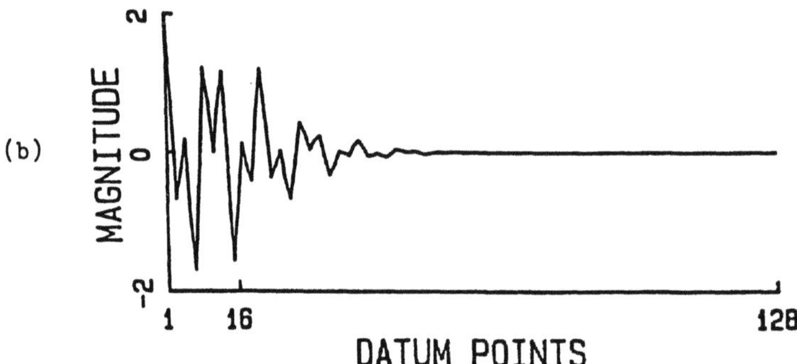

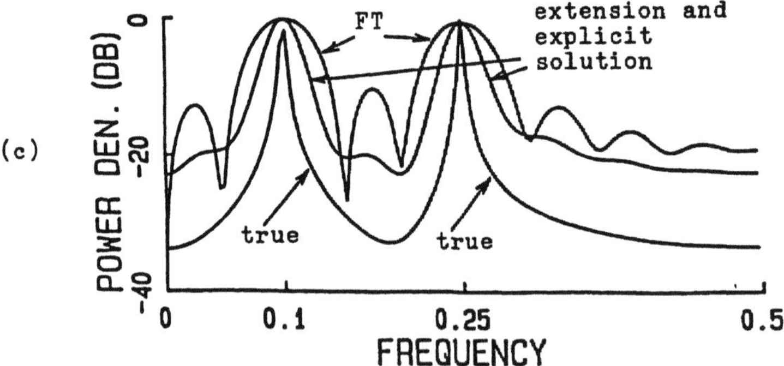

Fig. 3.6. The true series, extrapolated series and spectra in Example 2. $f_1 = 0.25$, $f_2 = 0.1$, otherwise same as in Fig. 3.5

Example 3. The real-valued series $x(n)$ is a realization of a stochastic process of length 256 (taken from a visibility function (real part) output from an aperture synthesis radio telescope), as shown in Fig. 3.7a. This series does not look like a minimum-phase one. Figures 3.7c, b are from the blind use of the explicit solution and data extension ($M = 15$).

On the other hand, for Fig. 3.8 the series has been converted to a minimum-phase one before proceeding to this MEM2 treatment.

Example 4. The real-valued $x(n)$ is a realization of another stochastic process of length 256 (from the same source as in Example 3), as shown in Fig. 3.9a. Here the data are processed as in Example 3.

The results from the blind use of the explicit solution and data extension are shown in Figs. 3.9c, b. Shown in Figs. 3.10a, c, b are the results from the use of these formulae after converting $x(n)$ to a minimum-phase series.

The last two Examples (3, 4) serve the purpose of showing the danger in the blind use of the explicit solution and data extension, and the applicability of the formulae when the series is indeed real, causal, and of minimum-phase.

From Figs. 3.7b, c it can be seen that in Example 3 the data extension and the estimated MEM2 spectrum by the blind use of the formulae are apparently wrong: The data extension goes wild, and the spectra calculated from $C(k)$ by (3.3.8) and from the data extension are remarkably different. On the other hand, from other figures for these two examples it can be seen that the MEM2 spectra calculated by the explicit solution and by the data extension are consistent for the minimum-phase series, and almost consistent for the almost minimum-phase series (the negligible difference cannot be shown in the plotting in Fig. 3.9c). So we may say that if the MEM2 spectra calculated by the two methods are consistent, then the results by the blind use of the formulae may be acceptable. In practice this self-consistency may be used as a criterion for judging whether the given data segment is part of a minimum-phase series.

3.4 Equivalents and Signal Model

In this section we present five equivalents to MEM2 and a signal model. They are: the ACF extension subject to the nonnegativity constraint, the Principle of MCE, exponential process (signal model), Bayesian method, and MLM. This section is more or less parallel to Sect. 2.3 in its contents, so the concepts and results in that section will be cited without explanation.

3.4.1 ACF Extension Subject to the Nonnegativity Constraint

It is well known that the extension of the ACF must ensure the nonnegativity of the ACF, i.e., its FT, the spectrum, must be nonnegative. For a complex-valued signal, the extrapolated ACF in each step must be within or on a

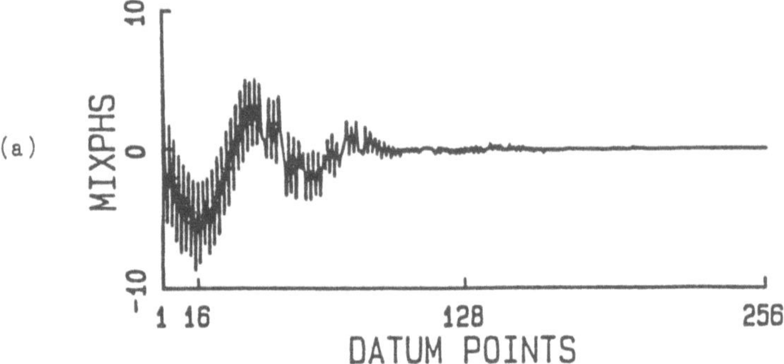

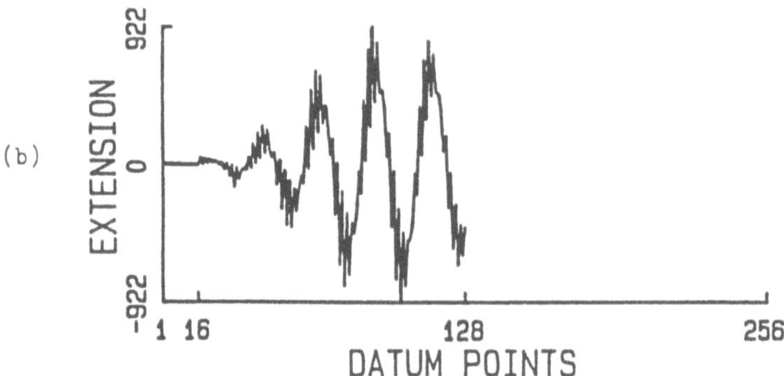

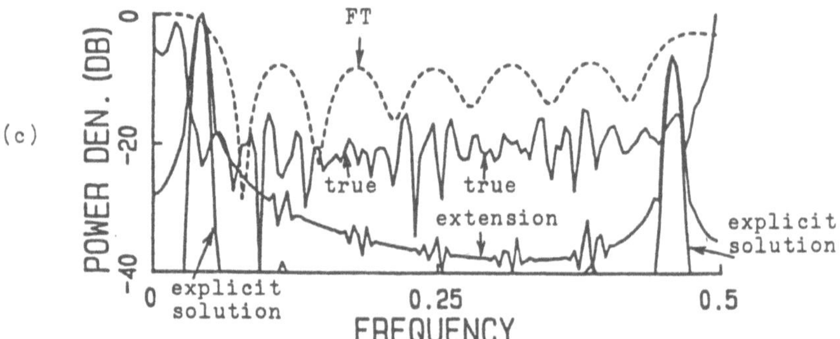

Fig. 3.7. The mixed-phase series, extrapolated series and spectra (from the blind use of the MEM2 formulae) in Example 3. (**a**) The original mixed-phase series. (**b**) The extrapolated series ($M = 15$). (**c**) Spectra: *true* spectrum, spectra by the *FT* method, calculated by $C(k)$ (*explicit solution*), and by data *extension*

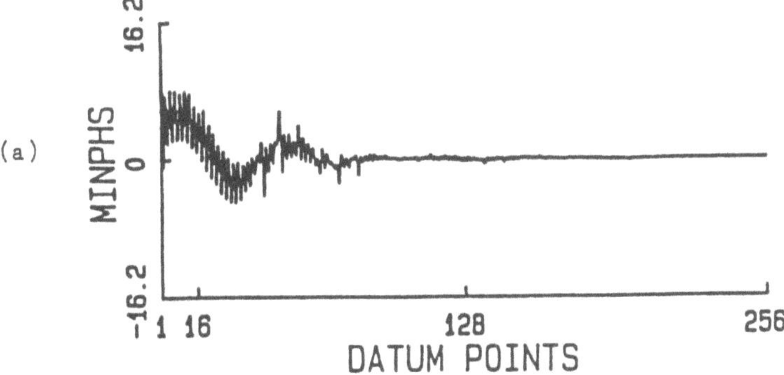

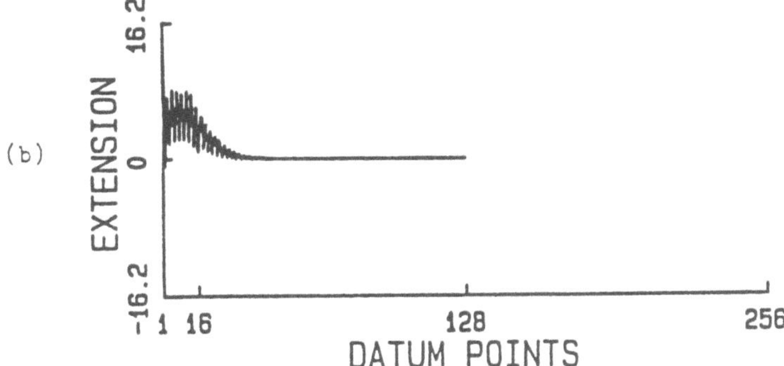

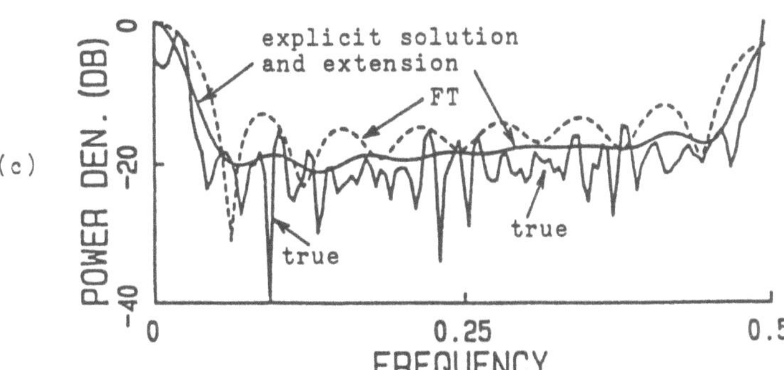

Fig. 3.8. Also for Example 3. (**a**) The equivalent minimum-phase series. (**b**) The extrapolated series ($M = 15$). (**c**) The spectra

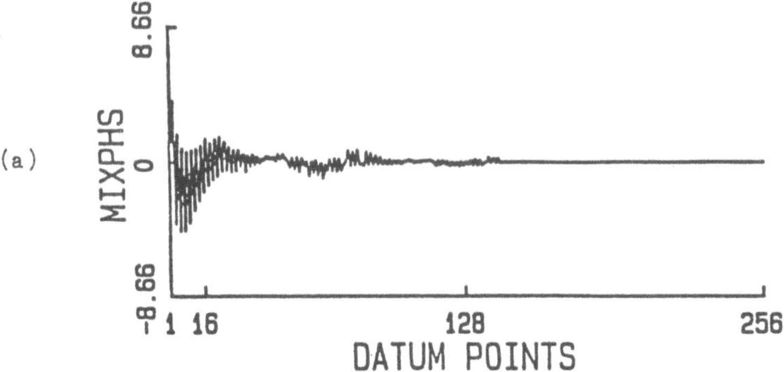

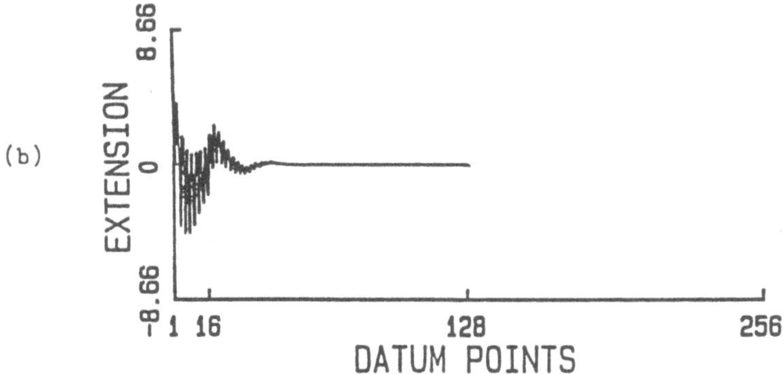

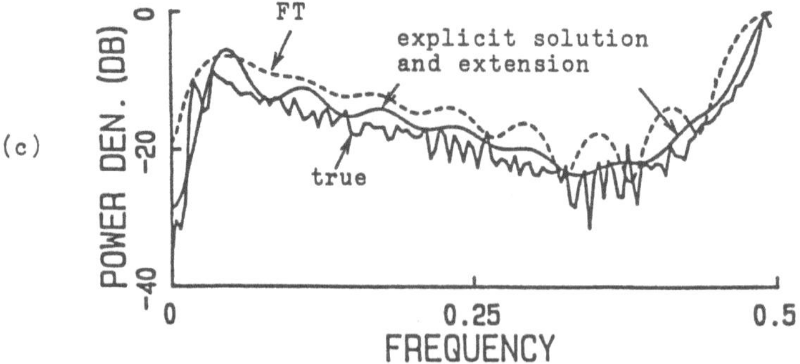

Fig. 3.9. For Example 4. (**a**) The original almost minimum-phase series. (**b**) The extrapolated series ($M = 15$). (**c**) The spectra

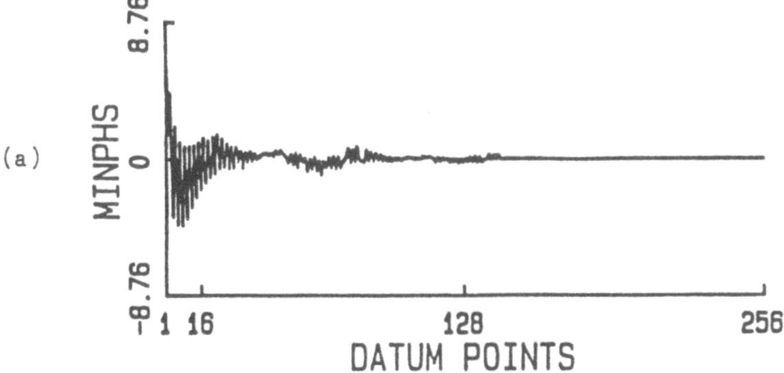

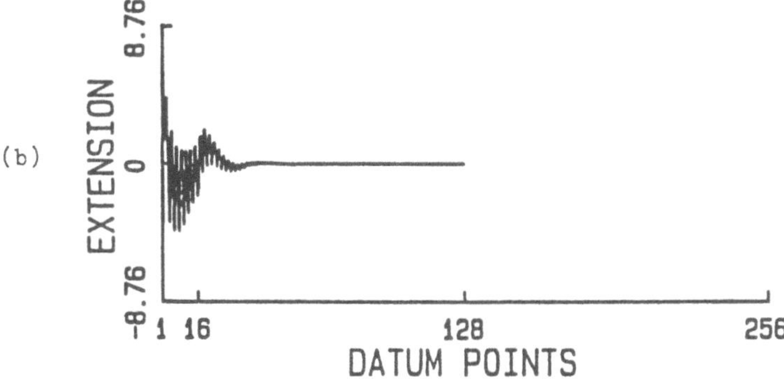

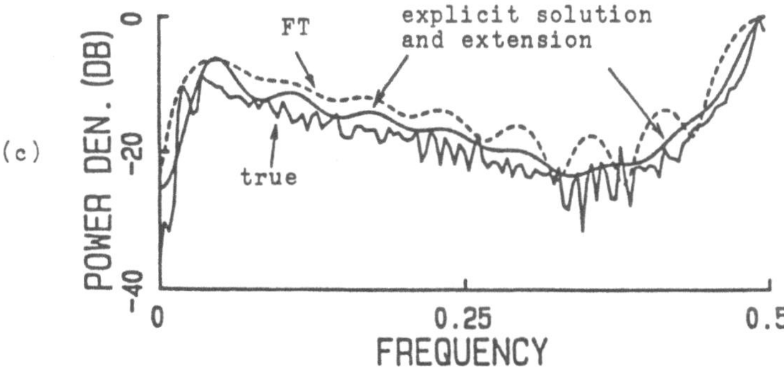

Fig. 3.10. Also for Example 4. (**a**) The equivalent minimum-phase series. (**b**) The extrapolated series ($M = 15$). (**c**) The spectra

circle on the complex plane (Fig. 2.4). In MEM1 the value at the center of the circle is taken as the extrapolated ACF. In MEM2, of course, the value at some point within or on the circle is taken. Clearly, this point cannot be the center of the circle since the data extensions in MEM1 and MEM2 are different. Generally speaking, this point is not on the circle either since, as pointed out in Sect. 2.3.1, once the extrapolated ACF lies on the circle, the successive circles determining the ranges of the permissible ACFs in the extension each shrink to a single point, hence there is no more "flexibility" in the extension. Then, where within the circle is this point representing the MEM2 extrapolated ACF? This is still an open question in theory, and is not successfully answered by investigating numerical examples.

3.4.2 Principle of MCE

In MEM1 the signal in the time domain is treated as a stochastic process while the spectrum as the expectation. Equation (2.3.20) is the cross-entropy expression in the frequency domain. If the prior estimate of the spectrum, $S_q(f)$, is a constant, say, one, then the cross-entropy H_c and (plain) $H1$ are related by

$$H_c = -H1 + [R(0) - 1] \,, \tag{3.4.1}$$

which shows the equivalence between the Principle of MCE and (plain) MEM1.

Denote the prior spectrum by $m(f)$ (equivalent to $S_q(f)$), and the posterior spectrum to be estimated by $S(f)$ (equivalent to $S_p(f)$). Then from the TP constraints,

$$\int_{-1/2}^{1/2} S(f)\mathrm{d}f = R(0) \,, \qquad \int_{-1/2}^{1/2} m(f)\mathrm{d}f = M(0) \tag{3.4.2}$$

($M(n)$ is the ACF corresponding to $m(f)$), the normalized spectra are known to be $S(f)/R(0)$ (posterior) and $m(f)/M(0)$ (prior).

Now we view the frequency f as a random variable, while the normalized spectra as its p.d.f.'s. This means that in measuring the Fourier components of the signal, the probabilities that the frequency is found to be in $f \sim f + \mathrm{d}f$ are $[S(f)/R(0)]\mathrm{d}f$ (posterior) and $[m(f)/M(0)]\mathrm{d}f$ (prior). The ACF $R(n)$ is viewed as the expectation of $R(0)\exp(2\pi \mathrm{i} f n)$:

$$R(n) = \int_{-1/2}^{1/2} [S(f)/R(0)]\,[R(0)\exp(2\pi \mathrm{i} f n)]\mathrm{d}f \,.$$

The cross-entropy is

$$H_c = \int_{-1/2}^{1/2} [S(f)/R(0)] \log\left(\frac{S(f)/R(0)}{m(f)/M(0)}\right) \mathrm{d}f$$

$$= \frac{1}{R(0)} \int S(f) \log \left[S(f)/m(f) \right] \mathrm{d}f + \frac{1}{R(0)} \int \log \left[M(0)/R(0) \right] \mathrm{d}f \,.$$

$$(3.4.3)$$

That is,

$$H_c = -\frac{1}{R(0)} H2 + \frac{1}{R(0)} \log \left[M(0)/R(0) \right] \,. \tag{3.4.4}$$

From (3.4.4) we see that given $R(0)$ and $M(0)$, minimizing H_c is equivalent to maximizing $H2$. Thus, the Principle of MCE is equivalent to proper MEM2, and plain MEM2 with constant $m(f)$ is their special cases.

In MEM2 the (normalized) spectrum itself is taken as the p.d.f. of the random variable f. Because f is a scalar and the mathematical operations are carried out only in the frequency domain, the derivation procedure of H_c is much simpler than that in MEM1.

3.4.3 Exponential Process (Signal Model)

The exponential model of signal, (3.3.9), in MEM2 has been obtained in Sect. 3.3; that is [3.10]

$$X(z) = \exp \left[C(0) + 2 \sum_{k=1}^{M} C(k) z^{-k} \right] \,. \tag{3.4.5}$$

The generator of the stochastic process $x(k)$ may be depicted by Fig. 2.5. This is a filter with the transfer function $H(z) = X(z)$, driven by white Gaussian noise of unit power. The output may be said to be an *exponential* (EXP) *process*, and denoted by EXP(M). M is the order of the process, and $C(k)$ are the EXP parameters.

Expanding the right-hand side of (3.4.5) in a Taylor series in z^{-1}, we see that the output $x(n)$ may be viewed as an MA process of infinite order. On the other hand, reforming the right-hand side to a fraction with the numerator being one, and then expanding the denominator in a Taylor series in z^{-1}, we see that $x(n)$ may be viewed as an AR process of infinite order. If you do think it is worth doing, you may reform (3.4.5) appropriately so that $x(n)$ may be viewed as an ARMA process of infinite order as well.

Because the impulse response of the filter $h(k)$ is of infinite duration and it is hard to calculate the coefficients for z^{-k} in the Taylor series from $C(k)$, in practice the realization of filtering in the frequency domain, but not in the time domain, is recommended. Specifically, to generate a segment of $x(k)$ of length N, take the discrete FT of the input sequence of length N, multiply it by $H(e^{2\pi \mathrm{i} j/N})$, and finally perform IFT, then a realization of the stochastic process is obtained. The larger N is, the closer the spectrum of the output is to $S(f)$.

Given the partial ACF or a time series, the EXP parameters must be calculated by $C(k) = \frac{1}{2}\lambda_k^*$ after estimating λ_k^* iteratively. Only when the given data $x(k), k = 0, \ldots, M$ is a segment of a real, causal and minimum-phase series, can $C(k)$, $k = 0, \ldots, M$ be calculated by the explicit solution.

3.4.4 Bayesian Method

Given the partial noisy ACF (visibility), by analogy with (2.3.33), the objective function to be maximized in MEM2 is

$$G = -\sum_{j=0}^{N-1} S(j) \log S(j) - \lambda \sum_{k \in K} |D(k) - R(k)|^2 / \sigma_k^2 \, , \tag{3.4.6}$$

where the relative weight of the maximization of the entropy to the data fit depends on the Lagrange multiplier λ.

First of all, we argue that if we set

$$p(S) \propto \exp \left[-\beta \sum_{j=0}^{N-1} S(j) \log S(j) \right] \, , \quad (S(j) \geq 0) \, , \tag{3.4.7}$$

then we have the following equivalence relationships:

Maximizing $p(S|D)$

$$\Longleftrightarrow \text{Max} \quad \log(D|S) + \log p(S)$$

$$\Longleftrightarrow \text{Max} \quad -\frac{1}{2} \sum_{k \in K} |D(k) - R(k)|^2 / \sigma_k^2 - \beta \sum_{j=0}^{N-1} S(j) \log S(j)$$

$$\Longleftrightarrow \text{Max} \quad -\sum_{j=0}^{N-1} S(j) \log S(j) - \alpha \sum_{k \in K} |D(k) - R(k)|^2 / \sigma_k^2 \, .$$

Here $\alpha = 1/(2\beta)$. The relative weight of the prior p.d.f. to the conditional p.d.f. depends on α.

Comparing the last expression with (3.4.6), we see that MEM2 is a special case of the Bayesian method [3.11]. It is special in that the prior p.d.f., $p(S)$, is assigned as in (3.4.7). The parameter $\alpha = 1/(2\beta)$ corresponds to λ.

In the remaining of this subsection we derive (3.4.7) from the stipulation of the Poisson distribution. The terminology of image processing will be used.

Consider the quantized $S(j)$ and its mean $m(j)$,

$$S(j) = n_j q \, , \quad m(j) = \lambda_j q \, , \tag{3.4.8}$$

where $n_j, \lambda_j = 0, 1, \ldots;$ q (equivalent to ΔB) is a small intensity unit (Sect. 3.1); λ_j is the mean of n_j.

We stipulate that $\{n_j\}$ has a Poisson distribution when all the pixels are independent of each other:

$$p(\{n_j\}) = \prod_{j=0}^{N-1} \frac{\lambda_j^{n_j}}{n_j!} e^{-\lambda_j} \, . \tag{3.4.9}$$

By Stirling's formula, $\log N! \sim N(\log N - 1)$ as $N \to \infty$, from (3.4.9) we have approximately

$$\log p(\{n_j\}) = \sum_{j=0}^{N-1} [n_j \log \lambda_j - n_j(\log n_j - 1) - \lambda_j]$$

$$= -\sum_{j=0}^{n-1} n_j \log(n_j/\lambda_j) + \sum_{j=0}^{N-1} n_j - \sum_{j=0}^{N-1} \lambda_j \,,$$

where the last two terms roughly cancel each other. From the above expression, in consideration of (3.4.8), we get

$$\log p(S) = -\frac{1}{q} \sum_{j=0}^{N-1} S(j) \log[S(j)/m(j)] \,. \tag{3.4.10}$$

This is the expression of the logarithm of the distribution $p(S)$ derived on the basis of the Poisson distribution stipulation. So we see that $m(j)$ is actually the prior estimate of $S(j)$. Let $m(j) = 1$, then (3.4.10) becomes

$$\log p(S) = -\frac{1}{q} \sum_{j=0}^{N-1} S(j) \log S(j) \,. \tag{3.4.11}$$

The unit q in the quantization of $S(j)$ may be indefinitely small, so we stipulate that $S(j)$ in (3.4.11) may assume continuous values. Having chosen q, dropping out the factor $1/q$, and adding a factor β in (3.4.11), we get

$$\log p(S) \propto -\beta \sum_{j=0}^{n-1} S(j) \log S(j) \,, \tag{3.4.12}$$

which is equivalent to (3.4.7).

3.4.5 MLM

In Sect. 3.1 it was MLM that we used to derive the $H2$ expression, so naturally MLM and MEM2 are equivalent. Note that here MLM is in the sense of the maximum multiplicity method, that is, maximizing the multiplicity W or, equivalently, its logarithm $\log W$.

Now we conclude the description of the equivalents and the signal model for MEM2. Compared with Sect. 2.3, this section is shorter. This reflects that compared with MEM1, MEM2 is simpler in some ways while more complex in others so that it is not mature yet. That this section is shorter can also be attributed partially to the omission of most of what was introduced in Sect. 2.3.

3.5 $R - \lambda$ Procedure

The $R - \lambda$ procedure is an algorithm for spectral estimation in MEM2
[3.6, 3.12]. The input data are the partial noiseless ACF (given or estimated
from the time series). This procedure is similar to the LM (Lim-Malik) algorithm introduced in Sect. 2.4. To save space, therefore, what was already
described in that section will be mostly omitted. On the other hand, techniques introduced here, such as the use of optimal scale factor and memory
coefficient, may also be employed for the LM algorithm. The $R - \lambda$ procedure
is used mainly for the study of the MEM2 properties. Most importantly, this
is the only algorithm available in the 2-D case for noiseless data.

In this section an alternative statement of the MEM2 problem is presented, then a computer flowchart for the $R - \lambda$ procedure is given, some
issues in practical computation are discussed, and, finally, a numerical example is demonstrated. More examples will be shown in Sects. 4.4, 4.6.

3.5.1 Statements of the MEM2 Problem

The statement of the 1-D MEM2 problem is originally like this (Sect. 3.2):

Given the ACF $R(n)$, $n \in D$, estimate the spectrum $S(f)$ such that

$$H2 = - \int_{-1/2}^{1/2} S(f) \log S(f) \mathrm{d}f$$

is maximized, and the constraints

$$R(n) = \mathrm{IFT}[S(f)] = \int_{-1/2}^{1/2} S(f) \exp(2\pi \mathrm{i} f n) \mathrm{d}f \,, \quad n \in D$$

are satisfied.

By investigating the structure of the solution (3.2.5) and the Lagrange
multiplier expression (3.2.7), with the discretized frequency and simplified
notation, then the MEM2 problem may be alternatively stated as follows:

Given the ACF $R(n)$, $n \in D$, estimate the spectrum $S(j)$ such that
(I) The constraints

$$R(n) = \mathrm{IFT}[S(j)] \,, \quad n \in D \tag{3.5.1}$$

are satisfied;
(II) The Lagrange multipliers

$$\lambda_n^* \begin{cases} \text{to be determined}\,, & n \in D\,, \\ = 0\,, & \text{otherwise}\,, \end{cases} \tag{3.5.2}$$

where

$$\lambda_n^* = \lambda_{-n} = \mathrm{IFT}[\log S(j)] = \mathrm{IFT}[\, \log \mathrm{FT}[R(n)]\,] \,. \tag{3.5.3}$$

The spectrum

$$S(j) = \exp\{\mathrm{FT}[\lambda_n^*]\} \ . \tag{3.5.4}$$

Obviously, the above statement is applicable, after doubling the subscripts, to the 2-D case. The data support may be of any shape, but must contain the origin for the TP constraint. For simplicity, we will use the 1-D notation throughout this section.

3.5.2 $R - \lambda$ Procedure

The $R - \lambda$ procedure, an iterative algorithm based on the last statement of the MEM2 problem, is best depicted by a computer flowchart in Fig. 3.11. The flowchart is self-explanatory. This is a practical flowchart taking into consideration the issues in the implementation of the algorithm.

The input data are the given ACF $R(n)$, $n \in D$. By setting

$$\lambda_n^{*(1)} = \begin{cases} \log R(0) \ , & n = 0 \ , \\ 0 \ , & \text{otherwise} \ , \end{cases}$$

the TP constraint is satisfied initially:

$$R^{(1)}(n) = \mathrm{IFT}[S^{(1)}(j)] = \mathrm{IFT}[R(0)] = R(0)\delta_n \ .$$

In the kth iteration, the calculated ACF, $R^{(k)}(n)$, is compared with the given $R(n)$ over the support D. If the difference between them is small enough, then $S^{(k)}(j)$ can be output as the MEM2 spectrum. The extrapolated ACF can also be obtained if $R^{(k)}(n)$ are output as well. If the difference is not small enough yet, the procedure should continue. That is, (1) correct the calculated $R^{(k)}(n)$ with the given $R(n)$ for $n \in D$; (2) calculate and truncate λ_n^* for $n \notin D$ according to (3.5.2). With these updated λ_n^*, the next iteration can be started. In repeating (1) and (2), we perform the transform back and forth between the data ($R(n)$) domain and the cepstrum (λ_n) domain, and impose the constraint (data, and λ_n^* must be truncated) on the current estimate in each domain. Hence the name $R - \lambda$ *procedure*.

In the following we discuss issues in implementing the algorithm.

1. Convergence Criterion

The normalized *rms* error is used as the measure of the difference between the calculated and given ACFs. Specifically, we define

$$\mathrm{DIFF}[R^{(k)}(n), R(n)] = \left(\frac{\sum_{n \in D} |R^{(k)}(n) - R(n)|^2}{\sum_{n \in D} |R(n)|^2} \right)^{1/2} \ , \tag{3.5.5}$$

which is equivalent to (2.4.17). The criterion for convergence is that DIFF decreases and falls just below the prescribed limit ε: DIFF $< \varepsilon$. Many numerical examples have shown that $\varepsilon = 0.01$ gives satisfactory results. Further reducing ε does not improve results but increases computational time.

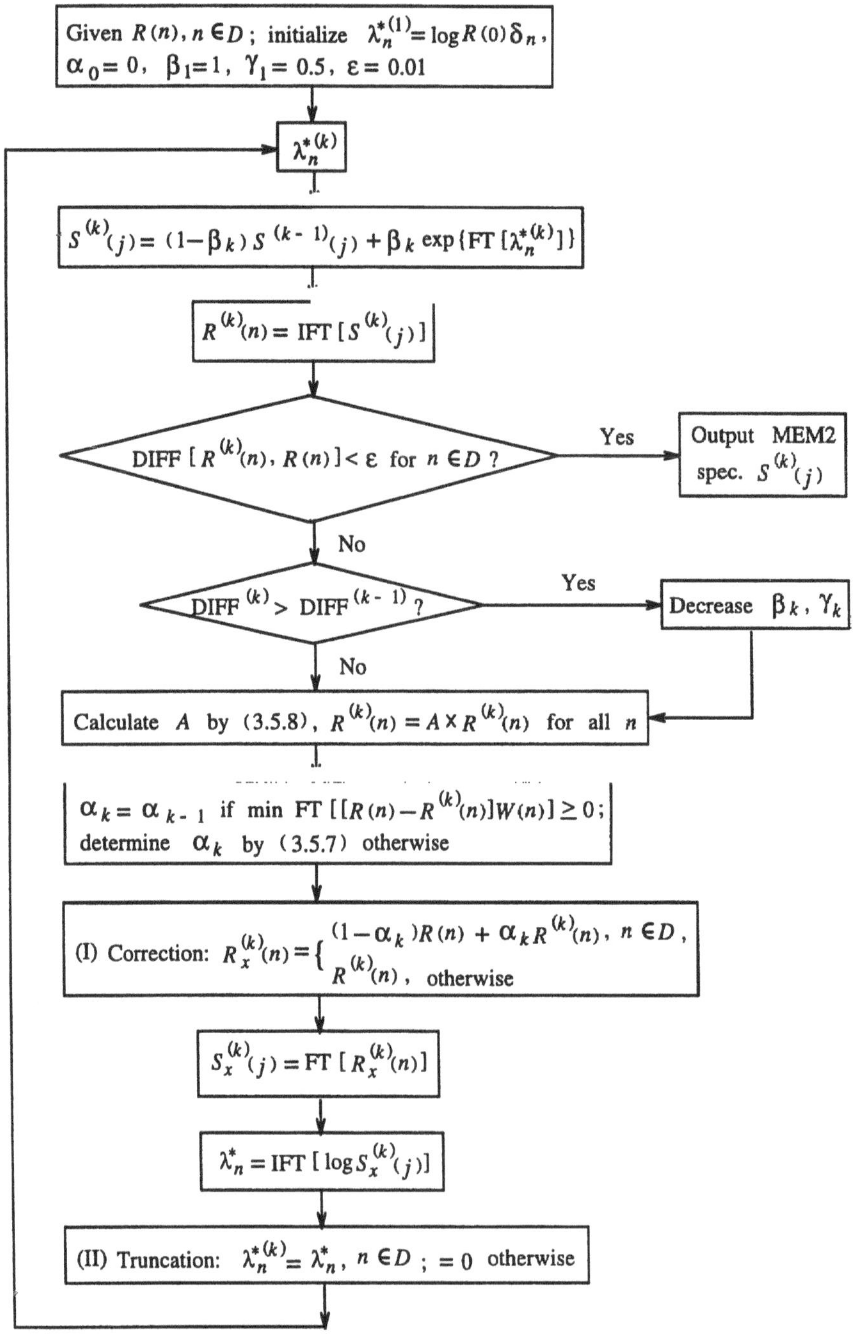

Fig. 3.11. The $R - \lambda$ procedure in MEM2. The superscript k, and subscript k of α, β, γ denote the kth iterate. The iteration starts with $k = 1$

2. FT Size

There is no point in using too large an FT size due to the modest extension of data in MEM2. In most cases we use the FT length as large as four times the given ACF size in each dimension.

3. Positive Definiteness of the ACF

The situation here is similar to that in the LM algorithm. $S(j)$ calculated from λ_n^* are always positive because of the exponential operation in (3.5.4). Therefore, the correction of $R^{(k)}(n)$ with $R(n)$ for $n \in D$ is the only possible way to compromise the PD of the ACF. Here there is no zero-crossing problem which was encountered in the LM algorithm.

In order to ensure the PD of the ACF, the calculated ACF is partially corrected. Specifically, in the iteration let

$$
R_x^{(k)}(n) = \begin{cases} (1 - \alpha_k)R(n) + \alpha_k R^{(k)}(n) , & n \in D , \\ R^{(k)}(n) , & \text{otherwise} \end{cases}
$$
$$
= R^{(k)}(n) + (1 - \alpha_k)[R(n) - R^{(k)}(n)]W(n) ; \qquad (3.5.6)
$$

cf. (2.4.19, 2.4.20). Here the window $W(n) = 1$ for $n \in D$, $W(n) = 0$ otherwise; the factor α is determined by

$$
\alpha_k = \max \left\{ \alpha_{k-1}, 1 - \gamma_k \min_{\{J_\alpha^-\}} \frac{\mathrm{FT}[R^{(k)}(n)]}{|\mathrm{FT}[\,[R(n) - R^{(k)}(n)]W(n)\,]|} \right\} , \qquad (3.5.7)
$$

cf. (2.4.24). γ_k is the convergence rate parameter, $0 < \gamma_k < 1$. The subset $\{J_\alpha^-\}$ is defined by

$$
\{J_\alpha^-\} = \{j : \mathrm{FT}[\,[R(n) - R^{(k)}(n)]W(n)\,] < 0\} .
$$

If $\{J_\alpha^-\}$ is an empty set, i.e.,

$$
\min_{\forall j} \mathrm{FT}[\,[R(n) - R^{(k)}(n)]W(n)\,] \geq 0 ,
$$

then α_k may take any value on $[0,1]$ in consideration of the PD. Equation (3.5.7) becomes

$$
\alpha_k = \max\{\alpha_{k-1}, 0\} = \alpha_{k-1} .
$$

The value of α can only increase or remain unchanged in the iteration.

4. Optimal Scale Factor

In the kth iteration, $R^{(k)}(n)$ can be scaled by a factor A before being corrected by $R(n)$ for $n \in D$. The scale factor is so chosen as to minimize the DIFF, that is, to achieve the best fit between $R^{(k)}(n)$ and $R(n)$ for $n \in D$. There are some points to be discussed in doing this:

(1) What is of interest to us is the shape of the spectrum, but not the individual values. Multiplying the ACF by A does not change the shape of the spectrum.

(2) By minimizing the DIFF, the TP ($R(0)$) constraint, along with the constraints on the other lags, is best satisfied. In this way convergence can be speeded up.

(3) Since the transform from $\lambda_n^{*(k)}$ to $R^{(k)}(n)$ is nonlinear, the scaled $R^{(k)}(n)$, $A \times R^{(k)}(n)$, do not correspond to the scaled $\lambda_n^{*(k)}$, $A \times \lambda_n^{*(k)}$, for which $\lambda_n^{*(k)} = 0$ for $n \notin D$ unless $A = 1$ (trivial). But λ_n^* must be truncated for the next iteration anyway.

We can expect A to become closer and closer to unity during the course of iteration in a monotonic or oscillating manner, so this scaling takes significant effect only in the early stage of the iteration.

The *optimal scale factor A* can be determined by differentiating $(\text{DIFF})^2$ with respect to A, where DIFF is defined by (3.5.5) with A prefixed to $R^{(k)}(n)$. Specifically, let

$$0 = \frac{\mathrm{d}(\text{DIFF})^2}{\mathrm{d}A} = \frac{\mathrm{d}}{\mathrm{d}A}\left(\frac{\sum |A \cdot R^{(k)}(n) - R(n)|^2}{\sum |R(n)|^2}\right),$$

(summations are over $n \in D$); that is

$$\begin{aligned}
0 &= \frac{\mathrm{d}}{\mathrm{d}A}\left[\sum |AR^{(k)}(n) - R(n)|^2\right] \\
&= 2A \sum |R^{(k)}(n)|^2 - 2 \sum \mathrm{Re}\{R(n)R^{(k)*}(n)\}.
\end{aligned}$$

Hence,

$$A = \frac{\sum \mathrm{Re}\{R(n)R^{(k)*}(n)\}}{\sum |R^{(k)}(n)|^2}. \tag{3.5.8}$$

The second derivative is

$$\frac{\mathrm{d}^2(\text{DIFF})^2}{\mathrm{d}A^2} = \frac{2 \sum |R^{(k)}(n)|^2}{\sum |R(n)|^2} > 0,$$

so that the factor A in (3.5.8) is indeed a minimum point.

From above we see that the formula for calculating A is fairly simple. In computation, the calculation of A requires little extra effort, while the number of iterations can be reduced significantly, especially when the initial value is improperly chosen. Sometimes this number can be reduced by one-third or so.

5. Memory Coefficient

To overcome the instability due to the exponential function in (3.5.4), i.e., peaks in the spectrum grow up very rapidly so arithmetic floating point overflow or divergence occurs in computation, we average successive iterates [3.13]. Specifically, the kth iterate $S^{(k)}(j)$ comes partially from the kth iterate $\lambda_n^{*(k)}$, and partially from the previous iterate $S^{(k-1)}(j)$, i.e.,

$$S^{(k)}(j) = (1 - \beta)S^{(k-1)}(j) + \beta \exp\{\text{FT}[\lambda_n^{*(k)}]\}, \tag{3.5.9}$$

where β is called the *memory coefficient*, $0 \leq \beta \leq 1$. If $\beta = 1$, then the first term on the right-hand side of (3.5.9) vanishes and the spectrum calculated from $\lambda_n^{*(k)}$ is assigned to $S^{(k)}(j)$ ("zero-memory"). On the contrary, if $\beta = 0$, then the second term vanishes and the spectrum obtained in the previous iteration is assigned to $S^{(k)}(j)$ ("full memory"). In this case the iteration stagnates and the value of DIFF does not change. In the computer program, the initial value of β is set to one. If the DIFF increases in iteration, then β will be reduced, say, to half.

The resultant spectrum does not change due to introducing β into the iteration. After the iteration has converged, $S^{(k)}(j)$, $S^{(k-1)}(j) \to S(j)$. Therefore, (3.5.9) becomes

$$S(j) = (1 - \beta)S(j) + \beta \exp\{\mathrm{FT}[\lambda_n^{*(k)}]\} \,.$$

That is,

$$S(j) = \exp\{\mathrm{FT}[\lambda_n^{*(k)}]\}.$$

This formula resembles (3.5.4).

A computer program can be written following the flowchart in Fig. 3.11. The two FTs for determining α_k and in evaluating $S_x^{(k)}(j)$ can be combined into one. Four calculations of FT/IFT are needed in one iteration. The FT and IFT can swap in position. Then, the initial $\lambda_n^{*(1)}$ must be changed accordingly. For flexibility, the initial γ, β and their rates of reduction can be changed in the computer program.

6. Convergence Problem

Now we come to the crucial question, the convergence problem of the $R - \lambda$ procedure. The MEM2 solution is unique (Sect. 4.3). But its existence is still an open question in theory. However, in hundreds of numerical examples with nonminimum-phase, noncausal real cosinusoids, or complex exponential or Gaussian functions as input data, we observed that with reasonably large zero-lag there was no difficulty in the converge of the $R - \lambda$ procedure, which suggests that no unduly restrictive conditions are required for the existence of the solution. MEM2 as well as MEM1 is a particular case of the positive definite extension of the ACF. By "reasonably large zero-lag" we mean that the zero-lag (which determines the positive background level in the spectrum) should be large enough to ensure the positiveness of the resultant spectrum under the modest data extension in MEM2. We are satisfied with this sweeping statement before a more precise one turns out. The exact zero-lag required depends on the specific ACF.

3.5.3 Example

Finally, we demonstrate a numerical example (Fig. 3.12). The sum of two complex Gaussian functions is used as the true ACF:

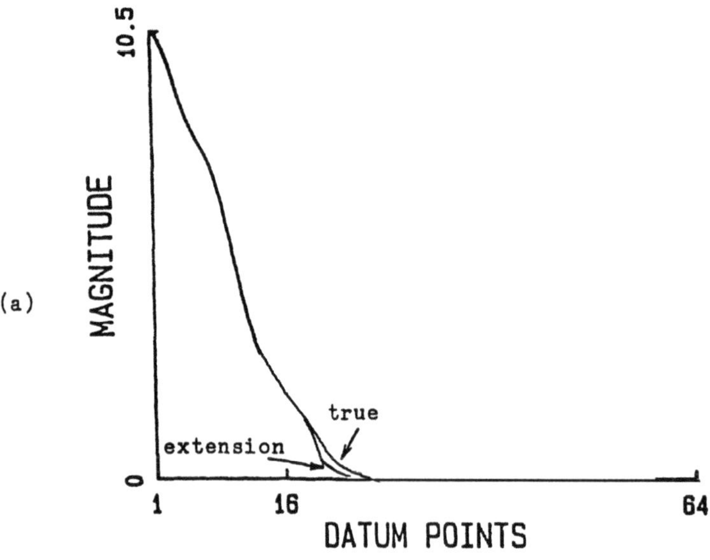

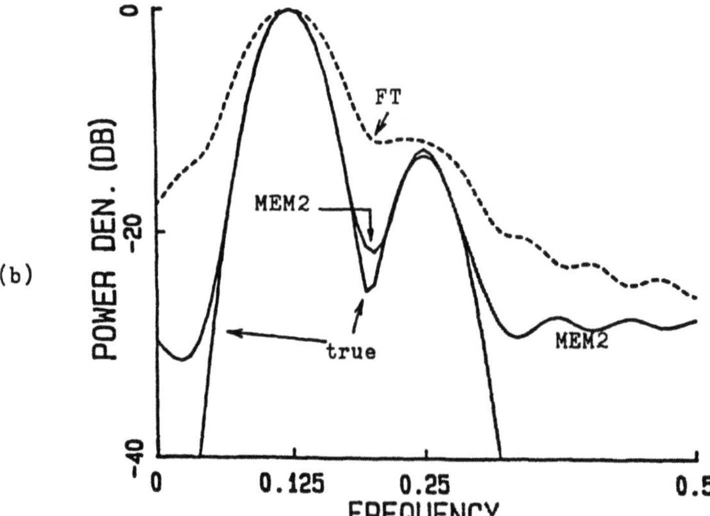

Fig. 3.12. Complex Gaussian function and its spectra. (a) The moduli of the *true* and extrapolated (*extension*) ACFs. (b) The *true*, *FT* (triangular window) and *MEM2* ($R - \lambda$ procedure) spectra. $M = 15$

$$R(n) = A_1 \exp\left(-\frac{n^2}{2\sigma_1^2}\right) \exp(2\pi i n_1 n/N)$$

$$+A_2 \exp\left(-\frac{n^2}{2\sigma_2^2}\right) \exp(2\pi i n_2 n/N), \quad n = \ldots, -1, 0, 1, \ldots .$$

$$(3.5.10)$$

The parameters are $A_1 = 10$, $A_2 = 0.5$, $n_1 = 16$, $n_2 = 32$, $\sigma_1 = \sigma_2 = 8$.

The moduli of the true and extrapolated ACF by MEM2 are shown in Fig. 3.12a. The first 16 true ACF values $R(n)$, $n = 0, \ldots, 15$ were used as the given data for the $R - \lambda$ procedure and used for calculating the FT spectrum (zero extension of data, triangular window). Accordingly, the FT length was chosen to be 128.

Shown in Fig. 3.12b are the true, FT and MEM2 (by the $R - \lambda$ procedure, 20 iterations) spectra. From the figure it can be seen that the lower peak (at $f_2 = 0.25$) degenerates to a suggestive shoulder in the FT spectrum, while it is distinct in the MEM2 spectrum. The MEM2 spectrum does not suffer from the sidelobe problem since the spectrum of a Gaussian function is a gradually rising and falling peak (contrasting sharply with the isolated spectral line(s) of a real cosinusoidal or complex exponential function).

3.6 Algorithms and Numerical Examples (I)

MEM2 is used mainly for 2-D image restoration. The emphasis has been placed on algorithms in research. We may say that the MEM2 research in the astronomical community is the most active. Most of the algorithms developed are applicable exclusively to astronomical image processing. Only a few of them may also be used in other areas such as medical imaging, crystallography and photo processing.

The algorithms to be introduced are those that played a key role in developing MEM2, and/or are in use currently, and/or may be referred to for tricks and approaches to certain problems in developing new algorithms.

Three sections are devoted to algorithms so that each one is not too lengthy. In this section the Frieden algorithm, Gull-Daniell algorithm and its revised version, and the simplified Newton-Raphson algorithm are described. They are used mainly for image restoration in astronomy. The next section is concerned with the Skilling-Bryan algorithm and differential equation approach. They are general-purpose algorithms, i.e., they can be used in astronomy or other areas. Described in the last section are algorithms developed in the past few years; namely, the MEM/MemSys5 package, the mem task in IRAF (Image Reduction and Analysis Facility), and several algorithms for restoring images with variable resolution. They are also general-purpose algorithms.

For all the algorithms mentioned above, the input data are noisy visibilities or degraded images.

We will describe in detail the basic ideas and main characteristics of these algorithms. The algorithms' publication dates, the authors' backgrounds and the specific problems to be approached were all quite different, therefore, a wide variety of terminology and notations were used in the literature. We will try our best to keep them what they were, but sometimes changes will be necessary and notes will be given. For simplicity, the 1-D notation is used throughout.

3.6.1 Frieden Algorithm

The Frieden algorithm [3.1] was introduced for optical image restoration. Denote the intensity distributions of the object and the measured degraded image by O_j $(j = 1, \ldots, J)$ and I_m $(m = 1, \ldots, M)$, respectively. The noise in measurement is n_m $(m = 1, \ldots, M)$. The units of the intensity and noise are ΔO and ΔN, respectively. The relationship is

$$I_m = \sum_{j=1}^{J} O_j S(m, j) + n_m \, , \quad m = 1, \ldots, M \, , \tag{3.6.1}$$

where $S(m, j)$ is the point spread function (PSF) of the image formation system. Let $J > M$, i.e., the number of pixels of the image to be restored is greater than that of datum points, which is desirable for resolution enhancement.

In the Frieden algorithm the object distribution O_j and the (biased) noise distribution N_m (see later for its definition) are estimated simultaneously.

Since $O_j \geq 0$, the entropy of the object is simply

$$H_0 = -\Delta O^{-1} \sum_{j=1}^{J} O_j \log O_j \, , \tag{3.6.2}$$

cf. (3.4.11). In contrast, since n_m may well assume a negative value, an expression similar to (3.6.2) for the noise is meaningless. To overcome this difficulty, we define the biased noise by

$$N_m = n_m + B \geq 0 \, , \tag{3.6.3}$$

where the noise bias is large enough to ensure the inequality holds. For example, the most negative value of the noise is estimated, and then its absolute value is taken as B. Now we are able to define the entropy of the (biased) noise, that is,

$$H_N = -\Delta N^{-1} \sum_{m=1}^{M} N_m \log N_m \, . \tag{3.6.4}$$

Based on (3.6.3), (3.6.1) becomes

$$I_m = \sum_{j=1}^{J} O_j S(m,j) + N_m - B , \quad m = 1,\dots,M . \tag{3.6.5}$$

$H_0 + H_N$ is the entropy of the object-noise combination. However, we change it to the following form:

$$H2 = -\sum_{j=1}^{J} O_j \log O_j - \rho \sum_{m=1}^{M} N_m \log N_m , \tag{3.6.6}$$

where the parameter $\rho = \Delta O / \Delta N$ represents the signal-to-noise uncertainty. $H2$ in (3.6.6) is the entropy to be maximized. The TP constraint is

$$P_0 = \sum_{j=1}^{J} O_j . \tag{3.6.7}$$

Introduce the Lagrange multipliers μ and λ_m for the TP constraint (3.6.7) and data constraints (3.6.5), respectively, then form the objective function G and let $\partial G/\partial O_j = 0$, $\partial G/\partial N_m = 0$. It is easy to find the MEM2 solution

$$\hat{O}_j = \exp\left[-1 - \mu - \sum_{m=1}^{M} \lambda_m S(m,j) \right] , \quad j = 1,\dots,J , \tag{3.6.8}$$

$$\hat{N}_m = \exp(-1 - \lambda_m/\rho) , \quad m = 1,\dots,M . \tag{3.6.9}$$

Substituting (3.6.8, 3.6.9) in (3.6.5, 3.6.7) yields $M + 1$ equations in $M + 1$ unknowns $\mu, \lambda_1, \dots, \lambda_M$, which can be solved by the Newton-Raphson method.

The parameters ρ and B are used to control the smoothness of the estimated object distribution $\hat{O}_j$ relative to the estimated noise distribution $\hat{N}_m$. The choice of ρ and B depends on the prior knowledge about the object and noise. For a wide range of ρ and B, the result is not sensitive to their values. We will not discuss this further.

The usefulness of the Frieden algorithm itself is limited. We may say that the most important achievement of this algorithm is that owing to its publication, the spatial or configurational entropy of the form

$$H2 = -\sum_{j} p_j \log p_j \tag{3.6.10}$$

was introduced into image restoration and MEM2 has gained popularity ever since. Its influence has reached far beyond the area of optical image restoration.

3.6.2 Gull-Daniell Algorithm

The Gull-Daniell (GD) algorithm [3.14] was introduced for image restoration for aperture synthesis radio telescopes. The original measured data are visibilities distributed on a grid in the polar coordinate system (for an East-West antenna array) or in an irregular manner (for an array containing at least one element in the North-South direction). For utilizing the FFT technique, the measured data are projected onto a rectangular grid in the Cartesian coordinate system before being used as the input to the algorithm. The antenna spacing coverage is normally sector-shaped (Fig. 3.13a). Signals from radio sources are usually very weak. So the problem facing us is image restoration with incomplete and noisy data.

Suppose that the input data are the visibilities P_k, $k \in D$ contaminated by white Gaussian noise with variance σ_k^2. The image (map) to be restored is B_j whose FT (i.e., the visibilities corresponding to B_j) is F_k. We would have $F_k = P_k$, $k \in D$ without noise. Now that the data are incomplete and noisy, the statistic

$$C = \sum_{k \in D} |P_k - F_k|^2 / \sigma_k^2 \tag{3.6.11}$$

should have a χ^2-distribution.

The entropy to be maximized is

$$H2 = -\sum_j B_j \log B_j \ . \tag{3.6.12}$$

In consideration of the noise, the data fit requires the χ^2-statistic C in (3.6.11) be equal to or less than its critical value C_e. For example, take $C_e = M$ (at 50% confidence level) or $C_e = M + 1.645\sqrt{2M}$ (at 95% confidence level), where M is the number of datum points. When M is very large, $M \gg 1.645\sqrt{2M}$, then the difference between the two C_e's is insignificant.

The required data fit $C \leq C_e$ is achieved by adjusting the Lagrange multiplier λ in iteration. Specifically, form the objective function

$$Q = H2 - \frac{\lambda}{2}C + \mu \sum_j B_j \ , \tag{3.6.13}$$

where the Lagrange multiplier μ is for the TP constraint. The necessity of including the term concerning the TP explicitly in the objective function is due to the fact that the data P_k usually contain no P_0 representing the TP (zero-baseline is unrealizable).

Let $\partial Q/\partial B_j = 0$; then it is easy to obtain the MEM2 solution of the form

$$B_j = A \exp \left\{ \frac{\lambda}{2} \frac{\partial C(B_j)}{\partial B_j} \right\} \ , \tag{3.6.14}$$

where the scale factor

$$A = \exp(-1 + \mu) \, , \tag{3.6.15}$$

$C(B_j)$ is a simplified notation of $C(\{B_j\})$ meaning that C is a function of all B_j.

The partial derivative in (3.6.14) can be expressed in terms of P_k, σ_k^2 and F_k. From

$$F_k = \mathrm{FT}[B_j] = \sum_{j=0}^{N-1} B_j \exp(-2\pi i j k/N) \, ,$$

it follows that

$$\begin{aligned}
\frac{\partial C(B_j)}{\partial B_j} &= \frac{\partial}{\partial B_j} \sum_{k \in D} (P_k - F_k)(P_k - F_k)^*/\sigma_k^2 \\
&= 2 \sum_{k \in D} (P_k - F_k) \left[\frac{\partial}{\partial B_j}(P_k - F_k) \right]^* /\sigma_k^2 \\
&= -2 \sum_{k \in D} \frac{P_k - F_k}{\sigma_k^2} \exp(+2\pi i j k/N) \\
&= -2N \, \mathrm{IFT} \left[\frac{P_k - F_k}{\sigma_k^2} W_k \right] \, , \tag{3.6.16}
\end{aligned}$$

where the window $W_k = 1$ for $k \in D$; $W_k = 0$ otherwise. In the above derivation the Hermitian property of the visibility (B_j are real) was utilized; the scale factor for the IFT was taken to be $1/N$.

Substituting (3.6.16) in (3.6.14) yields

$$B_j = A \exp \left\{ \lambda N \, \mathrm{IFT} \left[\frac{P_k - F_k}{\sigma_k^2} W_k \right] \right\} \, . \tag{3.6.17}$$

Equation (3.6.14) may be viewed as a solution of differential form, while (3.6.17) may be viewed as the one of integral form to the maximization. They are basic formulae for the GD algorithm.

The GD algorithm is an iterative procedure for solving (3.6.14) or equivalently (3.6.17), which is described in the following:

Given P_k and σ_k^2, for a particular λ (whose initial value is zero; i.e., starting with a uniform map) each new iterate is obtained by substituting B_j of the left-hand side into the right-hand side of (3.6.14). Symbolically,

$$B_j^{(n+1)}(GD) = A \exp \left(-\frac{\lambda}{2} \frac{\partial C(B_j^{(n)}(GD))}{\partial B_j} \right) \, , \tag{3.6.18}$$

that is,

$$B_j^{(n+1)}(GD) = A \exp \left\{ \lambda N \, \mathrm{IFT} \left[\frac{P_k - F_k^{(n)}}{\sigma_k^2} W_k \right] \right\} \, , \tag{3.6.19}$$

where $F_k^{(n)} = \mathrm{FT}[B_j^{(n)}(GD)]$, the superscript n denotes the nth iterate.

The solution to (3.6.14), B_j, exists and is unique for the appropriate choice of $\lambda > 0$ and σ_k^2. After convergence for a particular λ, C takes some value. Then increase λ, do iterations again until convergence is achieved, and a smaller C is obtained. In this way the iterations are repeated. When C falls just below its critical value C_e, then the data fit is considered to be appropriate, the value of λ is correct, and the desirable MEM2 image is obtained. Note the double iteration here. The inner iteration is for a fixed λ, while the outer iteration is for the correct data fit (adjusting λ).

The merits of the GD algorithm are the positiveness of successive iterates and rapid development of peak values due to the exponential function in (3.6.14). However, it is also due to the exponential function that the GD algorithm exhibits erratic and unstable behavior, especially at high values of λ. Sometimes peaks become higher and higher in the inner iteration for a large λ, and eventually arithmetic floating point overflow occurs in computation. To overcome this difficulty we must, as in the $R - \lambda$ procedure, average successive iterates to ensure convergence in the inner iteration by setting

$$B_j^{(n+1)}(GD) \;=\; (1-\beta)B_j^{(n)}(GD) + \beta A \exp\left(-\frac{\lambda}{2}\frac{\partial C(B_j^{(n)}(GD))}{\partial B_j}\right).$$

$$(3.6.20)$$

For a large λ, the memory coefficient β could be as small as a few percent; thus, the iteration effectively stops.

In order to overcome the drawbacks of instability and speed of convergence, two major techniques have been introduced to revise the GD algorithm. These constitute the subject of the next subsection.

3.6.3 Revised GD Algorithm

The two major techniques are presented in the following [3.6, 3.15]. They are: (1) Use of the second-order approximate solution; and (2) Use of an optimal scale factor.

1. Use of the Second-Order Approximate Solution

Formula (3.6.14) is the exact solution to the constrained maximization, but (3.6.18) is not (B_j's on the two sides are different). Apparently, the following solution is exact:

$$B_j^{(n+1)} = A \exp\left(-\frac{\lambda}{2}\frac{\partial C(B_j^{(n+1)})}{\partial B_j}\right).$$

$$(3.6.21)$$

Note $B_j^{(n+1)}$ on the right-hand side.

Expanding $\partial C(B_j^{(n+1)})/\partial B_j$ in a Taylor series yields

$$\frac{\partial C(B_j^{(n+1)})}{\partial B_j} = \frac{\partial C(B_j^{(n)})}{\partial B_j} + \sum_l \frac{\partial^2 C(B_j^{(n)})}{\partial B_j \partial B_l}[B_l^{(n+1)} - B_l^{(n)}] + 0. \quad (3.6.22)$$

The third and higher order terms vanish since C is a quadratic form in B_j. It is easy to see, by substituting (3.6.22) into (3.6.21), that (3.6.18) in the GD algorithm is the first-order approximate solution. The exact solution should be

$$B_j^{(n+1)} \;=\; B_j^{(n+1)}(GD)\exp\left\{-\frac{\lambda}{2}\sum_l \frac{\partial^2 C(B_j^{(n)})}{\partial B_j \partial B_l}[B_l^{(n+1)} - B_l^{(n)}]\right\}\;.$$

$$(3.6.23)$$

Utilizing (3.6.16), the second-order partial derivative is calculated to be

$$\begin{aligned}
\frac{\partial^2 C(B_j^{(n)})}{\partial B_j \partial B_l} &= 2\sum_{k\in D}\frac{1}{\sigma_k^2}\exp\left[\frac{2\pi\mathrm{i}}{N}k(j-l)\right]\\
&= 2Nw_{j-l}\;,
\end{aligned}$$

$$(3.6.24)$$

where

$$w_j = \mathrm{IFT}[W_k/\sigma_k^2]\;.$$

$$(3.6.24a)$$

Formula (3.6.23) can be written in the short form:

$$B_j^{(n+1)} = B_j^{(n+1)}(GD)\exp\{-\lambda N[B_j^{(n+1)} - B_j^{(n)}]*w_j\}\;.$$

$$(3.6.25)$$

Having calculated $B_j^{(n+1)}(GD)$ from $B_j^{(n)}$ by formula (3.6.18) or (3.6.19), equations (3.6.25) have N unknowns $B_j^{(n+1)}$. It is inevitable to invoke approximation for solving these equations because of the exponential function and convolution. Moreover, the solution method should be simple since (3.6.25) must be solved a number of times for a particular λ. There are three possible schemes, but only one of them to be described below has proved successful in practice.

For aperture synthesis radio telescopes, $\{w_j\}$ could be treated as a δ-function and the following would be a reasonable approximation for (3.6.25):

$$B_j^{(n+1)} = B_j^{(n+1)}(GD)\exp\{-\lambda N[B_j^{(n+1)} - B_j^{(n)}]w_0 q\}\;.$$

$$(3.6.26)$$

The factor q is introduced to improve the approximation due to using multiplication instead of convolution. q may be set to be the number of pixels in the mainlobe of antenna pattern, say $q = 1 \sim 2$.

Further approximation is necessary. Utilizing $\mathrm{e}^x \approx 1 + x$ to eliminate the exponential function in (3.6.26) yields, after a simple mathematical manipulation,

$$B_j^{(n+1)} = \frac{1 + \lambda N w_0 q B_j^{(n)}}{1 + \lambda N w_0 q B_j^{(n+1)}(GD)}B_j^{(n+1)}(GD)\;.$$

$$(3.6.27)$$

This is the formula to update the iterate in the revised GD algorithm. First, it is straightforward to evaluate $B_j^{(n+1)}$ from $B_j^{(n+1)}(GD)$. No extra FFT is needed. Second, the factor before $B_j^{(n+1)}(GD)$ on the right-hand side of (3.6.27) is less or greater than unity depending on whether $B_j^{(n+1)}(GD)$ is greater or less than $B_j^{(n)}$. Thus, a big jump from $B_j^{(n)}$ to $B_j^{(n+1)}(GD)$ will be compensated for by this factor, and high peaks will be damped down. Consequently, the stability is remarkably improved. As a matter of fact, $B_j^{(n+1)}$ is an interpolation between $B_j^{(n)}$ and $B_j^{(n+1)}(GD)$, $B_j^{(n+1)}$ being asymptotic to and upper bounded by $B_j^{(n)} + 1/(\lambda N w_0 q)$. As λ is increasing in the iteration, the limit on the increase from $B_j^{(n)}$ to $B_j^{(n+1)}$ (i.e., $1/(\lambda N w_0 q)$) will be decreasing.

Although the stability is improved by revising the algorithm, successive iterates still need averaging. Nevertheless, the memory coefficient can be larger so that the convergence rate is consequently higher. Numerical examples showed that β might be 0.25 or greater.

2. Use of an Optimal Scale Factor

Strictly speaking, the scale factor A in (3.6.15) should be adjusted in the iteration so that the TP meets the prescribed constraint. But in practice it is not necessary to do so. This is not only because the TP (zero-baseline visibility) is often unknown, but also because an appropriate scale factor can speed up the convergence as in the $R - \lambda$ procedure.

The TP is unknown; nevertheless its lower boundary is obvious:

$$\text{TP} = P_0 \geq \max_{k \in D} |P_k| \,,$$

which is the approximation of

$$\text{TP} = F_0 \geq \max_{k \in D} |F_k| \,.$$

Since λ starts with zero, initially let

$$A^{(0)} = B_j^{(0)} = \max_{k \in D} |P_k|/(\text{the number of pixels}) \,.$$

Suppose that after n iterations, $B_j^{(n+1)}$ for a particular λ have been calculated from $A^{(n)}$ and $F_k^{(n)} = \text{FT}[B_j^{(n)}]$. The scale factor for the $(n+1)$th iteration is $\alpha A^{(n)}$. The factor α is so chosen as to minimize the χ^2-statistic C in (3.6.11). Specifically, let

$$
\begin{aligned}
0 &= \partial C/\partial \alpha \\
&= \frac{\partial}{\partial \alpha} \sum_{k \in D} |P_k - \alpha F_k^{(n+1)}|^2/\sigma_k^2 \\
&= 2 \sum_{k \in D} (\alpha |F_k^{(n+1)}|^2 - \text{Re}\{F_k^{(n+1)} P_k^*\})/\sigma_k^2 \,.
\end{aligned}
$$

Then it is found that

$$\alpha = \frac{\sum_{k\in D} \text{Re}\{F_k^{(n+1)} P_k^*\}/\sigma_k^2}{\sum_{k\in D} |F_k^{(n+1)}|^2/\sigma_k^2} \; . \tag{3.6.28}$$

The second derivative is

$$\partial^2 C/\partial\alpha^2 = 2\sum_{k\in D} |F_k^{(n+1)}|/\sigma_k^2 > 0 \; ,$$

so α determined from (3.6.28) is indeed the minimum point of C.

After determining α, $B_j^{(n+1)}$ and $F_k^{(n+1)}$ are multiplied by α. Then calculate C and check whether $C \leq C_e$. The scale factor A is also updated for use in the next iteration:

$$A^{(n+1)} = \alpha A^{(n)} \; . \tag{3.6.29}$$

A is adjusted in this way for the optimization of the data fit. Hence A is called the *optimal scale factor*. As in the $R - \lambda$ procedure, α will approach unity rapidly. So this scale transformation takes significant effect only in the first few iterations.

The optimal scale factor and the second-order approximate solution techniques are the bases of the revised GD algorithm. Compared with the original version of the GD algorithm, the CPU time in computation can be reduced by $1/3 - 1/2$. A numerical example will be shown at the end of this section. The computer flowchart and a FORTRAN program list are presented in [3.6].

The revised GD algorithm was used in the FCP (Fleurs Control Program) package for image restoration at the Fleurs Radio Observatory. The second-order approximate solution was also used for nuclear magnetic resonance spectroscopy to effectively speed up the convergence [3.16].

Before concluding this subsection we discuss some points: (1) To introduce the prior estimate m_j of the image, let $A \propto m_j$ in (3.6.15) and change the initial value $A^{(0)}$ accordingly. (For the estimation of m_j, see the next subsection.) (2) $1/\sigma_k^2$ ($k \in D$) is called the accuracy. Usually, it is set to be the visibility of the antenna pattern of a synthesis telescope; that is, $\{w_j\}$ calculated by (3.6.24a) is taken as the antenna pattern. (3) The applicability of the revised GD algorithm. The first condition is that $\{w_j\}$ is approximately a δ-function. The second one is that the image is composed of sharp peaks. If these conditions are not satisfied, the approximation of (3.6.27) is not accurate, and the result is not good. Often aperture synthesis radio telescopes and images (radio maps) satisfy these conditions. (4) Radio maps are two-dimensional, so all the formulae should be modified accordingly.

3.6.4 Simplified Newton-Raphson Algorithm

The simplified Newton-Raphson (NR) algorithm was also developed for image restoration for aperture synthesis radio telescopes [3.4]. The input data are dirty map d (i.e., degraded image) and dirty beam w (i.e., antenna pattern or PSF). This algorithm is essentially an NR method for maximization with approximate inversion of a matrix of large size.

The relative entropy to be maximized is

$$H(b/m) = - \sum_{j=0}^{N-1} b_j \log(b_j/m_j) \, , \tag{3.6.30}$$

where $b = (b_0, \ldots, b_{N-1})^{\mathrm{T}}$ is the map (image) to be restored, whose prior estimate is $m = (m_0, \ldots, m_{N-1})^{\mathrm{T}}$.

The visibility and TP constraints are

$$\chi^2 = \sum_{k=0}^{N-1} W_k |V_k' - V_k|^2 \le \chi_c^2 \, , \tag{3.6.31}$$

$$F = \sum_{j=0}^{N-1} b_j = F_{\mathrm{ob}} \, , \tag{3.6.32}$$

where $V = (V_0, \ldots, V_{N-1})^{\mathrm{T}}$ is the observed visibility; $V' = (V_0', \ldots, V_{N-1}')^{\mathrm{T}}$ the visibility calculated from b; $W = (W_0, \ldots, W_{N-1})^{\mathrm{T}}$ the window used in calculating the dirty map. Let $\sum_{k=0}^{N-1} W_k = N$ so $w_0 = 1$. $W_k = 0$ at the points without observed data, so the summation is formally from $k = 0$ to $N - 1$. χ_c^2 is the critical value, say, $N\sigma^2$ (σ^2 is the average of σ_k^2). F_{ob} is the given TP.

Forming the objective function

$$J = H - \alpha\chi^2 - \beta F \, , \tag{3.6.33}$$

then searching for the maximum point of J by the NR method requires solving the equation $\nabla J = 0$. In the iteration the change of b needs calculating,

$$\Delta b = (-\nabla\nabla J)^{-1} \nabla J \, , \tag{3.6.34}$$

where the gradient

$$\nabla J = \nabla H - \alpha\nabla\chi^2 - \beta\nabla F \, , \tag{3.6.35}$$

where the components of the vectors are

$$(\nabla H)_j = \partial H/\partial b_j = -\log(b_j/m_j) - 1 \, , \tag{3.6.36a}$$

$$(\nabla\chi^2)_j = \partial\chi^2/\partial b_j = 2N \left(\sum_{k=0}^{N-1} p_{jk} b_k - d_j \right) \, , \tag{3.6.36b}$$

$$(\nabla F)_j = 1 \, . \tag{3.6.36c}$$

It is easy to prove (3.6.36a,c). Equation (3.6.36b) is derived as follows, cf. (3.6.16):

$$
\begin{aligned}
\partial\chi^2/\partial b_j &= 2N\{\mathrm{IFT}[W_k \cdot V_k'] - \mathrm{IFT}[W_k V_k]\} \\
&= 2N\{\mathrm{IFT}[W_k] * \mathrm{IFT}[V_k'] - d_j\} \\
&= 2N(w_j * b_j - d_j) \\
&= 2N\left(\sum_{k=0}^{N-1} p_{jk} b_k - d_j\right) ,
\end{aligned}
$$

where

$w_j = \mathrm{IFT}[W_k]$, $j = 0,\ldots,N-1$ is the dirty beam (PSF),

$d_j = \mathrm{IFT}[W_k V_k]$, $j = 0,\ldots,N-1$ is the dirty map (degraded image),

$p_{jk} = w_{|j-k|} = p_{kj}$, $(w_0 = 1)$.

Q.E.D.

The Hessian in (3.6.34) is

$$\nabla\nabla J = \nabla\nabla H - \alpha\nabla\nabla\chi^2 . \tag{3.6.37}$$

From (3.6.36a,b) it is easy to show the matrix elements to be

$$
\begin{aligned}
(\nabla\nabla H)_{jk} &= \partial^2 H/\partial b_j \partial b_k = -b_j^{-1}\delta_{jk} , \\
(\nabla\nabla\chi^2)_{jk} &= \partial^2\chi^2/\partial b_j \partial b_k = 2N p_{jk} ,
\end{aligned}
$$

and hence

$$(-\nabla\nabla J)_{jk} = b_j^{-1}\delta_{jk} + 2N\alpha p_{jk} . \tag{3.6.38}$$

The difficulty is to calculate the inversion of the $N \times N$ matrix $\nabla\nabla J$. In general, N is very large so exact inversion is prohibitive in practice. The solution is to neglect the nondiagonal elements of $\nabla\nabla J$. From (3.6.38) it can be seen that on the right-hand side only the second term contains the nondiagonal elements. Neglecting p_{jk}, $j \neq k$ is equivalent to ignoring the sidelobes of the dirty beam and treating it as a δ-function. The conditions for approximation here are the same as those for the revised GD algorithm. Having neglected p_{jk} for $j \neq k$, p_{jj} is increased to improve the approximation (using multiplication instead of convolution) by letting

$$p_{jj} = \sum_{k=0}^{N-1} p_{jk} = \sum_{j=0}^{N-1} w_j = q , \tag{3.6.39}$$

where the factor q is the same as that for the conversion between Jy/(beam area) and Jy/pixel for flux. (See later for the estimation of q.)

After the above simplification, an approximate formula is obtained:

$$(-\nabla\nabla J)_{jk}^{-1} = \begin{cases} \dfrac{1}{b_j^{-1} + 2N\alpha q} , & j = k , \\ 0 , & j \neq k . \end{cases} \tag{3.6.40}$$

So far we have obtained all the formulae for calculating the change Δb in the NR iteration: (3.6.34, 3.6.35, 3.6.40), etc.

In the iteration the parameters α and β, starting with zero, increase gradually until the constraints (3.6.31, 3.6.32) are finally satisfied. For particular α and β, the criterion for convergence is that the norm of ∇J should be much less than the norm of the vector $\mathbf{1} = (1,\ldots,1)^{\mathrm{T}}$, or

$$\|\nabla J\|^2 < \varepsilon \|\mathbf{1}\|^2,$$

where $\varepsilon \leq 0.01$ typically. The squared norm of a (covariant) vector $\boldsymbol{x} = (x_0,\ldots,x_{N-1})^{\mathrm{T}}$ is defined by

$$\begin{aligned}
\|\boldsymbol{x}\|^2 &= \boldsymbol{x}^{\mathrm{T}}(-\nabla\nabla J)^{-1}\boldsymbol{x} \\
&\approx \sum_{j=0}^{N-1} x_j^2/(b_j^{-1} + 2N\alpha q) .
\end{aligned}$$

Here $(-\nabla\nabla J)_{jk}^{-1}$ are used as a metric for the (image) $\boldsymbol{b}$-space. In calculating the norm, greater weights are given to points of greater b_j. This weighting is weakened to some extent because of the term $2N\alpha q$, especially when α is large in the late stage of iteration.

Finally, we discuss the following three problems:

(1) *Estimation of q*. For lack of the zero-baseline visibility, $W_0 = 0$ in (3.6.31). As a result, q calculated by (3.6.39) will be zero. The solution is to fit the mainlobe of the dirty beam (PSF) by a Gaussian function (or simply use the mainlobe itself), then sum up w_j in the (fitted) mainlobe to get q. Too large a value of q indicates a significant deviation of the dirty beam from the δ-function, and consequently (3.6.40) is a poor approximation. In practice $q < 50$ will do, otherwise the convergence will be very slow.

(2) *Calculation of χ^2 in the Image Space*. In the iteration the χ^2-statistic can be evaluated by (3.6.31). It is also possible to calculate χ^2 in the image space from the given data, the dirty map $\boldsymbol{d}$ and beam $\boldsymbol{w}$. The approximate formula is

$$\chi^2 \approx NE/q_1 , \tag{3.6.41a}$$

where

$$E = \sum_{j=0}^{N-1}\left|\sum_{k=0}^{N-1} p_{jk}b_k - d_j\right|^2 , \qquad q_1 = \sum_{j=0}^{N-1}|w_j|^2 . \tag{3.6.41b}$$

Proof. From Parseval's theorem for FT, it follows that

$$E = \frac{1}{N}\sum_{k=0}^{N-1} |W_k|^2|V_k' - V_k|^2 , \tag{3.6.42a}$$

$$\sum_{j=0}^{N-1} |w_j|^2 = \frac{1}{N}\sum_{k=0}^{N-1} |W_k|^2 . \tag{3.6.42b}$$

Suppose that $\{w_j\}$ is approximately a δ-function, then

$$W_k \begin{cases} \approx N/M\,, & k \in D\,, \\ = 0\,, & \text{otherwise}\,. \end{cases}$$

(D is the support of visibility, containing M points.)

From (3.6.31, 3.6.42b), we have, respectively,

$$\chi^2 \approx \sum_{k \in D} |V_k' - V_k|^2 N/M\,,$$

$$q_1 = \sum_{j=0}^{N-1} |w_j|^2 \approx N/M\,.$$

Then, from (3.6.42a), we get

$$E = \frac{1}{N} \sum_{k \in D} |V_k' - V_k|^2 N^2/M^2 \approx q_1 \chi^2/N\,,$$

and (3.6.41a) results.   Q.E.D.

(Our result here is slightly different from that in [3.4].)

q_1 may be estimated in a way by which q is estimated in (1), that is, only $|w_j|^2$ in the (fitted) mainlobe are taken into account.

(3) *Prior Estimate m_j.* In the usual case where only one degraded image (dirty map) is available, m_j may be assumed to be a constant (uniform prior). Besides, m_j may be estimated by one of the following four ways: (i) Segment the degraded image overlappingly. The pseudo-ensemble average calculated from these segments is taken as m_j. (ii) Convolve the image with a Gaussian function, or equivalently lowpass filter its visibility, to obtain a low resolution image, which is used as m_j after its zero and negative values have been clipped. (iii) Use a uniform $m_j^{(0)}$ to generate an MEM2 image, which is used as m_j after being convolved with a Gaussian function. (iv) Use a quick CLEAN to generate a small number of clean components, which are used as m_j after being convolved with a clean (usually Gaussian) beam.

The simplified NR algorithm is used in the AIPS (Astronomical Image Processing System) package for image processing in radio astronomy.

3.6.5 Numerical Example

The performance of the revised GD algorithm and simplified NR algorithm are similar. For a particular λ, or α and β, two calculations of FFT/IFFT are needed in each iteration. For processing a radio map, totally tens through one hundred FFT/IFFTs are needed. Restored MEM2 maps by the two algorithms are more or less the same in quality. Theoretically speaking, the MEM2 solution is unique. But in practice solutions from various algorithms are all approximate; hence they are different to some extent.

The result of processing the radio source 1637–771 by the revised GD algorithm is shown in Fig. 3.13. The visibility support (antenna spacing coverage) is shown in Fig. 3.13a by two shaded sectors. The dirty and MEM2 maps are shown, respectively, in Figs. 3.13b, c (ruled surface). Their central regions are shown, respectively, in Figs. 3.13d, e (contour plot).

It is evident from the ruled surface maps, b and c, that the noise has been remarkably suppressed in the MEM2 map; the weak source in the upper left corner is easily identified. The noise *rms* values are, respectively, 5.92 (dirty may) and 1.06 (MEM2 map). The *dynamic ranges* (*DRs*) are, respectively, 12.8 dB (dirty map) and 20.5 dB (MEM2 map). The definition is

$$\mathrm{DR} = -10\log_{10}(3 \times \text{noise } rms/\text{peak maximum}).$$

From the contour plots d and e, it can be seen that the resolution has been enhanced in the MEM2 map: the two peaks in the upper part are resolved. (A CLEAN map confirmed the existence of the two distinct peaks.)

In brief, MEM2 has shown its capability of suppressing noise and enhancing resolution at the same time; the compromise between them is controlled by the Lagrange multiplier λ. This good performance of MEM2 is achieved at the expense of much computational time. To generate the 256×256 point maps in Fig. 3.13, on a VAX-11/780 computer, about 3 minutes CPU time was needed for the dirty map, while about 40 minutes for the MEM2 map (totally 35 iterations). One more point should be noticed, that is, MEM2 tends to reduce small peaks' magnitudes relative to stronger peaks. This makes it difficult to estimate quantitatively the relative powers of sources in an MEM2 map.

3.7 Algorithms and Numerical Examples (II)

In this section we continue to introduce MEM2 algorithms for image restoration, also using 1-D notation.

3.7.1 Skilling-Bryan Algorithm

In the Skilling-Bryan (SB) algorithm the entropy to be maximized is [3.17]

$$S = -\sum_{j=0}^{N-1} f_j[\log(f_j/A) - 1], \qquad (3.7.1)$$

where $\boldsymbol{f} = (f_0, \ldots, f_{N-1})$ is the image to be restored. A is called the *default value* in the sense that the MEM2 solution will be $f_j = A$ if there are no data constraints (this is easily verified by letting $\partial S/\partial f_j = 0$). So we see that A, similar to m_j in the previous section, is the prior estimate of the image. A may be a constant, or assume different values for different j's.

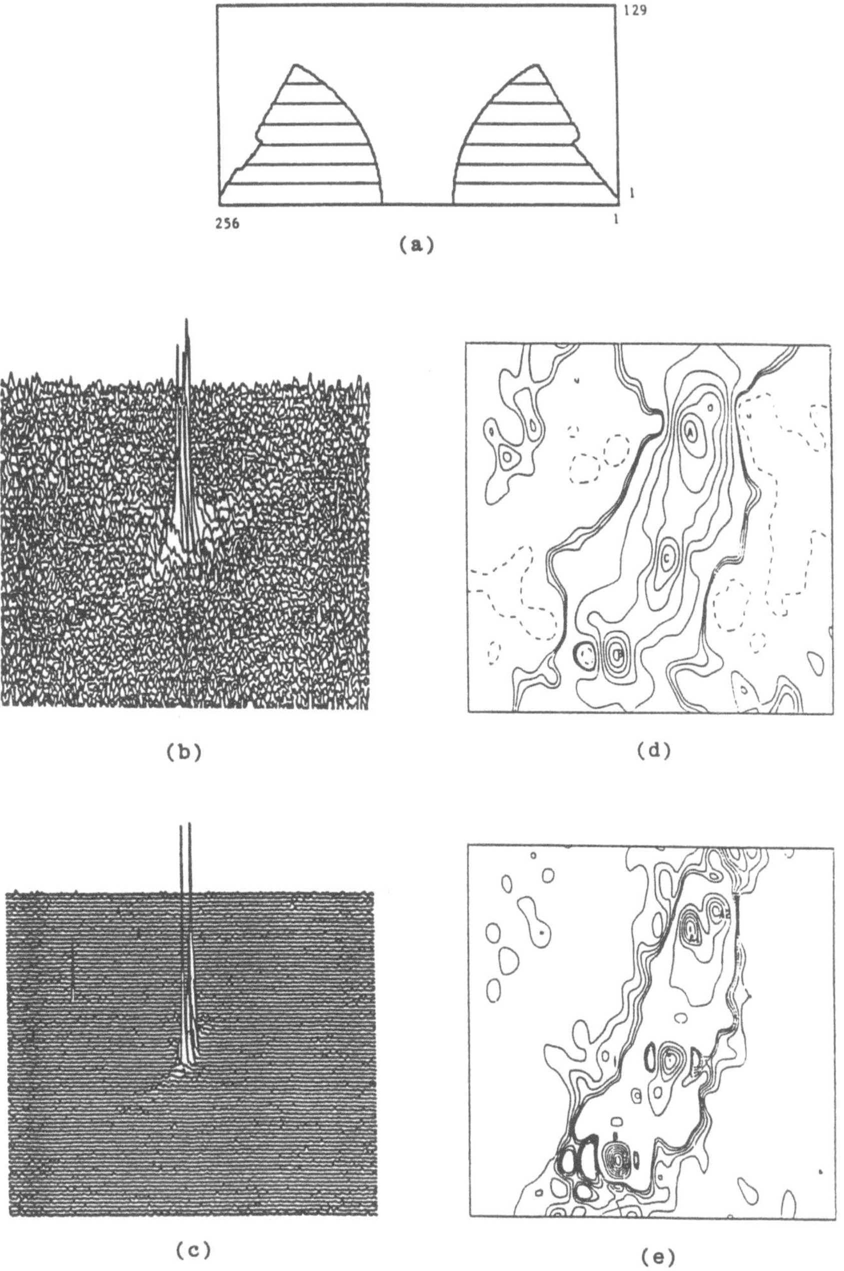

Fig. 3.13. The antenna spacing coverage and maps of the radio source 1637–771. (a) The antenna spacing coverage. Field= 0.8° × 0.8°, 256 × 256 points. (b) The dirty map. (c) The MEM2 map. (d) The central region of the dirty map. Contours: −50, −25, 6, 12, 18 : 346 : 50. (e) The central region of the MEM2 map. Contours: 4, 8, 12, 16, 20 : 398 : 50

The data fit requires the χ^2-statistic to be less than or equal to its critical value, that is,

$$C(f) = \chi^2 = \sum_{k \in D} |F_k - D_k|^2 / \sigma_k^2 \leq C_{\text{aim}} , \qquad (3.7.2)$$

where $F_k = \text{FT}[f_j]$, D_k is the measured visibility, σ_k^2 the variance of white Gaussian noise, D the data support, and C_{aim} the critical value.

The objective function is

$$Q = S - \lambda C . \qquad (3.7.3)$$

In the algorithms described in the previous section, the value of C (χ^2) is controlled by λ. For a particular λ, C is known after having found the maximum of Q in a search direction. The SB algorithm is characterized by that the problem of maximizing S subject to the constraint $C \leq C_{\text{aim}}$ is directly solved without using λ explicitly. More specifically, in each iteration in a three (or higher)-dimensional subspace, S and C are approximated, respectively, by quadratic forms $\tilde{S}$ and $\tilde{C}$; $\tilde{C}$ is reduced to a prescribed value $\tilde{C}_{\text{aim}}$ and $\tilde{S}$ attains the maximum under the constraint $\tilde{C} = \tilde{C}_{\text{aim}}$. The iteration stops when $\tilde{C}_{\text{aim}} = C_{\text{aim}}$. The iterative procedure is described in detail in the following:

1. Search Directions

Suppose that a subspace is spanned by r base vectors $e_1, \ldots, e_r$. Each e_μ is called a search direction. Usually, $3 \leq r \leq 10$.

Referring to the steepest ascent method, the gradient ∇Q should be one of the search directions. This direction is modified to $f(\nabla Q)$ to increase the weights for greater components f_j's. Here $f(\nabla Q)$ is defined as a "component-to-component" multiplication of two vectors, i.e.,

$$f(\nabla Q) = (f_0 \partial Q / \partial f_0, \ldots, f_{N-1} \partial Q / \partial f_{N-1}) .$$

Since ∇Q is a linear combination of ∇S and ∇C, we choose

$$e_1 = f(\nabla S) , \quad e_2 = f(\nabla C) .$$

In consideration that the gradient ∇Q is not constant but varying with f, $(\nabla \nabla Q) \nabla Q$ should have been a search direction. This direction is modified to $f(\nabla \nabla Q) f \nabla Q$ to take the weight f into account. Here $f(\nabla \nabla Q)$ is a matrix having $f_i \partial^2 Q / \partial f_i \partial f_j$ as its ith row, jth column element.

Furthermore, since $\nabla \nabla Q$ is a linear combination of $\nabla \nabla S$ and $\nabla \nabla C$, we have four search directions:

$$f(\nabla \nabla S) f(\nabla S) , \quad f(\nabla \nabla S) f(\nabla C) ,$$

$$f(\nabla \nabla C) f(\nabla S) , \quad f(\nabla \nabla C) f(\nabla C) .$$

But

$$(\nabla \nabla S)_{ij} = -\frac{1}{f_j} \delta_{ij} , \qquad (3.7.4)$$

so $f(\nabla\nabla S) = -I$. (I is an identity matrix.) As a result, the first two directions above are collinear to e_1 and e_2, respectively, and hence discarded. The last two are

$$e_3 = f(\nabla\nabla C)f\nabla S , \quad e_4 = f(\nabla\nabla C)f(\nabla C) .$$

In much the same way, we may use

$$f(\nabla\nabla C)f(\nabla\nabla C)f\nabla S , \quad f(\nabla\nabla C)f(\nabla\nabla C)f(\nabla C) , \quad \cdots\cdots$$

to take into consideration the change of e_3, e_4, etc.

In practice, it suffices to use the four search directions $e_1 \sim e_4$. Instead of e_3 and e_4, we can even simply use their linear combination,

$$f(\nabla\nabla C)f(|\nabla S|^{-1}\nabla S) - f(\nabla\nabla C)f(|\nabla C|^{-1}\nabla C) ,$$

so that the number of search directions is reduced to three. Here $|\nabla S|$, $|\nabla C|$ are the lengths of the vectors ∇S and ∇C, respectively.

It is the subspace spanned by the three search directions derived above that is involved in the following discussion. These three directions are repeated below:

$$\left\{ \begin{array}{lcl} e_1 & = & f(\nabla S) , \\ e_2 & = & f(\nabla C) , \\ e_3 & = & |\nabla S|^{-1}f(\nabla\nabla C)f(\nabla S) - |\nabla C|^{-1}f(\nabla\nabla C)f(\nabla C) . \end{array} \right. \tag{3.7.5}$$

As a digression we point out that the choice of search direction in the above can be understood from the point of view of the Newton method for maximization. The change in the Newton iteration is

$$-(\nabla\nabla Q)^{-1}\nabla Q .$$

Expanding $(\nabla\nabla Q)^{-1}$ in a Taylor series in $(\nabla\nabla Q)$, the terms contain

$$I, \quad \nabla\nabla Q , \quad (\nabla\nabla Q)^2 , \quad \cdots\cdots .$$

So the search directions are

$$\nabla Q, \quad (\nabla\nabla Q)\nabla Q, \quad (\nabla\nabla Q)^2\nabla Q, \quad \cdots\cdots .$$

Thus, we see that the method by which e_μ are chosen is just the approximate Newton method using only the first several terms (weighted by f) of the Taylor series.

2. Entropic Metric

The metric of the image space is defined by a matrix whose elements are

$$g_{ij} = (-\nabla\nabla S)_{ij} = \left\{ \begin{array}{ll} 1/f^i , & i = j , \\ 0 , & i \neq j . \end{array} \right.$$

(Note the tensor notation, and the Einstein convention hereafter). g_{ij} is used to calculate the inner product of contravariant vectors (e.g., f),

$$|f|^2 = g_{ij}f^if^j = \sum_j f^j .$$

The metric

$$g^{ij} = \begin{cases} f^i, & i = j, \\ 0, & i \neq j, \end{cases}$$

is used to calculate the inner product of covariant vectors (e.g., ∇S),

$$|\nabla S|^2 = g^{ij}(\nabla S)_i(\nabla S)_j = \sum_j f^j(\partial S/\partial f^j)^2,$$

$$|\nabla C|^2 = g^{ij}(\nabla C)_i(\nabla C)_j = \sum_j f^j(\partial C/\partial f^j)^2.$$

3. Search for the Maximum

In order to search for the maximum of S subject to the constraint $C = \tilde{C}_{\mathrm{aim}}$, starting with the current iterate f, the components x^μ of the change δf in the directions e_μ ($\mu = 1, 2, 3$) need to be determined. Denote $x = (x^1, x^2, x^3)$.

S and C in the vicinity of the current f are approximated by the following quadratic forms:

$$\begin{cases} \tilde{S}(x) & = & S_0 + S_\mu x^\mu - \frac{1}{2}g_{\mu\nu}x^\mu x^\nu, \\ \tilde{C}(x) & = & C_0 + C_\mu x^\mu + \frac{1}{2}M_{\mu\nu}x^\mu x^\nu, \end{cases} \tag{3.7.6}$$

where S_0 and C_0 are, respectively, the current S and C, $S(f)$ and $C(f)$, and

$$\begin{cases} S_\mu = e_\mu \cdot \nabla S, & g_{\mu\nu} = e_\mu \cdot e_\nu, \\ C_\mu = e_\mu \cdot \nabla C, & M_{\mu\nu} = \nabla\nabla C : e_\mu e_\nu, \end{cases} \tag{3.7.7}$$

where the vectors on the right-hand sides are determined by the current f, and the coefficients on the left-hand sides are calculated by inner production.

The iterate f is updated by

$$f^{(\mathrm{new})} = f + \delta f = f + x^\mu e_\mu, \tag{3.7.8}$$

where the squared length of δf is

$$l^2 = g_{\mu\nu}x^\mu x^\nu. \tag{3.7.9}$$

(1) *Form a Cartesian Coordinate System in the Subspace.* Normally e_1, e_2 and e_3 are not orthogonal to each other, and are linearly dependent sometimes. To simplify the algebra, a Cartesian coordinate system is formed by orthonormalizing the base vectors e_μ ($\mu = 1, 2, 3$) as follows: The matrices $(g_{\mu\nu})$ and $(M_{\mu\nu})$ are diagonalized simultaneously. If e_μ ($\mu = 1, 2, 3$) are linearly dependent, then one or more eigenvalues of $(g_{\mu\nu})$ will be close to zero in computation and are discarded. The corresponding eigenvectors are also discarded. The next step is to normalize the surviving eigenvectors (i.e., change their lengths to unity) by scale transformation. These eigenvectors are used to form the desirable Cartesian coordinate system, $g_{\mu\nu} = \delta_{\mu\nu}$. Now the distinction between covariant and contravariant indices disappears. Equations (3.7.6, 3.7.9) are simplified to

$$\begin{cases} \tilde{S}(\boldsymbol{x}) &= S_0 + S_\mu x_\mu - \frac{1}{2}x_\mu x_\mu \,, \\ \tilde{C}(\boldsymbol{x}) &= C_0 + C_\mu x_\mu + \frac{1}{2}\gamma(\mu)x_\mu x_\mu \,, \end{cases} \tag{3.7.10}$$

and

$$l^2 = x_\mu x_\mu \,, \tag{3.7.11}$$

where $\gamma(\mu)$ are the eigenvalues of $(M_{\mu\nu})$.

(2) *Basic Control.* In most applications, C is a downward convex function in $\boldsymbol{f}$, so that $\gamma(\mu) > 0$. From (3.7.10) it is easy to find the minimum point of $\tilde{C}(\boldsymbol{x})$ to be $-C_\mu/\gamma(\mu)$, and the minimum value to be

$$\tilde{C}_{\min} = C_0 - \frac{1}{2}\gamma^{-1}(\mu)C_\mu C_\mu \,. \tag{3.7.12}$$

$\tilde{C}_{\min}$ is the minimum of $\tilde{C}$ attainable in the subspace. The length of $\delta \boldsymbol{f}$, l, should be limited; otherwise the approximation of S and C by $\tilde{S}$ and $\tilde{C}$ in (3.7.6) is inappropriate. Too large an l may also result in zero or negative components of $\boldsymbol{f}^{(\text{new})}$. (When this happens, these components must be clipped.) Usually, the constraint is set to

$$|l|^2 \le l_0^2 = (0.1 \sim 0.5) \sum_j f_j \,.$$

However, with this constraint, the minimum in (3.7.12) may not be attained. So we set a more modest aim, that is, attempt to reduce C to

$$\tilde{C}_{\text{aim}} = \max \left\{ C_0 - \frac{1}{3}\gamma^{-1}(\mu)C_\mu C_\mu \,, \ C_{\text{aim}} \right\} \,. \tag{3.7.13}$$

Now we form the objective function

$$\tilde{Q} = \alpha \tilde{S} - \tilde{C} \,, \tag{3.7.14}$$

adjust the Lagrange multiplier α so that $\tilde{Q}$ attains its maximum (so does $\tilde{S}$) and the constraint

$$\tilde{C} = \tilde{C}_{\text{aim}} \tag{3.7.15}$$

is satisfied.

Here α needs to be adjusted a number of times. But the subspace is only 3-D and the function is a quadratic form, so it is easy to find the maximum of $\tilde{Q}$ for a particular α. Specifically, let $\partial \tilde{Q}/\partial x_\mu = 0$. Then it is easily shown, utilizing (3.7.10), that

$$x_\mu = (\alpha S_\mu - C_\mu)/[\gamma(\mu) + \alpha] \,. \tag{3.7.16}$$

For a particular α, substitute the above x_μ in (3.7.10) to calculate $\tilde{C}(\boldsymbol{x})$. Since the value of $\tilde{C}(\boldsymbol{x})$ increases (decreases) monotonically as α increases (decreases), gradually decrease (if the calculated $\tilde{C}(\boldsymbol{x}) > \tilde{C}_{\text{aim}}$) or increase (if $\tilde{C}(\boldsymbol{x}) < \tilde{C}_{\text{aim}}$) α and repeatedly calculate x_μ and $\tilde{C}(\boldsymbol{x})$ until the constraint (3.7.15) is satisfied. But it may be the case that l^2 defined in (3.7.11) reaches its prescribed limit l_0^2 before $\tilde{C}$ reduced to $\tilde{C}_{\mathrm{aim}}$. If this happens, the value of $\tilde{C}$ obtained at the limit $l = l_0$ is taken as $\tilde{C}_{\mathrm{aim}}$ (which is greater than expected in (3.7.13)). In other words, we are satisfied with whatsoever if only the value $\tilde{C}$ has been reduced as much as possible. A penalty term $-Pl^2$ may be added into the right-hand side of (3.7.14) lest l^2 exceed l_0^2 in the course of adjusting α.

Having determined x_μ for which $\tilde{C}(x) = \tilde{C}_{\mathrm{aim}}$ and S attains its maximum, move from the current f to the next $f^{(\mathrm{new})}$ in the image space according to (3.7.8). The procedure described in the above is repeated until achieving the aim $\tilde{C} = C_{\mathrm{aim}}$.

(3) *Test of MEM2 Solution.* In the iteration every iterate f is checked to see if it is really an MEM2 solution. This is done by a test of the parallelism between the vectors ∇S and ∇C. For an exact MEM2 solution, $\nabla Q = 0$ and hence $\nabla S // \nabla C$, cf. (3.7.3). The quantity,

$$\mathrm{TEST} = \frac{1}{2}|\,|\nabla S|^{-1}\nabla S - |\nabla C|^{-1}\nabla C\,|^2\,, \qquad (3.7.17)$$

should be zero. For an approximate MEM2 solution, $\mathrm{TEST} < 0.1$ is required.

Another way is to test the angle between ∇S and ∇C. For an exact MEM2 solution, $\theta =< \nabla S,\ \nabla C >= 0$, and $\mathrm{TEST} = 1 - \cos\theta = 0$. For an approximate one, it is required that $\theta \leq \frac{1}{2}$ radian, or equivalently, $\mathrm{TEST} = 1 - \cos\theta \leq 0.12$.

We have described the SB algorithm in detail. In practical computation the program requires as input, besides the degraded image f, the PSF for generating the "accuracy" $1/\sigma_k^2$. The number of search directions can be chosen to be $r = 3, 4$ or 6. Different types of default value A can be input. In addition to χ^2, other statistics may be used. It is even possible to choose the entropy expression.

In the SB algorithm, no assumptions such as approximate δ-function about the PSF and peaky image are made. Additionally, the outer iteration for λ is eliminated and several search directions are carefully chosen. This sophisticated scheme renders the algorithm effective, robust, and viable for general purpose.

By default $r = 3$. Six FFT/IFFTs are needed in one iteration. For processing an image, some 20 iterations, or roughly 100 FFT/IFFTs are needed. For processing a radio map of simple configuration, the SB, revised GD and simplified NR algorithms are almost the same as far as the numbers of FFT/IFFTs needed are concerned. This is because the latter two algorithms have taken advantage of the characteristic of data from aperture synthesis radio telescopes. The general purpose algorithm cannot show its superiority up. In contrast, for processing an image of complex configuration, the latter two are significantly inferior to the SB algorithm in the sense of more computational time and poorer quality of the restored image.

3.7.2 Differential Equation Approach

In the differential-equation algorithm the entropy to be maximized is [3.18]

$$H(\boldsymbol{f}) = -\sum_{i=1}^{n} f_i \log f_i \,, \tag{3.7.18}$$

and the measured data are a degraded image,

$$d_j = \sum_{i=1}^{n} A_{ji} f_i + e_j \,, \quad j = 1, \ldots, m \,, \tag{3.7.19}$$

where $\boldsymbol{f} = (f_1, \ldots, f_n)^{\mathrm{T}}$ is the image to be restored, $\boldsymbol{d} = (d_1, \ldots, d_m)^{\mathrm{T}}$ is the degraded image, the $m \times n$ matrix $A = [A_{ji}]$ represents the PSF, and $\boldsymbol{e} = (e_1, \ldots, e_m)^{\mathrm{T}}$ is the white noise vector with zero mean $\boldsymbol{0}$ and variance $\boldsymbol{\sigma}^2 = (\sigma_1^2, \ldots, \sigma_m^2)^{\mathrm{T}}$. Equation (3.7.19) can be written in a short form as

$$\boldsymbol{d} = A\boldsymbol{f} + \boldsymbol{e} \,. \tag{3.7.20}$$

Form the statistic

$$\begin{aligned}
Q(\boldsymbol{f}) &= \frac{1}{2} \sum_{j=1}^{m} \frac{1}{\sigma_j^2} \left(\sum_{i=1}^{n} A_{ji} f_i - d_j \right)^2 \\
&= \frac{1}{2} (A\boldsymbol{f} - \boldsymbol{d})^{\mathrm{T}} D (A\boldsymbol{f} - \boldsymbol{d}) \,, \tag{3.7.21}
\end{aligned}$$

where the diagonal matrix $D = \mathrm{diag}[1/\sigma_1^2, \ldots, 1/\sigma_m^2]$. According to the formula for the expectation of the χ^2-statistic, the data fit constraint is

$$Q(\boldsymbol{f}) = m/2 \,. \tag{3.7.22}$$

The TP constraint is

$$\sum_{i=1}^{n} f_i = t_c \,. \tag{3.7.23}$$

Forming the objective function

$$J(\boldsymbol{f}; \mu, \lambda) = H(\boldsymbol{f}) + \mu \sum_{i=1}^{n} f_i - \lambda Q(\boldsymbol{f}) \,, \tag{3.7.24}$$

then, from (3.7.18), we have

$$\nabla H = -(\log f_1 + 1, \ldots, \log f_n + 1)^{\mathrm{T}} \,.$$

From (3.7.21), it follows that

$$\nabla Q = \frac{1}{2} \times 2 \, A^{\mathrm{T}} D (A\boldsymbol{f} - \boldsymbol{d}) \,.$$

In addition, we have

$$\nabla \sum_{i=1}^{n} f_i = (1, \ldots, 1)^{\mathrm{T}} \overset{\triangle}{=} \boldsymbol{h} \ .$$

Utilizing the above three expressions, from the necessary condition for the maximization of J, we get

$$\begin{aligned} \boldsymbol{0} \ &= \ \nabla J = \nabla H + \mu \nabla \sum_{i=1}^{n} f_i - \lambda \nabla Q \\ &= \ -(\log f_1, \ldots, \log f_n)^{\mathrm{T}} + (\mu - 1)\boldsymbol{h} - \lambda A^{\mathrm{T}} D(A\boldsymbol{f} - \boldsymbol{d}) \ . \end{aligned} \qquad (3.7.25)$$

Since the Hessian

$$\nabla^2 J = \nabla \nabla J = -\mathrm{diag}[1/f_1, \ldots, 1/f_n] - \lambda A^{\mathrm{T}} D A, \quad (f_i > 0, \ \infty > \lambda \geq 0)$$

$$(3.7.26)$$

is a negative definite matrix, the solution of (3.7.25), $\boldsymbol{f}$, is indeed a maximum point.

We could solve the constrained maximization problem as before. Namely, find iteratively the maximum point $\boldsymbol{f}$ for particular λ and μ in (3.7.25); repeatedly adjust λ and μ so that the constraints (3.7.22, 3.7.23) are satisfied. Now we solve the problem by a different approach. That is, convert the solution of $\nabla J = \boldsymbol{0}$ to a Cauchy problem of differential equations; adjust λ so that (3.7.22) is satisfied, and then adjust μ so that (3.7.23) is also satisfied keeping the satisfaction of (3.7.22).

1. Cauchy Problem of Differential Equations

The solution of (3.7.25), $\boldsymbol{f}$, depends on λ and μ. For particular λ and μ, the solution $\boldsymbol{f} = \boldsymbol{f}(\lambda, \mu)$ is a point in the n-dimensional image space. As λ and μ are moving on the λ-μ plane, all the solutions form a hypersurface in the $(\lambda, \mu, \boldsymbol{f})$-space.

Clearly,

$$\nabla J(\boldsymbol{f}(\lambda, \mu); \mu, \lambda) \equiv 0 \ , \quad (\infty > \lambda \geq 0) \qquad (3.7.27)$$

is equivalent to the following equations:

$$\begin{cases} \dfrac{\mathrm{d}}{\mathrm{d}\lambda} \nabla J(\boldsymbol{f}(\lambda, \mu); \mu, \lambda) \equiv \boldsymbol{0} \ , \quad (\infty > \lambda \geq 0) \ , & (3.7.28\mathrm{a}) \\[2ex] \nabla J(\boldsymbol{f}(0, \mu); \mu, 0) = \boldsymbol{0} \ . & (3.7.28\mathrm{b}) \end{cases}$$

Utilizing ∇J in (3.7.25), from (3.7.28a) it follows that

$$\begin{aligned} \boldsymbol{0} \ &= \ \frac{\mathrm{d}}{\mathrm{d}\lambda}[\nabla H + (\mu - 1)\boldsymbol{h} - \lambda \nabla Q] \\ &= \ (\nabla^2 H - \lambda \nabla^2 Q)\frac{\mathrm{d}\boldsymbol{f}}{\mathrm{d}\lambda} - \nabla Q \\ &= \ \nabla^2 J \frac{\mathrm{d}\boldsymbol{f}}{\mathrm{d}\lambda} - \nabla Q \ . \end{aligned}$$

The solution of (3.7.28b) is

$$\boldsymbol{f}(\lambda,\mu)|_{\lambda=0} = \exp(\mu - 1)\boldsymbol{h} \ .$$

Combining the two expressions above yields the following Cauchy problem of differential equations:

$$\begin{cases} \nabla^2 J \dfrac{\mathrm{d}\boldsymbol{f}}{\mathrm{d}\lambda} = \nabla Q \ , \quad (\lambda_M > \lambda \geq 0) \ , & (3.7.29a) \\[2mm] \boldsymbol{f}|_{\lambda=0} = \exp(\mu - 1)\boldsymbol{h} \ , & (3.7.29b) \end{cases}$$

which is equivalent to (3.7.28a,b). Here the range of λ is limited deliberately for the purpose of avoiding unnecessary mathematical argument. Imposing a sufficiently large but finite upper bound λ_M on λ means assuming that the desirable data fit can be achieved using a finite λ. This will not give rise to any problem in practice.

It can be proved that the solution to the Cauchy problem (3.7.29a,b) exists and is unique. In practice, we assume that the initial value of Q, $Q(\boldsymbol{f}(0,\mu)) > m/2$, i.e., a uniform image cannot be the MEM2 solution satisfying the constraint (3.7.22). Thus, $\nabla Q(\boldsymbol{f}(\lambda,\mu))$ cannot be $\boldsymbol{0}$; $Q(\boldsymbol{f}(\lambda,\mu))$ decreases monotonically as λ increases from 0. (We have utilized this fact quite a number of times before.)

2. Solution of the Cauchy Problem

The Cauchy problem (3.7.29a,b) can only be solved by an approximate method.

(1) *Initial Values.* First of all, we introduce the following notations:

$$\alpha \ = \ \exp(\mu - 1) \ ,$$

$$\boldsymbol{r} \ = \ \left(\sum_{i=1}^{n} A_{1i}/\sigma_1, \ldots, \sum_{i=1}^{n} A_{mi}/\sigma_m \right)^{\mathrm{T}} \ ,$$

$$\boldsymbol{s} \ = \ (d_1/\sigma_1, \ldots, d_m/\sigma_m)^{\mathrm{T}} \ ,$$

$$F \ = \ \mathrm{diag}[1/f_1, \ldots, 1/f_n] \ .$$

From (3.7.21, 3.7.29b), the initial value of Q is

$$\begin{aligned} Q(\alpha\boldsymbol{h}) \ &= \ \frac{1}{2} \sum_{j=1}^{m} \frac{1}{\sigma_j^2} \left(\alpha \sum_{i=1}^{n} A_{ji} - d_j \right)^2 \\ &= \ a\alpha^2 + b\alpha + c \ , \end{aligned} \qquad (3.7.30)$$

where

$$a = \frac{1}{2}||\boldsymbol{r}||^2 \ , \quad b = -(\boldsymbol{r},\boldsymbol{s}) \ , \quad c = \frac{1}{2}||\boldsymbol{s}||^2 \ .$$

Since the squared inner product $(\boldsymbol{r},\boldsymbol{s})^2 \leq ||\boldsymbol{r}||^2||\boldsymbol{s}||^2$, i.e., $b^2 \leq 4ac$, and $a > 0$, $Q(\alpha\boldsymbol{h})$ has a minimum value when $\alpha = -b/(2a)$. In general, $A_{ji} \geq 0$

when the noise is not too strong, $(\boldsymbol{r},\,\boldsymbol{s}) > 0$, i.e., $-b/(2a) > 0$, then we can choose

$$\mu_0 = 1 + \log\left(-\frac{b}{2a}\right), \quad \boldsymbol{f}(0,\mu_0) = -\frac{b}{2a} \tag{3.7.31}$$

in order that $Q(\alpha\boldsymbol{h})$ may be minimized and as close to $m/2$ as possible.

Alternatively, taking the TP constraint into consideration, μ_0 can be chosen in a simple way such that the initial value of $\boldsymbol{f}$ is

$$\boldsymbol{f}(0,\mu_0) = (t_c/n)\boldsymbol{h} . \tag{3.7.32}$$

(2) *Iteration.* Let

$$L = A^{\mathrm{T}}DA, \quad \boldsymbol{p} = A^{\mathrm{T}}D\boldsymbol{d} .$$

Then, (3.7.26) becomes

$$\nabla^2 J = -F - \lambda L ,$$

and

$$\nabla Q = A^{\mathrm{T}}D(A\boldsymbol{f} - \boldsymbol{d}) = L\boldsymbol{f} - \boldsymbol{p} .$$

Approximating $\mathrm{d}\boldsymbol{f}/\mathrm{d}\lambda$ by $\delta\boldsymbol{f}/\delta\lambda = (\boldsymbol{f}^{(k+1)} - \boldsymbol{f}^{(k)})/(\lambda_{k+1} - \lambda_k)$ in (3.7.29a), we obtain the following formulae for iteration:

$$\begin{cases} \boldsymbol{f}^{(0)} = \exp(\mu_0 - 1)\boldsymbol{h} , \quad (\lambda_0 = 0) , \\ (F^{(k)} + \lambda_k L)(\boldsymbol{f}^{(k+1)} - \boldsymbol{f}^{(k)}) = -(\lambda_{k+1} - \lambda_k)(L\boldsymbol{f}^{(k)} - \boldsymbol{p}), \quad (k \geq 0), \end{cases}$$
$$\tag{3.7.33}$$

where the superscript k and subscript k of λ denote the kth iterates. $|\lambda_{k+1} - \lambda_k|$ is sufficiently small for all k, so that $\boldsymbol{f}^{(k)} \approx \boldsymbol{f}(\lambda_k,\mu_0)$.

Change (3.7.33) to the form

$$\boldsymbol{f}^{(0)} = \exp(\mu_0 - 1)\boldsymbol{h} , \quad (\lambda_0 = 0) , \tag{3.7.34a}$$

$$(F^{(k)} + \lambda_k L)\boldsymbol{f}^{(k+1)} = (2\lambda_k - \lambda_{k+1})L\boldsymbol{f}^{(k)} + \boldsymbol{h} + (\lambda_{k+1} - \lambda_k)\boldsymbol{p}, \quad (k \geq 0),$$
$$\tag{3.7.34b}$$

$(F^{(k)}\boldsymbol{f}^{(k)} = \boldsymbol{h})$, and let

$$\begin{aligned} P &= F^{(k)} + \lambda_k L , \\ \boldsymbol{b} &= (2\lambda_k - \lambda_{k+1})L\boldsymbol{f}^{(k)} + \boldsymbol{h} + (\lambda_{k+1} - \lambda_k)\boldsymbol{p} , \quad (k \geq 0) , \\ \boldsymbol{x} &= \boldsymbol{f}^{(k+1)} \approx \boldsymbol{f}(\lambda_{k+1},\mu_0) ; \end{aligned}$$

then (3.7.34b) can be written in a short form:

$$P\boldsymbol{x} = \boldsymbol{b} . \tag{3.7.35}$$

This is a system of linear equations of large size. In spite of this, (3.7.35) can be solved efficiently by the Gauss-Seidel iteration since the coefficient matrix P is positive definite (and symmetrical). Specifically, the iteration is represented by

$$\begin{cases} x_i^{(0)} = f_i^{(k)}, & i = 1,\ldots,n, \\ x_i^{(m)} = \dfrac{1}{P_{ii}} \left(b_i - \sum_{j<i} P_{ij} x_j^{(m)} - \sum_{j>i} P_{ij} x_j^{(m-1)} \right), & i = 1,\ldots,n, \\ & (m \geq 1). \end{cases}$$

(3.7.36)

Since $|\lambda_{k+1} - \lambda_k|$ is small, $\boldsymbol{f}^{(k+1)}$ is close to $\boldsymbol{f}^{(k)}$. Starting with the initial value $x_i^{(0)}$ above, the iteration converges fast. Experiments have shown that in most cases only one iteration is enough, and hence the formula for calculating $\boldsymbol{f}^{(k+1)}$ from $\boldsymbol{f}^{(k)}$ is simply

$$f_i^{(k+1)} = \frac{1}{P_{ii}} \left(b_i - \sum_{j<i} P_{ij} f_j^{(k+1)} - \sum_{j>i} P_{ij} f_j^{(k)} \right), \quad i = 1,\ldots,n. \quad (3.7.37)$$

The calculation of $\boldsymbol{b}$ involves multiplication of the matrix L and the vector $\boldsymbol{f}^{(k)}$. Let $\boldsymbol{g}^{(k)} = L\boldsymbol{f}^{(k)}$. At the beginning of the iteration, $k = 0$,

$$\boldsymbol{g}^{(0)} = L\boldsymbol{f}^{(0)} = \exp(\mu_0 - 1) \left(\sum_{i=1}^{n} L_{1i}, \ldots, \sum_{i=1}^{n} L_{ni} \right)^{\mathrm{T}}.$$

Later, $\boldsymbol{g}^{(k)}$, $k = 1,2,\ldots$ may be calculated more efficiently by a recursive scheme:

$$\begin{aligned} \boldsymbol{g}^{(k+1)} &= L\boldsymbol{f}^{(k+1)} \\ &= \left(2 - \frac{\lambda_{k+1}}{\lambda_k} \right) \boldsymbol{g}^{(k)} + \frac{1}{\lambda_k} \left(\frac{f_1^{(k)} - f_1^{(k+1)}}{f_1^{(k)}}, \ldots, \frac{f_n^{(k)} - f_n^{(k+1)}}{f_n^{(k)}} \right)^{\mathrm{T}} \\ &\quad + \left(\frac{\lambda_{k+1}}{\lambda_k} - 1 \right) \boldsymbol{p}, \quad (k \geq 1), \end{aligned}$$

(3.7.38)

which is easily derived from (3.7.34b) and $F^{(k)} = \mathrm{diag}[1/f_n^{(k)}, \ldots, 1/f_n^{(k)}]$. Now the matrix operations are eliminated, and only vector calculations are needed. The calculation of the initial $\boldsymbol{g}^{(1)}$ in (3.7.38) still requires matrix operation:

$$\boldsymbol{g}^{(1)} = L\boldsymbol{f}^{(1)}.$$

(3.7.39)

(3) The iterative algorithm for finding the MEM2 solution under the constraint (3.7.22) is summarized as follows:

(i) Initialize
$$k = 0, \quad \lambda_0 = 0, \quad f^{(0)} = \exp(\mu_0 - 1)h, \quad g^{(0)} = Lf^{(0)}, \quad Q^{(0)} = Q(f^{(0)}).$$
Choose μ_0, say by (3.7.31), such that $Q^{(0)}$ is as small as possible. Normally, $Q^{(0)} > m/2$.

(ii) Solve (3.7.34a,b) for $f^{(k+1)}$ by the Gauss-Seidel iteration.

(iii) Calculate $Q^{(k+1)} = Q(f^{(k+1)})$.

(iv) If $|Q^{(k+1)}/(m/2) - 1| < \varepsilon$, then stop and output the solution $f^{(k+1)}$; otherwise continue.

(v) Calculate $g^{(k+1)}$ by (3.7.39) if $k = 0$, or by (3.7.38) if $k \geq 1$.

(vi) If $Q^{(m+1)} > m/2$, set $\delta\lambda > 0$; otherwise set $\delta\lambda < 0$.

(vii) Let $k = k + 1$, $\lambda_k = \lambda_{k-1} + \delta\lambda$.

(viii) Go back to Step (ii) and repeat the iteration.

Here ε is a prescribed small positive number, say, 0.1. The change $|\delta\lambda|$ is chosen empirically. $|\delta\lambda|$ may be increased gradually to speed up the convergence.

3. Adjusting μ to Satisfy the TP Constraint

The iterative procedure described above may be viewed as adjusting λ along the straight line $\mu = \mu_0$ on the λ-μ plane so that the MEM2 solution $f(\lambda_0, \mu_0)$ satisfies the constraint $Q(f(\lambda_0, \mu_0)) = m/2$. For this solution, the TP takes some value t_0. If we do not uphold the prescribed constraint (3.7.23), then t_0 may be taken as t_c, and $f(\lambda_0, \mu_0)$ as the final solution. If this is not the case, then we must, starting with $f(\lambda_0, \mu_0)$, modify $f(\lambda, \mu)$ along a differentiable curve $\lambda = \lambda(\mu)$ passing through the point $(\lambda(\mu_0), \mu_0)$ on the λ-μ plane so that (3.7.23) is satisfied and meanwhile the constraint (3.7.22) remains in effect. This means that the curve must satisfy the following equations:

$$\frac{\mathrm{d}Q(f(\lambda(\mu),\mu))}{\mathrm{d}\mu} = 0 , \tag{3.7.40}$$

$$\nabla J(f(\lambda(\mu),\mu); \mu, \lambda(\mu)) = \mathbf{0} . \tag{3.7.41}$$

From (3.7.40), we have

$$[\nabla Q(f(\lambda(\mu),\mu))]^{\mathrm{T}} \frac{\mathrm{d}f(\lambda(\mu),\mu)}{\mathrm{d}\mu} = 0 . \tag{3.7.42}$$

Utilizing ∇J in (3.7.25), from (3.7.41) it follows that

$$\begin{aligned}
\mathbf{0} &= \frac{\mathrm{d}\nabla J}{\mathrm{d}\mu} = \frac{\mathrm{d}}{\mathrm{d}\mu}[\nabla H + (\mu - 1)h - \lambda\nabla Q] \\
&= (\nabla^2 H - \lambda\nabla^2 Q)\frac{\mathrm{d}f}{\mathrm{d}\mu} + h - \frac{\mathrm{d}\lambda(\mu)}{\mathrm{d}\mu}\nabla Q ,
\end{aligned}$$

that is,

$$\nabla^2 J \frac{\mathrm{d}f}{\mathrm{d}\mu} + h - \frac{\mathrm{d}\lambda(\mu)}{\mathrm{d}\mu}\nabla Q = \mathbf{0} . \tag{3.7.43}$$

Hence we obtain

$$\frac{\mathrm{d}\boldsymbol{f}(\lambda(\mu),\mu)}{\mathrm{d}\mu} = [\nabla^2 J(\boldsymbol{f}(\lambda(\mu),\mu);\lambda(\mu))]^{-1}\left[\frac{\mathrm{d}\lambda(\mu)}{\mathrm{d}\mu}\nabla Q(\boldsymbol{f}(\lambda(\mu),\mu)) - \boldsymbol{h}\right].$$

$$(3.7.44)$$

Substituting (3.7.44) in (3.7.42) yields

$$[\nabla Q]^{\mathrm{T}}[\nabla^2 J]^{-1}\left[\frac{\mathrm{d}\lambda(\mu)}{\mathrm{d}\mu}\nabla Q - \boldsymbol{h}\right] = 0\ .$$

In practical application, we assume that $\nabla Q(\boldsymbol{f}(\lambda,\mu),\mu)$ is never $\boldsymbol{0}$, then

$$\frac{\mathrm{d}\lambda(\mu)}{\mathrm{d}\mu} = \frac{[\nabla Q]^{\mathrm{T}}[\nabla^2 J]^{-1}\boldsymbol{h}}{[\nabla Q]^{\mathrm{T}}[\nabla^2 J]^{-1}[\nabla Q]}\ . \tag{3.7.45}$$

The initial condition is

$$\lambda(\mu_0) = \lambda_0\ . \tag{3.7.46}$$

To derive the equation concerning the TP

$$t(\mu) = \sum_{i=1}^{n} f_i(\lambda(\mu),\mu)\ ,$$

premultiplying (3.7.44) by $\boldsymbol{h}^{\mathrm{T}}$ yields

$$\begin{aligned}
\frac{\mathrm{d}t(\mu)}{\mathrm{d}\mu} &= \boldsymbol{h}^{\mathrm{T}}[\nabla^2 J]^{-1}\nabla Q\frac{\mathrm{d}\lambda(\mu)}{\mathrm{d}\mu} - \boldsymbol{h}^{\mathrm{T}}[\nabla^2 J]^{-1}\boldsymbol{h} \\
&= \frac{\{[\nabla Q]^{\mathrm{T}}[\nabla^2 J]^{-1}\boldsymbol{h}\}^2}{[\nabla Q]^{\mathrm{T}}[\nabla^2 J]^{-1}[\nabla Q]} - \boldsymbol{h}^{\mathrm{T}}[\nabla^2 J]^{-1}\boldsymbol{h}\ .
\end{aligned} \tag{3.7.47}$$

Note that (3.7.45) and the symmetry of the matrix $\nabla^2 J$ were utilized in the above.

Finally, we obtain the Cauchy problem for adjusting μ so that the TP constraint (3.7.23) is satisfied:

Equations: (3.7.45, 3.7.43, 3.7.47);

Initial conditions:

$$\boldsymbol{f}(\lambda(\mu),\mu)|_{\mu=\mu_0} = \boldsymbol{f}(\lambda_0,\mu_0)\ ,$$

$$\lambda(\mu)|_{\mu=\mu_0} = \lambda_0\ ,$$

$$t(\mu)|_{\mu=\mu_0} = t_0 = \boldsymbol{h}^{\mathrm{T}}\boldsymbol{f}(\lambda_0,\mu_0)\ .$$

The iterative scheme for solving the above Cauchy problem is as follows:

Given initial values: $\mu_0,\ \lambda_0,\ \boldsymbol{f}^{(0)} = \boldsymbol{f}(\lambda_0,\mu_0),\ t_0 = \boldsymbol{h}^{\mathrm{T}}\boldsymbol{f}^{(0)}$;

Iterative formulae:

$$\frac{\lambda_{k+1} - \lambda_k}{\mu_{k+1} - \mu_k} = \frac{[\nabla Q(\boldsymbol{f}^{(k)})]^{\mathrm{T}}[\nabla^2 J(\boldsymbol{f}^{(k)};\lambda_k)]^{-1}\boldsymbol{h}}{[\nabla Q(\boldsymbol{f}^{(k)})]^{\mathrm{T}}[\nabla^2 J(\boldsymbol{f}^{(k)};\lambda_k)]^{-1}[\nabla Q(\boldsymbol{f}^{(k)})]}\ ,$$
$$(\text{ by } (3.7.45))\ ,$$

$$[\nabla^2(\boldsymbol{f}^{(k)};\lambda_k)](\boldsymbol{f}^{(k+1)} - \boldsymbol{f}^{(k)})$$
$$= (\lambda_{k+1} - \lambda_k)[\nabla Q(\boldsymbol{f}^{(k)})] - (\mu_{k+1} - \mu_k)\boldsymbol{h}\,, \quad (\text{ by } (3.7.43))\,,$$

$$\frac{t_{k+1} - t_k}{\mu_{k+1} - \mu_k} = \frac{\{[\nabla Q(\boldsymbol{f}^{(k)})]^{\mathrm{T}}[\nabla^2 J(\boldsymbol{f}^{(k)};\lambda_k)]^{-1}\boldsymbol{h}\}^2}{[\nabla Q(\boldsymbol{f}^{(k)})]^{\mathrm{T}}[\nabla^2 J(\boldsymbol{f}^{(k)};\lambda_k)]^{-1}[\nabla Q(\boldsymbol{f}^{(k)})]}$$
$$- \boldsymbol{h}^{\mathrm{T}}[\nabla^2 J(\boldsymbol{f}^{(k)};\lambda_k)]^{-1}\boldsymbol{h}\,, \quad (k \geq 0)\,, \quad (\text{ by } (3.7.47))\,.$$

Four numerical examples of image restoration by MEM will be demonstrated in the next section, and none is presented in this section.

3.8 Algorithms and Numerical Examples (III)

In this section we describe the most recently developed MEM algorithms for image restoration. Emphasis will be put on their practical, rather than theoretical, aspects.

To begin with, we introduce a more sophisticated entropy expression and image formation model to be used.

1. Entropy Expression

The more sophisticated entropy expression of an image is of the form

$$S(\boldsymbol{h}) = \sum_j [h_j - m_j - h_j \log(h_j/m_j)]\,, \tag{3.8.1}$$

where $\boldsymbol{h} = \{h_j\}$ represents the image to be restored, $\boldsymbol{m} = \{m_j\}$ the prior estimate or model, and the summation is over the whole image.

Comparing (3.8.1) with (3.7.1) we note an extra term of $-m_j$ in the summation. This term was introduced after a quite complicated argument [3.19], which is outlined as follows:

Given the model $\boldsymbol{m}$, for the restored image $\boldsymbol{h}$ which is selected from among all the feasible solutions to satisfy some axioms (subset independence, coordinate invariance, etc.), the entropy expression is required to be of the form

$$S(\boldsymbol{h}) = S(\boldsymbol{h}, \boldsymbol{m}) = \sum_j [h_j - h_j \log(h_j/m_j)]\,. \tag{3.8.2}$$

This is equivalent to (3.7.1), which was derived on the basis of some arguments we already understood.

In the case where the model is not given but should be inferred from the data, based on a set of same axioms, the function to be maximized is found to be

$$S(\boldsymbol{m}) = S(\boldsymbol{m}, \boldsymbol{h}) = \sum_j [-m_j - h_j \log(h_j/m_j)]\,. \tag{3.8.3}$$

Now combining (3.8.2, 3.8.3) gives

$$S(\boldsymbol{h}, \boldsymbol{m}) = \sum_j [h_j - m_j - h_j \log(h_j/m_j)] . \tag{3.8.4}$$

This can be used to select the "best" image-model given the data. From the point of view of mathematical operation, adding in the summation of (3.8.2) the term $-m_j$, which is constant for the given model $\boldsymbol{m}$, has no effect on the maximization with respect to $\boldsymbol{h}$. The solution (maximum point of $\boldsymbol{h}$) does not change. However, the maximum value of S does. The absolute maximum of S is always zero and attained at $\boldsymbol{h} = \boldsymbol{m}$. This is beneficial since the iteration in image restoration always starts at $\boldsymbol{h} = \boldsymbol{m}$ and, therefore, the change of the model $\boldsymbol{m}$ does not change the starting value of S.

2. Image Formation Model

The image formation model presented in Appendix B has been used in the previous sections concerning image restoration. In this model, no correlations between pixels of an image are taken into consideration. In the real world, however, such correlations may well exist. These correlations do not manifest themselves in the entropy expression. This problem is resolved by introducing a hidden space in which the entropy S is defined for the hidden image $\boldsymbol{h}$ as shown in (3.8.1). This hidden image is then convolved with an Intrinsic Correlation Function (ICF) to generate a visible image that is what we really see [3.20].

Thus, the image formation is modeled like this:

hidden image $*$ ICF $=$ visible image,

visible image $*$ PSF $=$ blurred image, and

blurred image $+$ noise $=$ degraded image.

The above formulae may be combined in one expression:

hidden image $*$ ICF $*$ PSF $+$ noise $=$ degraded image.

Normally, the ICF is a Gaussian function. Its standard deviation σ is a scale on which the neighboring pixels are correlated. However, for an image having complex configuration, a single ICF is not sufficient in modeling this image since its correlation scales vary with the position in the image. In this case, multiple ICFs should be employed for the best result in image restoration. We will see this technique later in this section.

Now we are ready to describe a number of more sophisticated algorithms in image restoration by MEM.

3.8.1 MEM/MemSys5 Package

Basically, the package MemSys5 is a set of routines for optimization (or maximization). It can be used for different purposes with appropriate interface programs. MEM is one of such interface programs designed especially for restoring images in optical astronomy as well as ordinary photos. They were written in the FORTRAN language. While MEM is free of charge, MemSys5 is a commercial package. From the point of view of algorithm, only MemSys5 will be described in detail [3.20, 3.21].

Image restoration is treated as a statistical inference problem in the MemSys5 package; that is, given the model image, estimating the hidden image from the degraded image (data). Not surprisingly, the whole theory is based on the Bayesian method (Sect. 3.4.4).

Given the model image m, the density of joint probability that the hidden image and degraded image (data) have, respectively, the intensity distributions h and d is

$$P(h, d|m) = P(h|d, m)P(d|m) = P(d|h, m)P(h|m) .$$

This equation gives Bayes' theorem which is applicable in our case:

$$P(h|d, m) = \frac{P(d|h, m)P(h|m)}{P(d|m)} . \tag{3.8.5}$$

Given the model m, the likelihood $P(d|h, m)$ can be determined from the image formation model, while the prior probability density $P(h|m)$ is assigned by MEM. The denominator $P(d|m)$, which is equivalent to $p(D)$ in Sects. 2.3.4, 3.4.4 where $p(D)$ was viewed simply as a normalization factor, is called the *evidence* and can be evaluated. Then the expression of the posterior probability density $P(h|d, m)$ can be worked out.

The most probable image $\hat{h}$, taken as the MEM solution, is obtained by maximizing the posterior probability density $P(h|d, m)$.

1. Entropy and Likelihood

The entropy is of the form in (3.8.1), namely,

$$S(h) = \sum_i [h_i - m_i - h_i \log(h_i/m_i)] . \tag{3.8.6}$$

The (logarithmic) likelihood function under the assumption of zero-mean white Gaussian noise with variance σ_j^2 is:

$$L(h) = \frac{1}{2} \sum_j [d_j - d_j'(h)]^2/\sigma_j^2 = \frac{1}{2}\chi^2(h) , \tag{3.8.7}$$

where $d' = \{d_j'(h)\}$ is the "mock" data calculated from the estimated h.

After some mathematical manipulation [3.22], (3.8.5) gives

$$P(h|d, m) \propto \exp[Q(h)] , \tag{3.8.8}$$

where

$$Q(h) = \alpha S(h) - L(h) . \tag{3.8.9}$$

The MEM solution or the most probable image is found by maximizing $P(h|d, m)$ in (3.8.8) or, equivalently, $Q(h)$ in (3.8.9) for an appropriate α with $S(h)$ and $L(h)$ defined in (3.8.6, 3.8.7), respectively.

2. Evidence

The appropriate value of the parameter α, servicing for the purpose of regularization, is determined by maximizing the evidence $P(d|m)$, the probability density of the data d given the model m. More specifically, the evidence can be expressed in the form of $P(d|\alpha)$ [3.22]. In the iteration, $P(d|\alpha)$ is calculated for each image converged for a particular α. The iteration stops at the appropriate α for which $P(d|\alpha)$ attains its maximum. This value of α is considered to be the best.

3. Optimization Algorithm

The mock data are

$$d' = \text{PSF} * \text{ICF} * h = \text{PSF} * v \, ,$$

where v stands for the visible image. Put in the matrix notation, the above expression becomes

$$d' = RCh = Rv \, , \tag{3.8.10}$$

where R and C are the PSF and ICF matrices, respectively, and the visible image $v = \text{ICF} * h = Ch$.

Utilizing (3.8.9, 3.8.6, 3.8.7, 3.8.10), let $\partial Q/\partial h_i = 0$; then we have, after some mathematical manipulation,

$$h_i = m_i \exp\left\{\frac{1}{\alpha}\sum_j \frac{1}{\sigma_j^2}[d_j - d'_j(h)](RC)_{ji}\right\} \, . \tag{3.8.11}$$

Let

$$w = \alpha^{-1}[\sigma_j^{-1}](d - RCh) \, , \tag{3.8.12}$$

where $[\sigma_j^{-1}]$ denotes the diagonal matrix with σ_j^{-1} as its elements. Then (3.8.11) can be written in the form

$$h = m \exp(C^{\mathrm{T}}R^{\mathrm{T}}[\sigma_j^{-1}]w) \, . \tag{3.8.13}$$

Substituting h above in (3.8.9), $Q(h)$ becomes a function in the parameter w, $Q(w)$.

The Newton method is used to find the maximum point w of the objective function $Q(w)$ for a fixed α. The scaled residual w and the hidden image h are related by (3.8.12, 3.8.13).

The introduction of the parameter w facilitates the implementation of the subpixelization (subsampling) technique in MEM image restoration.

Now a big issue remains in implementing the Newton method, namely, in order to determine the search direction, how does one solve a set of linear equations of large size in the matrix form of $Ax = b$.

The direct inversion of the matrix is impracticable. The approximation like that in the simplified Newton-Raphson method described in Sect. 3.6.4 is inadequate. An accurate solution is necessary.

It is easy to verify that the maximum point of x of the function

$$f(x) = 2bx - x^{\mathrm{T}}Ax \qquad (3.8.14)$$

satisfies the equation $Ax = b$. In light of this, the above maximization problem is solved iteratively by the conjugate gradient method, efficiently in terms of memory size and CPU time on a computer, to obtain an accurate solution to $Ax = b$.

4. Iteration Scheme

A double iteration scheme is used. In the inner iteration for a fixed α, the maximum point of w, and hence that of h, is found for the objective function Q in (3.8.9). The values of entropy, evidence, etc. are calculated. The parameter α is updated and another iteration is started. This procedure is repeated until α reaches an appropriate value for which $P(d|\alpha)$ attains its maximum. No total power (flux) constraint is imposed.

The convergence criterion, maximum evidence, makes this "classic" MaxEnt distinct from other MEM algorithms that use the χ^2 criterion for convergence. The critical value of α determined by maximum evidence is smaller than that determined by the χ^2 criterion. (Note that now α is associated with the entropy but not with the likelihood.) This means that the iteration goes further in the classic MaxEnt. The consequence is higher resolution, less bias and higher noise level in the restored image, and more computational time.

5. The Interface MEM

An interface program called MEM has been written especially for restoring optical images from CCD (Charge Coupled Device) detectors. The basic inputs to MEM/MemSys5 are: degraded image (data), model image (usually a flat image or output image from the previous run of the program), PSF, noise parameters (CCD read-out noise and the analogue-to-digital conversion ratio), and other parameters and selections as described in the following:

The program can use solely Poisson statistics to handle the "Poisson noise only" case.

The program can use a subpixelization technique, which means that the restored image can be subsampled on a finer grid (i.e., having smaller pixel size) than the input images' grid [3.23].

The program can use multi-data, which means that more than one degraded image can be used as constraints for generating a single restored image.

The program can use a multichannel technique to improve resolution and linearity of photometry. More specifically, in the image formation model, a number of hidden images are used. They are associated with ICFs of different widths. The restored images from such channels with different resolutions are summed up to form a single output image.

While running the program, the total number of transforms (including convolutions and correlations) will be printed out. This number times two gives the number of FFTs.

The program can be run interactively or automatically. The execution of the program can be stopped, and the environment and intermediate images can be saved on disk for rerunning the program later.

3.8.2 MEM Task in IRAF

The MEM task (program) was first written in IRAF for image restoration in optical astronomy. Then it was further developed especially for restoring images from the Hubble Space Telescope [3.24, 3.25]. It was written in the SPP language. It is included in the subpackage STSDAS of IRAF. It is a public domain program and available by ftp.

1. Entropy Expression, Image Formation Model, and Constraints

The entropy expression (3.8.1) is used, namely,

$$S(\boldsymbol{h}) = \sum_j [h_j - m_j - h_j \log(h_j/m_j)] \, . \tag{3.8.15}$$

The image formation model described at the beginning of this section is employed. In contrast to MEM/MemSys5, only one channel (i.e., one ICF) may be used.

Both the data constraint and TP constraint are imposed. They are

$$E = \chi^2 = \sum_j \left| \sum_k p_{jk} h_k - d_j \right|^2 / v_j = E_{\mathrm{c}} \tag{3.8.16}$$

and

$$F = \sum_j h_j = F_{\mathrm{ob}} \, , \tag{3.8.17}$$

where $\boldsymbol{h} = \{h_j\}$ represents the hidden image to be restored, $\boldsymbol{m} = \{m_j\}$ the model or prior estimate of the image, $\boldsymbol{d} = \{d_j\}$ the degraded image (data), $\{v_j\}$ the noise variance, (p_{jk}) the PSF; the subscript c stands for critical value, and ob for observational value.

The objective function to be maximized is

$$J = S - \frac{1}{2}\alpha E - \beta F \, ,$$

where α and β are the Lagrange multipliers.

2. Optimization Algorithm

The preconditioned conjugate gradient method or the Newton method (accurate, not approximate) is used to find the maximum point of $\boldsymbol{h}$ for the objective function J.

Preconditioned Conjugate Gradient Method. In the ordinary conjugate gradient method, the search direction for the maximum point in each iteration (except the first one) is the current gradient's component conjugate (or, loosely speaking, orthogonal) to the search direction in the preceding iteration which is no longer useful in the current iteration. In this way the search direction in the current iteration is better than the current gradient. But this gradient itself is not a very good direction in searching for the maximum point. Further improvement is possible. Recalling that the search direction was determined easily in the approximate Newton(-Raphson) method (Sect. 3.6.4), we can use it instead of the gradient in the current iteration for a better result.

This conjugate method based on the approximate Newton method is commonly known as the conjugate gradient method preconditioned by the Jacobian method.

Newton Method. This method is the same as that in MEM/MemSys5 for optimization, except that a set of linear equations of large size in the form of $A\boldsymbol{x} = \boldsymbol{b}$ is solved by the approximate Newton method, but not by the conjugate gradient method.

3. Model Updating Technique

In all the methods introduced for optimization, including the accurate and approximate Newton method and the preconditioned conjugate gradient method, the linear equations to be solved (accurately or approximately) is of the specific form

$$(-\nabla\nabla J)\Delta\boldsymbol{h} = \nabla J\,, \tag{3.8.18}$$

where the matrix elements

$$(-\nabla\nabla J)_{jk} = h_j^{-1}\delta_{jk} + \alpha\sum_n p_{nj}p_{nk}/v_n\,.$$

The nondiagonal elements ($j \neq k$) are proportional to the multiplier α. The smaller α is, the more diagonally dominant the matrix $-\nabla\nabla J$ is, and the easier it is to solve the equations. In the extreme case, $\alpha = 0$, the matrix is a diagonal one, and the equations can be solved with little effort.

In the course of iteration, in order to reduce the value of E, α gradually steps up as schematically depicted by the lines located in the upper part of Fig. 3.14. Consequently, the solution of (3.8.18) is getting more and more difficult since the matrix is becoming less and less diagonally dominant.

The model updating (MU) technique comes to help [3.26]. At the beginning of iteration, as usual, $\alpha = 0$ and $\boldsymbol{h} = \boldsymbol{m}$. α is increased to some value to start the first outer iteration, and an ME solution is obtained at the end. Then this ME solution is used as the model to start the second outer iteration. Accordingly, α is increased to some value from zero but not from the value used in the first outer iteration. The later outer iterations proceed in the same manner. In this way α will jump up and down, but not progressively step up, in the course of iteration as depicted by the thick lines located in the lower part of Fig. 3.14. As a consequence, the values of α are kept to a minimum. Obviously, this will result in a more rapid convergence of iteration.

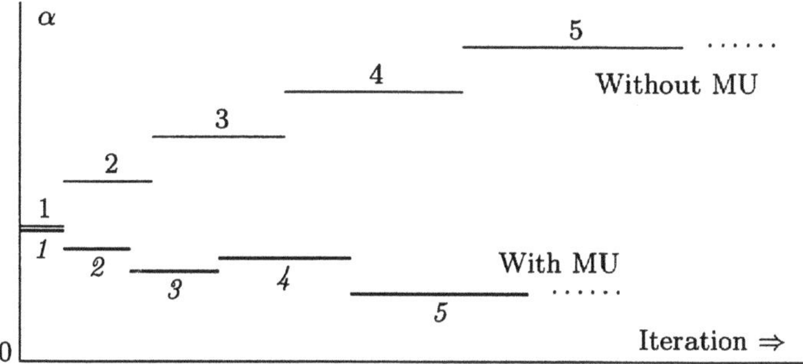

Fig. 3.14. The change of α in iteration. The outer iteration numbers are as annotated. Each outer iteration has a fixed α, and contains a number of inner iterations

4. Iteration Scheme

As described above, a double iteration scheme is used. The inner iteration is for finding the maximum point of h for fixed α and β. In the outer iteration, α is increased from zero (with model updated) to some value, and β is also changed if necessary, so that the initial magnitude of the gradient ∇J falls within a reasonable range $[0.15, 0.35]$. This magnitude is brought down to a small value, say, 0.05, (ideally zero) after a number of inner iterations, and then a new outer iteration starts. The iteration stops when the constraints (3.8.16, 3.8.17) are satisfied with some tolerances.

5. Functions and Usage

The task can handle the "Poisson noise only" case, in which the Poisson statistics are used in data constraint. A quantity similar to χ^2 defined in (3.8.16) is used as a measure for data fit.

The task employs the techniques of model updating and subpixelization. By default, in search for the maximum point of the objective function, the accurate Newton method is used in the "Poisson noise only" case and the preconditioned conjugate gradient method is used otherwise. Once the search direction has been determined, an optimal step is taken by the 1-D accurate search or parabolic approximation.

The number of FFTs in one inner iteration is 4 if the preconditioned conjugate gradient method is used. The numbers of iterations and convolutions/correlations are printed out so that the total number of FFTs can be calculated (2 FFTs per convolution/correlation).

The total number of maximally allowed inner iterations can be set to a desirable value. The output intermediate image before convergence (χ^2 is not small enough) can be used to restart the program.

Every effort has been made to create a user-friendly interface. The number of enquired parameters is kept to a minimum. The default values of optional parameters are carefully chosen and normally satisfactory. The diagnostic messages are informative and may be brief or verbose. Last but not least, a help file is well written and of great value in assisting the user to run the task.

3.8.3 Restoration with Variable Resolution

An image may have structures of different scales at different locations. It would be ideal if this knowledge can be incorporated in image restoration in such a way that objects are smoothed according to their structural scales. That is, the resolution varies over the whole restored image.

There are two key issues in the implementation of this idea. They are: How to determine the resolution map representing the structural scales over an image, and how to smooth objects accordingly.

Variable resolution can be achieved by the following approaches:

1. Multichannel Using ICFs of Gaussian-Type

Recalling the image formation model, the hidden image is convolved with an ICF to generate a visible image. Obviously, a broader (narrower) ICF will result in lower (higher) resolution or a smoother (less smooth) visible image. Suppose, as usual, ICFs are Gaussian functions with standard deviation σ's. Then we easily see that the value of σ can be used to control the resolution or smoothness of the restored (visible) image.

In image restoration algorithms using the multichannel technique, a single or multiple hidden images are restored. Each channel has its own ICF having a particular σ, which is convolved with the single or one of the multiple restored hidden images to generate a visible image for that channel. The visible images from all channels which have different resolutions are summed up with appropriate weights, W_c, to form a single restored (visible) image, namely,

$$\text{Restored image} = \sum_{c=1}^{N} W_c \times \text{image of channel } c \,. \tag{3.8.19}$$

(1) In the package MEM/MemSys5 [3.20], if the multichannel technique is used, each channel has its own hidden image and ICF. The number of channels N may be empirically chosen to be 5. The ICFs are of Gaussian-type, their sigmas being $1, 2, 4, 8$ and 16 pixels, respectively.

The weighting is realized in the entropic regularization:

$$\text{Objective function} \;=\; \sum_{c=1}^{N} w_c \times \text{entropy of hidden image of channel } c$$
$$+ \text{ data constraint} \,.$$

The weight w_c is inversely proportional to that channel's sigma squared. In this way, channels with narrower ICFs are more heavily regularized.

All the weights $W_c = 1$ in (3.8.19) . Therefore, to any pixel in the restored image, none of the channels is excluded from making a contribution. There is no need to determine the resolution map.

(2) In the pixon method [3.27, 3.28], which restores images by MLM but not by MEM, all channels share the common hidden image (pseudo-image) but have different ICFs (pixons). The numbers of ICFs $N = 12$ by default. The ICFs are parabolic functions with the bases from 1 up to 30 pixels increasing logarithmically.

For each pixel in the restored image, one and only one channel is chosen among the N channels according to the resolution map (pixon map). That is, only a single W_c in (3.8.19) is set to unity and all the others are zero. The consequence is that errors in calculating the resolution map may lead to noticeable degradation in the restored image. For this reason it is fair to say that the multichannel technique in MEM5/MemSys5 is more robust.

The crucial pixon map is calculated in the following steps:

First, an image is restored by MLM using the degraded image (data) and PSF. This image is convolved with PSF to get noise-free data (an image blurred but without noise).

Second, the restoration is repeated with the noise-free data as weights, and the resultant image is convolved with PSF to get better estimated noise-free data.

Third, the difference between the data (blurred and noisy image) and noise-free data (blurred only) is taken as a noise map. The data is convolved with each of the 12 pixons (ICFs) having sizes ranging from 1 (smallest) up to 30 (largest) pixels. Each time after convolution, the difference between the resultant image blurred by a particular pixon and the data is taken as the change caused by this pixon. If this change at a pixel is significant, i.e., the change is greater than a few, say 5, times of the square root of the χ^2 local to this pixel in the noise map, then a pixon of this size is assigned to this pixel. At the end a pixon map is formed. Each pixel is associated with a pixon number in the range $[1, 12]$, equivalent to the channel index C in (3.8.19).

The convolution between the pseudo-image and the pixon numbered C will be assigned to the pixels which have the index C in the the pixon map.

2. Using Variable Weighting Between the Entropy and Data Constraint in the Objective Function

In the objective function to be maximized,

$$J = S - \lambda E \, ,$$

the multiplier λ is used to control the relative weight between the maximization of the entropy S and the global data fit represented by E. Compared with a small λ, a large λ gives more weight to the data fit and results in a restored image with higher resolution on the whole and lower SNR (signal-to-noise ratio). If the image has structures of different scales and different SNRs in different regions, a single λ cannot give satisfactory results.

The solution is to use a number of multipliers λ. In a region where high resolution is required, a large λ is used and therefore more weight is given to the data fit. In contrast, in a region of smooth or having low SNR, a small λ is used and hence more weight is given to the entropy for stronger regularization. Symbolically, the objective function is now

$$J = S - \sum_{(\text{all regions } k)} \lambda_k E_k \ . \tag{3.8.20}$$

In the program FMAPEVAR (Fast Maximum A Posterior with Entropy prior and VARiable resolution) [3.29, 3.30], a hyper-parameter Δa playing a role equivalent to λ_k in (3.8.20) serves this purpose.

The resolution map (distribution of Δa) may be generated by inspection. A few regions are identified and appropriate values of Δa are assigned to them. Another way is to convolve the image restored by MLM with a Gaussian function. The resultant image is appropriately scaled to obtain the resolution map.

The research on the division of an image into different regions and the determination of suitable Δa for them by segmentation technique in image processing is in progress.

3. Multichannel Using Variable Entropic Regularization

In the Multiscale Maximum Entropy Method [3.31], the multiscale entropy of an image is defined as

$$S(I) \ = \ \frac{1}{\sigma_I} \sum_{(\text{all scales } k)} \sum_j [1 - M(k,j)]\sigma_k\{w_k(j) - m_k$$

$$-|w_k(j)| \log[|w_k(j)|/m_k]\} \ ,$$

where σ_I is the standard deviation of the noise in the image I; $w_k(j)$, σ_k, and m_k are the wavelet coefficients (positive or negative), the standard deviation of the noise, and the model at scale k, respectively; and j is the pixel number. The function $M(k,j) = 1$ if $w_k(j) \geq \beta\sigma_k$, $M(k,j) = 0$ otherwise. Usually, $\beta = 3$.

The entropy of the image is the sum of the entropy at all the scales. At each scale k, if SNR is large at pixel j, then no entropic regularization will be imposed ($1 - M(k,j) = 0$).

So we see that multichannel is accomplished by wavelet transform having multiscale, while variable weighting is achieved at each channel (scale) by the introduction of the function $M(k,j)$, which is called the *multiresolution support*.

The objective function to be minimized is

$$J = \frac{1}{2}\chi^2 - \alpha S \ .$$

The Lagrange multiplier α is empirically set to $0.5 \times$ maximum of PSF. This minimization problem is solved by the gradient method.

4. Discussion

(1) In carrying out the smoothing operation in methods 1 and 2 described above, having a correct resolution map is crucial. But to generate a correct resolution map is not an easy task. It was observed that even with a sophisticated scheme as that employed in the pixon technique, a misleading resolution map might be generated and some structures were lost in the restored image. This issue needs to be pursued further.

(2) Computer program restoring images with variable resolution, especially those using a large number of ICFs, need very much memory and CPU time. In running the pixon program, for example, more than 60 images may be held in the core memory with the number of pixons equal to 17 (the maximum pixon size being 30 by default). The CPU time ranges from a few tens of minutes for restoring a 128×128 image, up to a few tens of hours for restoring a 512×512 image on a SUN SPARCstation 2 with 32 MB RAM and 66 MB swap space on the disk.

3.8.4 Numerical Examples

In this section we present four examples of image restoration by MEM. The mem task in IRAF described in the second subsection was used.

Example 1. Car Number Plate (Fig. 3.15)

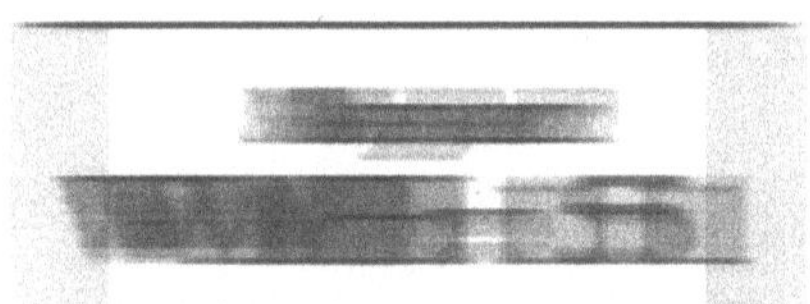

Fig. 3.15. The images of a car number plate for Example 1

A simulated car number plate blurred by linear motion is shown in the left of Fig. 3.15. The PSF is a horizontal line of 19 pixels in length. No noise is added. The number cannot be read.

Shown in the right of the figure is the plate's image restored by MEM. It is easy to read the numbers. It is also fairly easy to recognize the name of the state based on some geographic knowledge about the United States of America.

Example 2. A Portrait (Fig. 3.16)

Shown in the left of Fig. 3.16 is a girl's portrait degraded by a defect camera (the Wide Field Camera of the Hubble Space Telescope before the first servicing mission, WFC I of HST). The image was blurred by the camera's aberration and contaminated by Gaussian and Poisson noises.

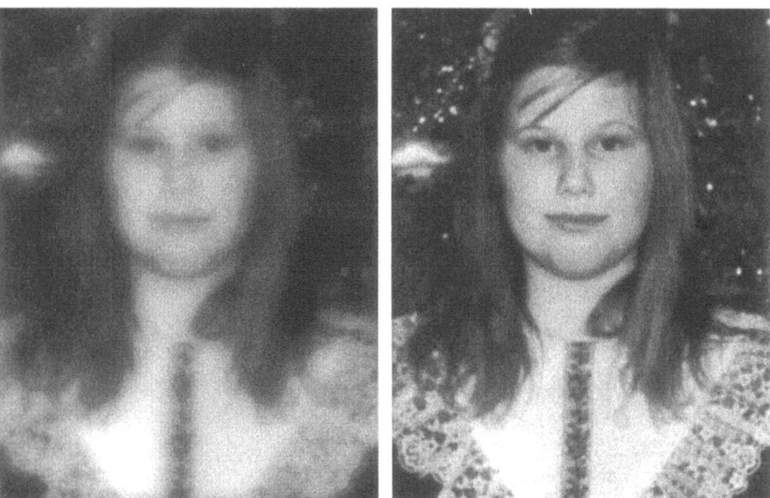

Fig. 3.16. The girl's portrait for Example 2

The restored image is shown in the right of the figure. It is obvious that the resolution has been improved remarkably. The pretty girl's face reappears clearly with the beautiful pattern of her lace collar below and prominent lights of a Christmas tree in the background.

Example 3. A Star Cluster (Fig. 3.17)

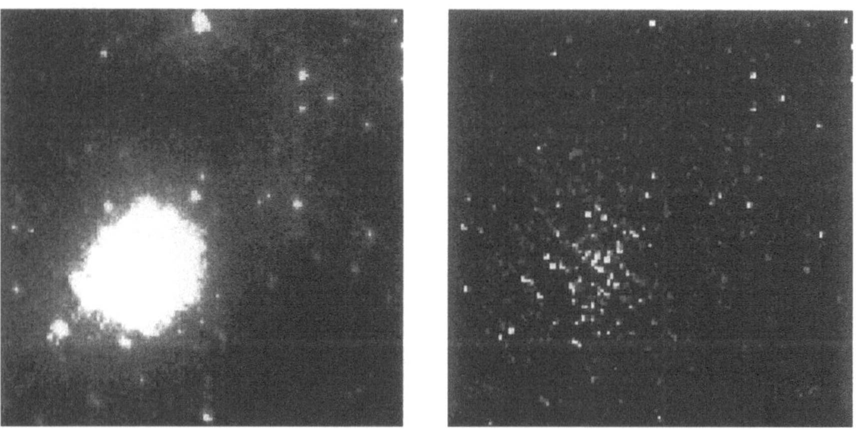

Fig. 3.17. The image of a star cluster for Example 3

Shown in the left of Fig. 3.17 is a real image of a star cluster taken by the WFC I of HST. The core of the cluster is composed of many stars and looks like a blob.

In the restored image shown in the right of the figure, the resolution is greatly enhanced. Individual stars in the core become visible.

Example 4. Image of Saturn (Fig. 3.18)

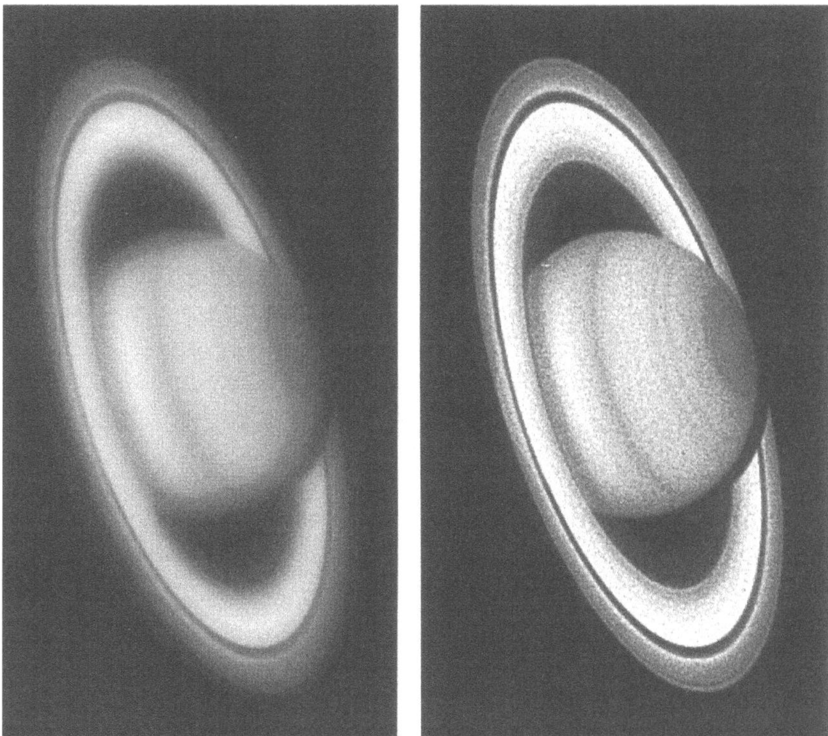

Fig. 3.18. The image of Saturn for Example 4

Shown in the left of Fig. 3.18 is a real image of the planet Saturn taken by the Planetary Camera of HST before the first servicing mission (known as PC I of HST). It is degraded due mainly to the aberration of the camera.

As shown in the right of the figure, the restored image has higher resolution. The rings are especially well resolved.

3.8.5 Other Algorithms

MEM2 image restoration is widely used in many areas. For further studies, the reader is referred to the following literature:

1. γ-ray astronomy [3.32],
2. X-ray astronomy [3.13],
3. Infrared astronomy [3.33],
4. Radar imaging [3.34, 3.35],
5. Polarized image reconstruction [3.5, 3.36],
6. Image reconstruction from projection data [3.37].

4. Analysis and Comparison
of the Maximum Entropy Method

This chapter is devoted to the analysis and comparison of the three schools of thought on MEM, namely, MEM1, MEM2 and Generalized MEM (GMEM), in basic idea, solution properties, etc. Emphasis will be put on the comparison of the MEM1 and MEM2 solution properties. Both experimental results and theoretical analysis are presented. For the purposes of completeness or further studies, some issues mentioned in the previous chapters will be discussed again in this chapter.

This chapter begins in Sect. 4.1 with a detailed introduction to GMEM, including its basic idea, mathematical expressions, and properties. Then Sect. 4.2 summarizes various expressions of entropy. Section 4.3 studies major properties of the MEM solution. Sections 4.4, 4.5 are devoted to the comparison between MEM1 and MEM2 in resolution enhancement and data extension (which are closely related to each other), respectively, presenting experimental results and theoretical analysis. The following Sects. 4.6, 4.7 are also devoted to the comparison but concerning the accuracy of peak location and relative power estimation in the spectra estimated by MEM1 and MEM2, respectively, for experimental results and theoretical analysis. Finally, Sect. 4.8 concludes this chapter with comments on the three schools of thought on MEM.

It should be pointed out that in MEM research, there are quite a few open questions in theory and much work needs to be done in experiment. Also, different opinions can be heard about some problems that have been studied in depth. We will try our best to make a comprehensive review and give a fair account. However, readers should not be constrained by the existing conclusions but give the reins to their imagination and creativity. At the same time it should be kept in mind that freedom of expression can certainly be exercised, but one should think twice before jumping to a conclusion – especially to a conclusion contradictory to the established one.

4.1 Generalized MEM

In the previous two chapters we described MEM1 and MEM2, which are related to the concept of information-theoretic entropy. In this section we introduce GMEM in detail, which has, nevertheless, nothing to do with information theory.

The idea goes like this: Now that in determining a solution we may maximize $H1$ or $H2$, why not maximize functions of other forms? Compared with using $H1$ or $H2$, the results may be similar or even better. Needless to say, these functions must satisfy some conditions and include $H1$ and $H2$ as special cases.

Högbom [4.1] suggested that these kinds of functions should also be called "entropy". We call the corresponding method GMEM. It should be noted, therefore, that "entropy" in the following may not be in the information-theoretic sense. The reasons for introducing GMEM in detail are: (1) GMEM is used by some researchers and mentioned in the literature sometimes, and (2) studying GMEM helps to understand and compare MEM1 and MEM2.

To express in the image processing language, when the measured data are not sufficient to uniquely determine the image intensity distribution $B(x,y)$, a remedy is to introduce a requirement of the form

$$F(B(x,y)) = \max \tag{4.1.1}$$

in image restoration. To facilitate mathematical operations, the function $F(B)$ takes the following convenient and simplified form:

$$H_G(B) = \int\int f(B(x,y))\mathrm{d}x\mathrm{d}y \ , \tag{4.1.2}$$

where the integral is taken over the whole image, and becomes the summation in the discrete case. $f(B)$ is a function of individual values of B. Hence, H_G cannot be used to express the information about spatial relations between features on the image. Nevertheless, it permits the use of information related to the statistics of the values of B.

4.1.1 Formulation of GMEM

The GMEM problem can be stated as follows [4.2, 4.3]:

Maximization : $H_G(B)$ in (4.1.2) ,

Constraints : $R(m,n) = \int\int B(x,y)\exp[-2\pi\mathrm{i}(mx+ny)]\mathrm{d}x\mathrm{d}y \ ,$

$$(m,n) \in K \ , \tag{4.1.3}$$

(the integral is over the whole image), where

$B(x,y)$ is the image intensity distribution, equivalent to the spectrum;
$R(m,n)$ is the visibility, equivalent to the ACF;
$R(m,n) = \mathrm{FT}[B(x,y)]$, $B(x,y) = \mathrm{IFT}[R(m,n)]$;
K is the support of the measured data.

Note the FT direction between R and B chosen according to the convention in image processing.

Form the objective functional:

$$
\begin{aligned}
G \;=\; & \int\!\!\int f(B)\mathrm{d}x\mathrm{d}y \\
& - \sum_{(m,n)\in K} \lambda_{mn} \left\{ \int\!\!\int B\exp[-2\pi\mathrm{i}(mx+ny)]\mathrm{d}x\mathrm{d}y - R(m,n) \right\} .
\end{aligned}
$$

Let the variation of G with respect to f be zero, i.e.,

$$
0 = \delta G = \int\!\!\int \left\{ f'(B) - \sum \lambda_{mn}\exp[-2\pi\mathrm{i}(mx+ny)] \right\} \delta B \mathrm{d}x\mathrm{d}y .
$$

Then, since δB is arbitrary, it follows that

$$
\begin{aligned}
f'(B(x,y)) \;&=\; \sum_{(m,n)\in K} \lambda_{mn}\exp[-2\pi\mathrm{i}(mx+ny)] \\
&=\; \sum_{(m,n)\in K} \sigma_{mn}\exp[2\pi\mathrm{i}(mx+ny)] ,
\end{aligned} \tag{4.1.4}
$$

$(\sigma_{mn} = \lambda^*_{mn})$. If σ_{mn} are viewed as the (inverse) Fourier coefficients of $f'(B)$, then $f'(B)$ is a band-limited function since the data are incomplete, i.e., the support K is only part of (m,n) domain: $\sigma_{mn} = 0$ for $(m,n) \notin K$. A nonlinear function g inverse to f' is necessary to restore B from $f'(B)$:

$$
B(x,y) = g(f'(B)) = g(\text{ band-limited function}) . \tag{4.1.5}
$$

Here $g = (f')^{-1}$ must be a nonlinear function because any linear function cannot change the band-limitedness of $f'(B)$, i.e., cannot extrapolate σ_{mn}.

The nonlinear transformation represented by (4.1.5) is shown schematically in Fig. 4.1. The function g is so chosen that B, compared with $f'(B)$, has sharp peaks and a flat background. The requirements to g are:

(1) The first derivatives

$$
f'(B) \to \infty \quad \text{as} \quad B \to +0 . \tag{4.1.6a}
$$

This ensures the positivity of B.

(2) The second derivatives

$$
f''(B) < 0 . \tag{4.1.6b}
$$

This is a necessary condition for the solution to be unique.

(3) The third derivatives

$$
f'''(B) > 0 . \tag{4.1.6c}
$$

Under this condition the ripples of $f'(B)$ are reduced so that the background is flattened, and the peaks are sharpened.

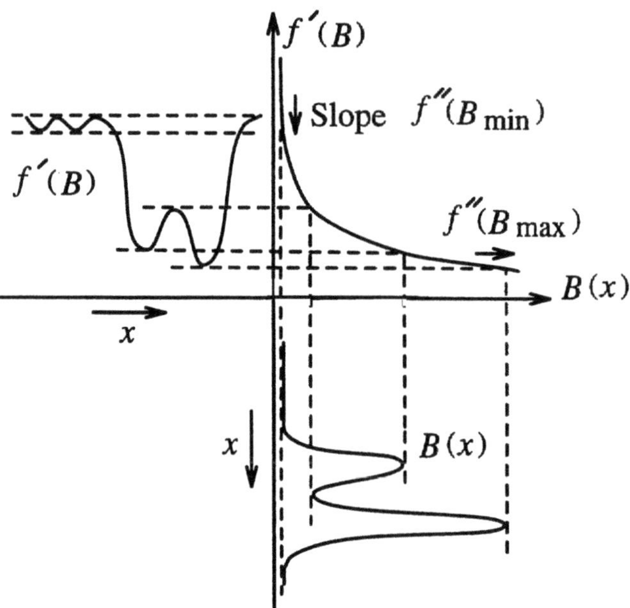

Fig. 4.1. Image $B(x)$ is a nonlinear transform of the band-limited function $f'(B)$ [4.3]

4.1.2 "Entropy" Expressions in GMEM

A family of functions satisfying the above three requirements are functions $f(B)$ for which

$$f''(B) \propto -B^{-n}, \ n \geq 1, \quad \text{(positive proportional factor)}. \tag{4.1.7}$$

The verification is straightforward.

Different values of n give different forms of "entropy" as listed below (up to $n = 3$):

n	$f(B)$	"Entropy"
1	$-B\log B$	$H2$
3/2	$B^{1/2}$	$H_{1/2}$
2	$\log B$	$H1$
3	$-B^{-1}$	H_{-1}

$H1$ and $H2$ in the list are the same as defined before. It is easy to show that $f(B)$ in the list satisfies (4.1.7).

The degree of nonlinearity of the transformation g for a fixed n is measured by a parameter R defined as

$$R = f''(B_{\min})/f''(B_{\max}) = (B_{\max}/B_{\min})^n , \tag{4.1.8}$$

where $B_{\min}$ and $B_{\max}$ are the minimum and maximum values of the image, respectively. Note that (4.1.7) was utilized in deriving the second equality in (4.1.8).

Thus, the behavior of the transformation g relies on two parameters: n (the form of "entropy") and R (the degree of nonlinearity). In Fig. 4.1 the curve representing the transformation g (in the first quadrant) can be viewed as a characteristic curve representing the input-output relationship of a nonlinear amplifier. $|f''(B)|$ can be viewed as the nonlinear "amplification factor" depending on the value of B.

Based on a semi-quantitative analysis of Fig. 4.1, the definition of R in (4.1.8), and results of numerical examples, the following conclusions can be inferred:

(1) For a fixed n, i.e., a particular form of "entropy", as R increases, the degree of nonlinearity becomes higher, the span of the transformation curve becomes wider, and hence the peak sharpening, i.e., resolution enhancement, becomes more evident. When R tends to very high values, the extreme nonlinearity may lead to superresolution (i.e., false peaks). Decreasing R has the opposite effect. In the extreme case where $R \to 1$, $B_{\max} \approx B_{\min}$, that is, the image is composed of small peaks on a very high background, the segment of the curve transforming the peaks becomes approximately a segment of a straight line, g can be viewed roughly as a linear transformation, and hence the method degenerates to a linear one and gains no benefit in resolution enhancement. Indeed, this conclusion can be shown quantitatively in MEM1 and MEM2. In particular, if the zero-lag ACF $R(0)$ is very large, then the resultant spectrum (image) has a very high flat background, and MEM1, MEM2 and the FT method are comparable in resolvability. Numerical examples will be presented in Sect. 4.4 while a theoretical argument will be given in Sect. 4.5.

(2) For a fixed R, as n increases, the "entropy" becomes "harder", the transformation curve tends to touch the horizontal and vertical axes, and hence the method becomes more effective in resolution enhancement. For instance, MEM1 ($n = 2$) is more effective than MEM2 ($n = 1$). When $n = 0$, the transformation curve becomes a straight line, and the method degenerates to a linear one. The smallest value possible is $n = 1$. So $H2$ is the "softest entropy".

(3) The parameter R can be used to control resolution enhancement. This technique is called "floating". In the implementation it is achieved by adding a constant C to the image so that

$$R = [(B + C)_{\max}/(B + C)_{\min}]^n$$

meets a prescribed value in the iteration. Adding C to the whole image is equivalent to changing the total intensity of the image or the zero-lag ACF. Note that B is changing in the iteration; so is C. Note also that in this scheme, the positivity of solution ensures $(B + C)$, but not B itself, to be positive. R should be within an intermediate range $10 \sim 10^3$. Typically, $R = 100$.

(4) If the function $f'(B(x,y))$ in (4.1.4) is expanded in a Taylor series at its minimum, then

$$f'(B(x,y)) \approx \sigma_0 + ax^2 + by^2 \, , \quad (a, b > 0) \, .$$

(To simplify the expression, the origin of (x,y) has been translated to the minimum point, and the coordinate axes have been rotated so that the crossed term xy vanishes.) For MEM1, $f_1 = \log B$, $f_1' = B^{-1}$. For MEM2, $f_2 = -B \log B$, $f_2' = -1 - \log B$. From (4.1.5) we have

$$\text{MEM1}: \quad B = g(f_1'(B)) = (f_1')^{-1} \approx (\sigma_0 + ax^2 + by^2)^{-1} \, ,$$

$$\text{MEM2}: \quad B = g(f_2'(B)) = \exp(-1 - f_2') \approx (-1 - \sigma_0 - ax^2 - by^2) \, .$$

So we see that roughly speaking, peaks in the MEM1 and MEM2 solutions are Lorentzian-shaped and Gaussian-shaped, respectively. They are used to fit the actual peaks.

(5) The image discussed above is composed of peaks on a low background. This is the case of emission image (spectrum). For an absorption image which has a high and flat background with valleys, an easy approach is to use a function $C - B$ instead of B. In this way the absorption image is converted to an emission one. The constant C should be greater than the value of the background of B.

(6) None of the specific forms of "entropy" has general superiority over the others. GMEM, including MEM1 and MEM2, is no more than model-fitting by nonlinear transformation of a band-limited function, assuming that the image has sharp peaks on a flat background. The concepts of information theory, maximum likelihood method, etc., are all irrelevant. (This is purely a pragmatic point of view.)

4.1.3 Properties of GMEM

From the conditions the nonlinear transformation g should meet in GMEM, properties common to the MEM spectra or images (denoted by S) estimated by use of different "entropy" expressions can be inferred.

1. Uniqueness and Positivity of Solution

We have pointed out that $f''(S) < 0$ ensures the uniqueness of solution. For the proof, see Sect. 4.3.

We have also pointed out that $f'(S) \to \infty$ as $S \to +0$ ensures the positivity of solution. The positiveness of the solutions of MEM1, MEM2 and $\text{MEM}_{1/2}$ can be seen immediately from their entropy expressions $H1$, $H2$ (both containing the logarithmic function), and $H_{1/2}$ (containing the square root). For MEM_{-1}, the function $f(S) = -S^{-1}$ in H_{-1} cannot ensure the positiveness of S. However, the values of S will never go negative or zero as long as its initial values are positive due to $f'(S) \to \infty$ as $S \to +0$. So we see that in GMEM the ACF extensions all satisfy the positivity (or nonnegativity) constraint. Except for MEM1, it is not clear which points within the region (an interval or circle) of permissible values are taken as the extrapolated ACF.

2. Flat Spectrum

GMEM gives as flat a spectrum as possible under the constraints. This can be understood by noticing the following facts:

(1) If the measured data contain only the zero-lag ACF (TP, total power), then maximizing $H1$, $H2$, $H_{1/2}$ or H_{-1} under this constraint will result in a flat spectrum. (The verification is easy and therefore not presented here.)

(2) Suppose that $S(j) > S(k)$. Then for the two to approach equality while keeping the TP constant, let $\Delta S(j) = -\varepsilon$, $\Delta S(k) = +\varepsilon$, $(0 < \varepsilon \ll 1)$. The "entropy" will consequently increase. That is to say, maximizing "entropy" will tend to even up the two peaks.

Take $H2$ as an example. The change of $H2$ is

$$
\begin{aligned}
\Delta H2 &= \Delta\left[-\sum_m S(m)\log S(m)\right] \\
&= -[1 + \log S(j)]\Delta S(j) - [1 + \log S(k)]\Delta S(k) \\
&= \varepsilon \log[S(j)/S(k)] > 0 \,.
\end{aligned}
$$

Simple calculation gives the following similar results:

$$
\begin{aligned}
\Delta H1 &= \varepsilon\left[\frac{1}{S(k)} - \frac{1}{S(j)}\right] > 0 \,, \\
\Delta H_{1/2} &= \frac{1}{2}\varepsilon\left[\frac{1}{S^{1/2}(k)} - \frac{1}{S^{1/2}(j)}\right] > 0 \,, \\
\Delta H_{-1} &= \varepsilon\left[\frac{1}{S^2(k)} - \frac{1}{S^2(j)}\right] > 0 \,.
\end{aligned}
$$

Presented above is the simple case: Only two elements undergo small changes at a time. The analysis will not be so straightforward in the complex case where many elements change at a time.

3. Resolution Enhancement and Background Flattening

These properties are the consequence of $f'''(S) > 0$ and the particular shape of the transformation curve in Fig. 4.1. Another way of understanding these properties is from the point of view of data extension. The unmeasured high order components are "recovered", though perhaps in a wrong way, by the nonlinear transformation. So the resolution is enhanced.

4. Noise Suppression

For noisy data, it is a common practice to use a statistic, say, χ^2 to indicate the data fit. A Lagrange multiplier λ is used to control the value of χ^2, i.e., the data fit in maximizing the "entropy". In other words, λ is used to control the balance between resolution enhancement and noise suppression, so that results satisfactory in these two respects are obtained. But the situation in practice is usually much more complex. This balance or compromise relies on the estimation of noise, the purposes of data processing, personal experience, etc. Use of an adjustable parameter λ provides flexibility, but also causes difficulty as to choice of a "correct" value of the Lagrange multiplier.

4.2 Expressions of Entropy

We have so far introduced various expressions of entropy. They are summarized in Table 4.1. For simplicity the 1-D notation in spectral analysis of time series is used. $H1$ and $H2$ are omitted from the GMEM column to save space. Note that the more sophisticated entropy expression (3.8.1) used in image restoration is not included in the table.

$H1$ is originally defined in the time domain by the joint p.d.f. $p(\boldsymbol{x})$ of the time series. It has derived expressions in the time domain and frequency domain, which are functions of the ACF matrix R and/or the spectrum S. In contrast, $H2$ and H_G have only original definitions in the frequency domain. Their derived expressions in the time domain are not known yet.

The prior estimates $m(f)$, $m(j)$ and m_j are generally assumed to be uniform (constant 1) in spectral analysis, but may be nonuniform in image restoration (e.g., degraded images of low resolution). Usually $H1$ is used in spectral analysis, and is then simplified to the common form (in plain MEM1). Note that R_m in the table is the ACF matrix corresponding to the prior spectrum $m(f)$.

Now let us look into the entropy expressions in the frequency domain. As far as dissimilarity is concerned, $H1$ is derived from the original definition in the time domain, while $H2$ is derived by MLM or applying the concept of entropy to a set of proportions. $H1$ and $H2$ correspond to different prior p.d.f.'s assigned in the Bayesian method. On the other hand, from the point of view of similarity, proper MEM1 and MEM2 are both equivalent to the Principle of MCE. $H1$ and $H2$ are both included in H_G, which is obtained from some requirements to the nonlinear transformation. Maximizing any one of the H_G (including $H1$ and $H2$) expressions is equivalent to extrapolating the ACF in a manner subject to the nonnegativity constraint. $H1$ and $H2$ may be compared in some more aspects.

4.3 Solution's Properties

This section is concerned with the properties of the MEM solution, including its existence, uniqueness, consistency, and statistical properties.

4.3.1 Existence

Only in the following cases has the existence of the MEM solution been theoretically proved:

Table 4.1. 1-D entropy expressions

Signal $x = (\ldots, x_{-1}, x_0, x_1, \ldots)$ (time series, spatial sequence, ...)

ACF $R(n), n = 0, \pm 1, \ldots$ (ACF, visibility, ...)

Spectrum $S(f)$ or $S(j)$ (power spectrum, image, ...)

Frequency f or $j \cdot 1/N$ (temporal, spatial, ...)

MEM1	MEM2	GMEM

Time domain

Original form

$$H1 = - \int p(x) \log[p(x)/m(x)]\mathrm{d}x$$

$$[\, H1 = - \int p(x) \log p(x)\mathrm{d}x \,]$$

Derived form

$$H1 = \log[\det(R)] - \log[\det(R_m)] - \sum_{k=0}^{M} S(k)/m(k) + (M+1)$$

$$[\, H1 = \log[\det(R)] \,]$$

Freq. domain

Derived form	Original form	Original form

Continuous:

$$H1 =$$

$$\int [\log \tfrac{S(f)}{m(f)} - \tfrac{S(f)}{m(f)} + 1]\mathrm{d}f$$

$$[\, H1 = \int \log S(f)\mathrm{d}f \,]$$

$$H2 =$$

$$- \int S(f) \log[\tfrac{S(f)}{m(f)}]\mathrm{d}f$$

$$(\int S(f)\mathrm{d}f = \mathrm{const})$$

$$H_{1/2} = \int S^{1/2}(f)\mathrm{d}f$$

$$H_1 = - \int S^{-1}(f)\mathrm{d}f$$

Discrete:

$$H1 = \sum [\log \tfrac{S(j)}{m(j)} - \tfrac{S(j)}{m(j)} + 1]$$

$$[\, H1 = \sum \log S(j) \,]$$

$$(1)\ H2 = - \sum p_j \log(\tfrac{p_j}{m_j})$$

$$(p_j = S(j)/\sum S(j))$$

$$(2)\ H2 = - \sum S(j) \log[\tfrac{S(j)}{m(j)}]$$

$$(\sum S(j) = \mathrm{const})$$

$$H_{1/2} = \sum S^{1/2}(j)$$

$$H_1 = - \sum S^{-1}(j)$$

---

1. For Noiseless Data

(1) 1-D MEM1, given partial ACF $R(n)$, $|n| \leq M$.

(2) 2-D MEM1, given partial ACF $R(n_1, n_2)$, $|n_1| \leq M_1$, $|n_2| \leq M_2$.

(3) 1-D MEM2, given a segment of the series $x(n)$, $n = 0, \ldots, M$, $x(n)$ being real, causal, and of minimum-phase.

2. For Noisy Data

(4) 1-D or 2-D, MEM1 or MEM2, given partial ACF, and the data fit is controlled by the χ^2-statistic.

In Cases (1, 3), the existence of a solution is evident since the explicit solutions have been found. In Case (2) the existence can be proved in two steps. First, prove that the 2-D MEM1 spectrum is equivalent to the spectrum of a 2-D discrete Gaussian-Markovian random field. Second, prove that there exists a unique Gaussian-Markovian random field whose ACF is equal to the given $R(n_1, n_2)$ for $|n_1| \leq M_1$, $|n_2| \leq M_2$. The proof is rather complicated and hence will not be presented here. The interested reader is referred to [4.4].

We have mentioned the existence of a solution in Case (4) for quite a number of times. An intuitive way of its proof illustrated graphically is to be presented in the next subsection concerning the uniqueness. Here we only outline an analytic proof for 1-D or 2-D MEM2 in the 1-D notation [4.5].

Recalling that in the differential equation approach for MEM2 image restoration (Sect. 3.7.2), the problem is formulated as follows:

$$\max_{f} \; J(f; \lambda, \mu) \;\; = \;\; H(f) + \mu \sum_{i=1}^{n} f_i - \lambda Q(f) \quad ((3.7.24)) \, ,$$

$$\text{Constraints}: \qquad Q(f) = \frac{1}{2} \, , \quad (\text{data fit}, (3.7.22) \text{ normalized by } m) \, ,$$

$$\sum_{i=1}^{n} f_i = t_c \, , \quad (\text{TP}, (3.7.23)) \, .$$

If we can first find a solution to the maximization problem subject only to the data fit constraint, then there exists a way to adjust μ so that the solution $f(\lambda, \mu)$ satisfies one more (TP) constraint. Therefore, we need only to prove the existence of the solution to the conditional maximization in (3.7.24, 3.7.22) (after the normalization of $Q(f)$ by m).

We can assume

$$\text{MIN} = \min_{f_i \geq 0, \forall i} Q(f) \leq \frac{1}{2} \, ,$$

otherwise there would be no point in formulating the problem in the first place. We can prove that along the solution curve $f(\lambda, \mu)$ of (3.7.24),

$$Q(f(\lambda, \mu)) - \text{MIN} \leq O(\frac{1}{\lambda}) \quad \text{as} \quad \lambda \to \infty \, .$$

Consequently, if MIN $< \frac{1}{2}$, there must exist a unique λ^* such that

$$Q(\boldsymbol{f}(\lambda^*, \mu)) = \frac{1}{2} \ .$$

That is to say, $Q(\boldsymbol{f}(\lambda^*, \mu))$ is the solution to (3.7.24, 3.7.22).

If MIN $= \frac{1}{2}$, then we can prove that $\lim_{\lambda \to \infty} \boldsymbol{f}(\lambda, \mu)$ exists and uniquely solves (3.7.24, 3.7.22). This completes our proof for the existence in Case (4).

Apart from the four cases listed above, the existence of solution is still an open question in theory, although a wide variety of practical algorithms have been designed.

4.3.2 Uniqueness

To show the uniqueness of the MEM solution, we present here simplified and intuitive versions of the proof for two cases. Note that by the uniqueness of the solution we mean that the (local) maximal point of the entropy function under the constraints must be a global maximum point, i.e., the maximal point must be unique.

1. For Noiseless Data

Denote the image (or spectrum) by $B(\boldsymbol{x})$, where $\boldsymbol{x}$ is a position vector, e.g., $\boldsymbol{x} = x$ in the 1-D case while $\boldsymbol{x} = (x, y)$ in the 2-D case. Partial FT components (ACF, visibility, etc.) of B are given on the support D. $f''(B) < 0$ (Sect. 4.1). The problem is to determine the MEM solution by maximizing [4.2]

$$H = \int f(B(\boldsymbol{x})) \mathrm{d}\boldsymbol{x} \ . \tag{4.3.1}$$

Suppose that $B_1(\boldsymbol{x})$ and $B_2(\boldsymbol{x})$ are the solutions. Let B be an interpolation of B_1 and B_2,

$$B(\alpha) = (1 - \alpha)B_1 + \alpha B_2 \ , \tag{4.3.2}$$

where $0 \le \alpha \le 1$. Since B_1 and B_2 satisfy the constraints on D:

$$\mathrm{FT}[B_1] = \mathrm{FT}[B_2] \quad \text{on} \quad D \ ,$$

so B also satisfies the constraints on D:

$$\mathrm{FT}[B] = (1 - \alpha)\mathrm{FT}[B_1] + \alpha\mathrm{FT}[B_2] = \mathrm{FT}[B_1] = \mathrm{FT}[B_2] \quad \text{on} \quad D \ .$$

The entropy of $B(\alpha)$, $H(B(\alpha)) = H(\alpha)$, is a function of α, and

$$H(\alpha) = \int f(B)\mathrm{d}\boldsymbol{x} = \int f((1 - \alpha)B_1 + \alpha B_2)\mathrm{d}\boldsymbol{x} \ ,$$

$$\frac{\mathrm{d}^2}{\mathrm{d}\alpha^2}H(\alpha) = \int f''((1 - \alpha)B_1 + \alpha B_2)(-B_1 + B_2)^2\mathrm{d}\boldsymbol{x}$$

$$= \int f''(B(\alpha))(B_1 - B_2)^2\mathrm{d}\boldsymbol{x} < 0 \tag{4.3.3}$$

since $f''(B) < 0$ (if $B_1 \ne B_2$).

From (4.3.2) we know $B(0) = B_1$, $B(1) = B_2$. If $B_1 \neq B_2$, i.e., B_1 and B_2 are two distinct points in the (infinitely dimensional) image space, then from (4.3.3) we see that the entropy $H(\alpha) = H(B(\alpha))$ is an upward convex function of α (Fig. 4.2). This implies that at least one of B_1 and B_2 cannot be the maximal point. Therefore, we conclude that $B_1 = B_2$ must hold. Q.E.D.

Since the solution is unique, usually we do not check the sufficient condition for the maximization of entropy, and only the necessary condition is utilized in finding the MEM solution.

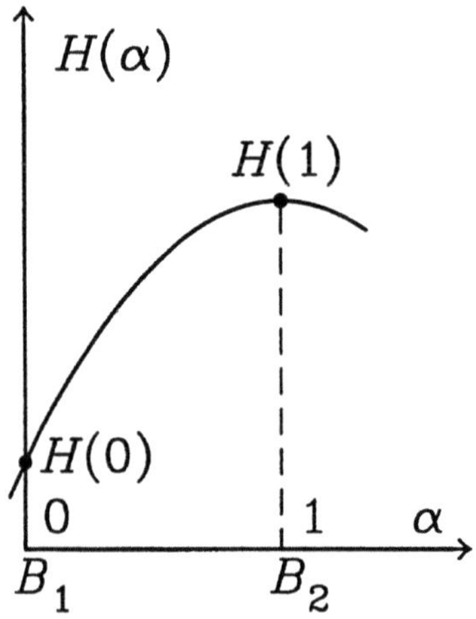

Fig. 4.2. Entropy $H(\alpha)$ is an upward convex function. $(B_1 \neq B_2)$

2. For Noisy Data

$$\text{Entropy}: \quad H2 = -\sum_j B_j \log B_j \,,$$

$$\text{Constraint}: \quad \chi^2 = \sum_{k \in K} |F_k - D_k|^2 / \sigma_k^2 = C_e \,,$$

where $F_k = \text{FT}[B_j]$, D_k is the measured data, σ_k^2 the noise variance, K the data support, and C_e the critical value (Sects. 3.6, 3.7) [4.6].

$H2$ is an upward convex function of B_j. χ^2 is a quadratic function of B_j, and the contours of χ^2 form a family of ellipses. Shown schematically in Fig. 4.3 is the case of three pixels only. The interior of the equilateral triangle represents the image space. The distances from an interior point to the three sides represent B_j $(j = 1, 2, 3)$. Their sum is a constant, the TP.

Form the objective function

$$Q(\{B_j\}) = H2 - \lambda \chi^2 \,.$$

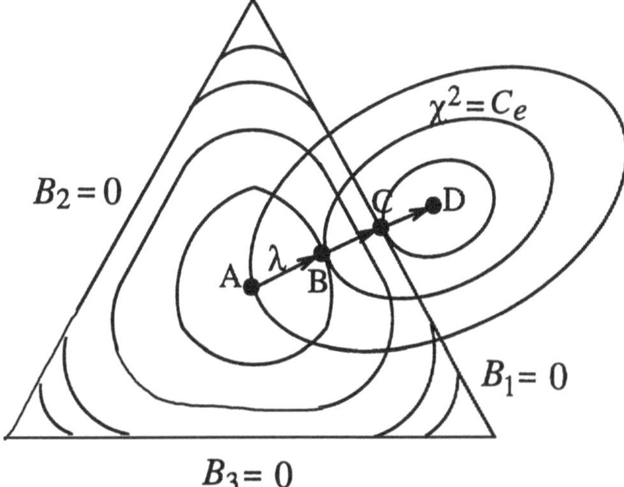

Fig. 4.3. Schematic diagram of the MEM2 solution for noisy data. A family of closed curves centered at A inside the triangle represent the $H2$ contours. A family of ellipses centered at D represent the χ^2 contours

Since $H2$ is an upward convex function (having a unique maximal point) while χ^2 is a downward convex function (having a unique minimal point), Q is an upward convex function for $\lambda \geq 0$ and therefore has a unique maximal point (maximum point). When $\lambda = 0$, B_j in the solution are all equal and χ^2 attains its maximum value, which corresponds to the center of the triangle, point A. As λ increases, the point representing the solution leaves A and moves towards the center of the ellipses, point D; the maximum value of the entropy decreases and the value of χ^2 also decreases. At a point B, the $H2$ contour and the ellipse $\chi^2 = C_e$ are tangential to each other, $\nabla H2 // \nabla \chi^2$, hence B represents the MEM solution satisfying the constraint. The corresponding λ is thought to be correct.

When the noise is sufficiently strong, the solution corresponding to the exact data fit will have some negative B_j. To signify this fact, in Fig. 4.3 the center of the ellipses, D, (corresponding to $\chi^2 = 0$, $\lambda = \infty$) is deliberately put outside the triangle. The maximum value of λ and the minimum value of χ^2 permissible correspond to the tangential point C of a χ^2 contour and a side of the triangle. This restriction on λ is removed if the noise is weak, and hence D falls inside the triangle.

Clearly, the above argument is applicable to other forms of "entropy" H_G (including $H1$) since they are also upward convex functions.

4.3.3 Consistency

The consistency requirement to the solution was mentioned in Sect. 1.3. The consistency of the MEM solution stems from the composition law in deriving the entropy function.

In the introduction of the Principle of MCE, it has been shown that the cross-entropy must take the form of $\int p\log(p/q)\mathrm{d}x$ in order that the solution may satisfy the consistency requirement [4.7]. Since we are concerned only with the time domain and frequency domain, only cross-entropies defined in these two domains can satisfy the consistency requirement. The cross-entropy defined in the time domain is equivalent to $-H1$ while that defined in the frequency domain is equivalent to $-H2$. Minimizing these two cross-entropies is equivalent to maximizing $H1$ and $H2$. Therefore, we can draw the conclusion that only MEM1 and MEM2, among GMEM, can satisfy the consistency requirement to the solution.

The consistency of solution ensures that the final result (solution) depends only on the total data available but not on the order in which the data are used. This is why we may first derive an entropy expression without any measured data constraints, and add any data constraints later on in maximizing the entropy.

In what follows we demonstrate the consistency of the MEM solutions through specific examples in simple and representative cases. In addition, a counterexample in which the solution is not consistent is presented (Sect. 1.3 and [4.8]).

1. Example of MEM2

In statistical mechanics, data D_1 is the ensemble average of a function $A(x)$,

$$< A >= \sum_j A_j p_j = \sum_j A(x_j)p_j \ , \tag{4.3.4}$$

and data D_2 is the ensemble average of another function $B(x)$,

$$< B >= \sum_j B_j p_j = \sum_j B(x_j)p_j \ , \tag{4.3.5}$$

where p_j is the probability that the random variable x takes the value x_j.

The normalization condition is

$$\sum_j p_j = 1 \ . \tag{4.3.6}$$

Under the three constraints above, the entropy $H2$ is maximized,

$$H2 = -\sum_j p_j \log(p_j/m_j) \ . \tag{4.3.7}$$

(i) Use D_1 to determine the solution S_1 first. Let $m_j = 1$ without prior knowledge about p_j. Introduce the Lagrange multipliers $\alpha^{(1)}$ and $\lambda^{(1)}$ for the constraints (4.3.6, 4.3.4), respectively; then it is easy to find the MEM1 solution S_1 to be

$$p_j^{(1)} = e^{-\alpha^{(1)} - \lambda^{(1)} A_j} , \tag{4.3.8}$$

where $\alpha^{(1)}$ and $\lambda^{(1)}$ are determined by (4.3.6, 4.3.4).

Then, modify S_1 by D_2. By modification we mean that the solution is forced to meet one more constraint (4.3.5), and at the same time the constraints (4.3.6, 4.3.4) necessarily remain in effect. Now $p_j^{(1)}$ should be taken as the prior estimate m_j of p_j. Introduce $\alpha^{(12)}$, $\lambda_1^{(12)}$ and $\lambda_2^{(12)}$ for (4.3.6, 4.3.4, 4.3.5), respectively; the MEM solution $S_{1,2}$ can be easily shown to be

$$
\begin{aligned}
p_j^{(12)} &= m_j e^{-\alpha^{(12)} - \lambda_1^{(12)} A_j - \lambda_2^{(12)} B_j} \\
&= e^{-(\alpha^{(1)} + \alpha^{(12)}) - (\lambda^{(1)} + \lambda_1^{(12)}) A_j - \lambda_2^{(12)} B_j} ,
\end{aligned}
\tag{4.3.9}
$$

where $\alpha^{(12)}$, $\lambda_1^{(12)}$ and $\lambda_2^{(12)}$ are determined by (4.3.4–4.3.6). Note that (4.3.8) was utilized in the above.

(ii) Use D_2 to determine the solution S_2, i.e., $p_j^{(2)}$, first; then modify S_2 by D_1 to obtain $S_{2,1}$, i.e., $p_j^{(21)}$.

Since (4.3.4, 4.3.5) are of the same form, it is expected that the property of the final solution does not change by swapping D_1 and D_2 in order of use, that is, the solutions $S_{1,2}$ and $S_{2,1}$ behave in the same manner. So we will not discuss $S_{2,1}$.

(iii) Use D_1 and D_2 at once to obtain the final solution S_{1+2}:

$$p_j^{(1+2)} = e^{-\alpha^{(1+2)} - \lambda_1^{(1+2)} A_j - \lambda_2^{(1+2)} B_j} , \tag{4.3.10}$$

where $\alpha^{(1+2)}$, $\lambda_1^{(1+2)}$ and $\lambda_2^{(1+2)}$ are determined by (4.3.4–4.3.6).

Now we compare S_{1+2} with S_{12}. Substituting $p_j^{(1+2)}$ and $p_j^{(12)}$, respectively, in (4.3.4–4.3.6) to solve for $\alpha^{(1+2)}$, $\lambda_1^{(1+2)}$, $\lambda_2^{(1+2)}$ and $\alpha^{(12)}$, $\lambda_1^{(12)}$, $\lambda_2^{(12)}$, we will definitely find

$$
\begin{cases}
\alpha^{(1+2)} &= \alpha^{(1)} + \alpha^{(12)} \\
\lambda_1^{(1+2)} &= \lambda^{(1)} + \lambda_1^{(12)} \\
\lambda_2^{(1+2)} &= \lambda_2^{(12)}
\end{cases}
. \tag{4.3.11}
$$

Consequently, we have

$$p_j^{(1+2)} \equiv p_j^{(12)}. \tag{4.3.12}$$

We see clearly that in proving $S_{1+2} = S_{1,2}$, the exponential function plays a crucial role. Its operational property

$$e^x e^y = e^{x+y}$$

renders the MEM2 solution unchanged in form.

If more data D_3, D_4, etc., are added, we can prove in the same way the relationships like (4.3.11, 4.3.12): $\alpha^{(1+2+3)} \equiv \alpha^{(1)} + \alpha^{(12)} + \alpha^{(123)}, \ldots,$ $p_j^{(1+2+3)} \equiv p_j^{(123)}$, etc.

2. Example of MEM2

In image restoration, the normalized intensities are

$$p_j = f_j \bigg/ \sum_j f_j \ ,$$

and the normalized visibilities are $r(n) = R(n)/R(0)$. The visibility constraint $R(n) = \mathrm{FT}[f_j] \ (|n| \leq M)$ becomes

$$r(n) = \mathrm{FT}[p_j] = \sum_{j=0}^{N-1} p_j \exp(-2\pi i j n/N) \ , \quad |n| \leq M \ .$$

The MEM2 solution with the prior estimate $m_j = 1$ is

$$p_j^{(1)} = \exp\left[- \sum_{n=-M}^{M} \lambda_n^{(1)}(-2\pi i j n/N) \right] \ .$$

If more data are added so that $|n| \leq M + L$, then the above $p_j^{(1)}$ is taken as m_j and the modified solution is

$$
\begin{aligned}
p_j^{(12)} &= m_j \exp\left[- \sum_{n=-(M+L)}^{M+L} \lambda_n^{(12)} \exp(-2\pi i j n/N) \right] \\
&= \exp\left[\sum_{n=-(M+L)}^{M+L} (\lambda_n^{(1)} + \lambda_n^{(1+2)}) \exp(-2\pi i j n/N) \right] ,
\end{aligned}
$$

where $\lambda_n^{(1)} = 0$ for $|n| > M$. It has the same form as the solution $p_j^{(1+2)}$ which is determined by using all $r(n)$, $|n| \leq M + L$ at once,

$$p_j^{(1+2)} = \exp\left[- \sum_{n=-(M+L)}^{M+L} \lambda_n^{(1+2)} \exp(-2\pi i j n/N) \right] \ .$$

Consequently, substituting $p_j^{(1+2)}$ and $p_j^{(12)}$, respectively, in the constraints $r(n) = \mathrm{FT}[p_j]$ for $|n| \leq M + L$ yields

$$
\begin{aligned}
\lambda_n^{(1+2)} &\equiv \lambda_n^{(1)} + \lambda_n^{(12)} \ , \\
p_j^{(1+2)} &\equiv p_j^{(12)} \ .
\end{aligned}
$$

Similar relationships hold if more data are added.

3. Example of MEM1

The entropy is

$$H1 = \sum_j \left(\log \frac{S(j)}{m(j)} - \frac{S(j)}{m(j)} + 1 \right) .$$

The ACF constraints are

$$R(n) = \text{IFT}[S(j)] = \frac{1}{N} \sum_{j=0}^{N-1} S(j) \exp(2\pi ijn/N) , \quad |n| \le M .$$

Taking the prior estimate $m_j = 1$, the MEM1 solution is (Sect. 2.3.2):

$$S^{(1)}(j) = \frac{1}{1 + \displaystyle\sum_{n=-M}^{M} \lambda_n^{(1)} \exp(-2\pi ijn/N)} .$$

If more data are added so that $|n| \le M+L$, then taking the above $S^{(1)}(j)$ as $m(j)$, the modified solution is

$$\begin{aligned}
S^{(12)}(j) &= \frac{1}{\dfrac{1}{m(j)} + \displaystyle\sum_{n=-(M+L)}^{M+L} \lambda_n^{(12)} \exp(-2\pi ijn/N)} \\
&= \frac{1}{1 + \displaystyle\sum_{n=-(M+L)}^{M+L} (\lambda_n^{(1)} + \lambda_n^{(12)}) \exp(-2\pi ijn/N)} ,
\end{aligned}$$

where $\lambda_n^{(1)} = 0$ for $|n| > M$. It has the same form as $S^{(1+2)}(j)$,

$$S^{(1+2)}(j) = \frac{1}{1 + \displaystyle\sum_{n=-(M+L)}^{M+L} \lambda_n^{(1+2)} \exp(-2\pi ijn/N)} .$$

Therefore, from the constraints $R(n) = \text{IFT}[S(j)]$, $|n| \le M + L$, it follows that

$$\begin{aligned}
\lambda_n^{(1+2)} &\equiv \lambda_n^{(1)} + \lambda_n^{(12)} , \\
S^{(1+2)}(j) &\equiv S^{(12)}(j) .
\end{aligned}$$

This time, in proving the consistency, the inverse operation plays a crucial role. Its property

$$(x^{-1})^{-1} = x$$

renders the MEM1 solution to keep the same form. It is also easy to see that consistency of the solution holds if the number of data groups is greater than two.

4. Counterexample of MEM1

Consider the "entropy" expression

$$H = \sum_j m_j \log(p_j/m_j) = -\sum_j m_j \log(m_j/p_j) \,, \qquad (4.3.13)$$

(Sect. 2.1.3). The constraints are still

$$< A > = \sum_j A_j p_j \,, \quad < B > = \sum_j B_j p_j \,, \quad \sum_j p_j = 1 \,.$$

The maximum "entropy" solution

$$p_j = \frac{m_j}{\alpha + \lambda_1 A_j + \lambda_2 B_j + \cdots}$$

seems to be reasonable. The prior estimate m_j manifests itself in the solution. That is, the solution will be $p_j \propto m_j$ without any measured data constraints. But this solution does not meet the requirement of consistency.

By similar calculation, as in the first example of MEM2 above, the following results will turn out:

$$(i) \qquad p_j^{(1)} = \frac{1}{\alpha^{(1)} + \lambda^{(1)} A_j} \,,$$

$$p_j^{(12)} = \frac{1}{(\alpha^{(1)} + \lambda^{(1)} A_j)(\alpha^{(12)} + \lambda_1^{(12)} A_j + \lambda_2^{(12)} B_j)} \,.$$

(ii)   (Omitted.)

$$(iii) \qquad p_j^{(1+2)} = \frac{1}{\alpha^{(1+2)} + \lambda_1^{(1+2)} A_j + \lambda_2^{(1+2)} B_j} \,.$$

If $p_j^{(1+2)} \equiv p_j^{(12)}$, then we would have

$$\begin{cases} \alpha^{(1+2)} & \equiv & \alpha^{(1)}\alpha^{(12)} \\ \lambda_1^{(1+2)} & \equiv & \lambda^{(1)}\alpha^{(12)} + \alpha^{(1)}\lambda_1^{(12)} \\ \lambda_2^{(1+2)} & \equiv & \alpha^{(1)}\lambda_2^{(12)} \\ 0 & \equiv & \lambda^{(1)}\lambda_1^{(12)} \\ 0 & \equiv & \lambda^{(1)}\lambda_2^{(12)} \end{cases} \,.$$

But the Lagrange multipliers by their nature cannot be zero. (A zero multiplier means removing its corresponding constraint.) Thus, we have shown $p_j^{(1+2)} \not\equiv p_j^{(12)}$ by reduction to absurdity.

The continuous form corresponding to (4.3.13) is

$$H = \int m(f) \log[S(f)/m(f)]\mathrm{d}f \,.$$

With the partial ACF as constraints, although the maximum "entropy" solution can show up the prior estimate of the spectrum, cf. (5) in Sect. 2.2.3, yet it can be shown in a similar way as above that this solution cannot meet the requirement of consistency.

4.3.4 Statistical Properties

In the case of noisy data, the given ACF or the ACF estimated from the given segment of a time series is contaminated by noise. Solutions obtained by using formulae derived for noiseless data must also be contaminated by noise (error). When a segment of a time series is given, the MEM1 solution can be found by estimating AR parameters directly from the data, and the MEM2 solution can also be determined directly from the data under certain conditions. Solutions obtained in this way are also contaminated by noise.

Because of the high degree of nonlinearity of MEM, it is very difficult to understand the propagation of noise from data to solutions. For 1-D MEM1, theoretical results are available. The mathematical calculus is rather complicated. Experimental results were also reported. For MEM2, a general explicit solution is not available even in the 1-D case. Hence theoretical analysis is not possible, and only experimental results were reported. Based on the above understanding, we will cite only the main results from theoretical analysis and the qualitative conclusions from experiments. Relevant literature will be referred to in due course.

1. 1-D MEM1. The ACF is estimated from a segment of a time series by the quasi-ensemble average method. The estimated spectrum has a multidimensional generalization of the "noncentral student's t" distribution. The variance of the estimated spectrum is then calculated [4.9]. In this respect, similar results can be found in [4.10].

2. 1-D MEM1. The data are as given above. The ACF is estimated by the same method. The confidence interval is calculated from the given confidence level [4.11].

3. 1-D MEM1. Asymptotic properties of solution [4.12, 4.13]. Let the length of the segment of a time series be N, and the maximum lag of the unbiased estimated ACF (i.e., the order of AR model) be M. The true spectrum is assumed to be reasonably smooth relative to the resolution.

(1) The MEM1 spectral estimate is asymptotically normal distributed.

(2) The MEM1 spectral estimate is asymptotically unbiased. That is, as $N, M \to \infty$, the ensemble average of the solution is equal to the true spectrum:

$$< S_{\mathrm{MEM1}}(f) > = S_{\mathrm{true}} .$$

(3) As $N, M \to \infty$, the number of degrees of freedom of the MEM1 estimate is

$$\nu_{\mathrm{MEM1}} = N/M .$$

The number of degrees of freedom is defined as

$$\nu = 2(\text{mean})^2/\text{variance}.$$

In comparison, the results for the FT spectral estimates are:

$$\nu_{\text{BT}} = 2N/M, \text{ (biased ACF estimate (2.4.2), triangular window)};$$

$$\nu_{\text{FT}} = N/M, \text{ (unbiased ACF estimate (2.4.1), rectangular window)}.$$

So we see that MEM1 is somewhat inferior to the FT method.

Presented above are the main results from theoretical analysis. The main results from experiments are reported in the following:

4. 1-D MEM1. Noise is added to the exact ACF of a real sinusoid. The results of experiments show that compared with the BT spectral estimates (the ACF is triangular windowed and then Fourier transformed), the MEM1 estimates seem to have very much larger percentage variations in the peak value, and somewhat larger percentage fluctuations in the background. However, the percentage changes in the area under the peak estimated by MEM1 (representing the estimated power) are comparable to those in the peak value estimated by the BT method [4.14].

5. 1-D MEM1. AR parameters are estimated from segments of complex exponentials (time series) by various algorithms, and the MEM1 spectra are calculated. Experimental results show that the LUD algorithm (its results are the same as those from the Marple algorithm when their orders are chosen to be equal) is superior to the Burg algorithm in the sense of smaller bias and variance [4.15].

6. MEM1. As the initial phase of a sinusoid (time series) randomly changes, the peak value, peak location, bias and variance of the MEM1 spectral estimate also change. For detailed experimental results, see [4.15–4.17].

7. 1-D MEM2. Biased ACFs (triangular window) are estimated from data segments of AR, MA and ARMA processes (time series), and then the spectra are calculated. Experimental results show that generally speaking, the FT method (ACF being further Papoulis windowed) has a smaller bias but a larger variance than MEM1 in the peak value; and MEM2 is in between the FT method and MEM1 [4.18]. (A Papoulis window takes nonnegative values in the frequency domain with minimum bias of the estimated spectrum.)

4.4 Resolution Enhancement and Data Extension (Experimental Results)

The analysis and comparison of the MEM solution are carried out experimentally and theoretically. The theoretical job is a difficult one because of the high degree of nonlinearity of MEM. This is true even in the few cases where explicit solutions are available. The mathematical manipulations are complicated. During the course of developing MEM, it was often the case that experimental results were reported first, then theoretical explanations followed. Accordingly, our strategy is to report experimental results first, and then to try our best to explain the main results [4.19, 4.20].

Our studies will be concentrated on (1) the resolvability of the MEM1 and MEM2 spectral estimation, and (2) the accuracy of peak location and relative power estimation in the estimates. Results from the FT method will be cited for the purposes of comparison. Four sections are devoted to these studies so that each section is of reasonable length. This section and the next are for the resolvability and its close relative, data extension, concerning, respectively, experimental results and theoretical analysis. Then, two following sections are for peak location and power estimation.

MEM algorithms utilized in experiments will be stated along with numerical examples. Unless otherwise specified, in the FT method the Bartlett (i.e., triangular) window is used; the results are signified by BT.

First of all, we would like to make a statement about the data used in the experiments. Unless otherwise specified, the data are taken from the ACF of a collection of complex exponentials buried in white Gaussian noise:

$$1\text{-D} : R(n) = \sum_{k=1}^{K} a_k^2 \exp(2\pi \mathrm{i} f_k n) + \sigma^2 \delta_n \, , \qquad (4.4.1a)$$

$$2\text{-D} : R(n_1, n_2) = \sum_{k=1}^{K} a_k^2 \exp[2\pi \mathrm{i}(f_{k_1} n_1 + f_{k_2} n_2)] + \sigma^2 \delta_{n_1, n_2} \, , \qquad (4.4.1b)$$

where K is the number of exponentials, a_k^2 is the power of the kth exponential, $f_k(f_{k_1}, f_{k_2})$ is the frequency (pair) determining the kth peak location, and σ^2 is the noise power.

Deterministic and noiseless signals are used to facilitate the analysis and comparison of the results. Of course, this incurs limitation, to some extent, to the generality of the conclusions drawn. The reason for using complex exponentials instead of real sinusoids is that the interference between the positive and negative frequency lobes should be precluded so that properties of spectral estimates can manifest themselves best.

The partial ACF is given over a support D, which is an interval $|n| \leq m$ (1-D) or a square symmetrical about and containing the origin, $|n_1|, |n_2| \leq m$ (2-D) (Fig. 4.4). The size of D will be specified by the number of points $(2M-1)$ (1-D) or $(2M-1) \times (2M-1)$ (2-D), or simply by M. $2M-1 = 2m+1$ is the number of points along each coordinate axis, and $M = m + 1$ is its nonnegative part.

The signal-to-noise ratio (SNR) is defined as the sum of power of each exponential divided by the noise power in the logarithmic scale. Symbolically, the SNR is defined by

$$\mathrm{SNR(dB)} = 10 \log_{10} \left(\sum_{k=1}^{K} a_k^2 / \sigma^2 \right) \, .$$

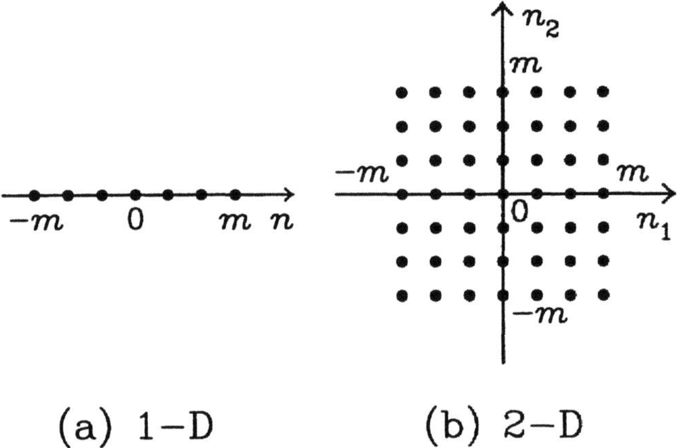

(a) 1–D   (b) 2–D

Fig. 4.4. The support D of the partial ACF. $M = m + 1$

At this point it is worthwhile to say a few words about the term "noise" since it is somewhat misleading. The original signal (exponentials) is assumed to be buried in the white noise. However, the exact ACF generated from them is noiseless. The point is that the signal and the white noise are independent and the ACF of the white noise is the δ-function. We assume that the ACF has been determined exactly over the finite support D. So the data, ACF, are incomplete but noiseless. Decreasing SNR just means increasing the zero-lag ACF component so that the background in the spectra will be raised.

4.4.1 Examples

Now we are ready to investigate the experimental results concerning resolvability and data extension in spectral estimation. The following four numerical examples may help us get a feeling first:

Example 1. (Fig. 4.5). The data are taken from the ACF of a sinusoid buried in white Gaussian noise:

$$R(n) = a^2 \cos(2\pi f n) + \sigma^2 \delta_n , \quad n = \ldots, -1, 0, 1, \ldots , \tag{4.4.2}$$

where $f = 0.25$ and SNR $= 10 \log_{10}(a^2/\sigma^2) = -3$dB. A part of the ACF is shown in Fig. 4.5a. Let $M = 16$, the data extrapolated by MEM2 are shown in Fig. 4.5b. The spectra are shown in Fig. 4.5c.

Algorithms: the explicit solution (solution of the normal equations) of MEM1, the implicit solution (CAM) of MEM2 (its results are the same as those from the $R - \lambda$ procedure).

Example 2. (Fig. 4.6). Same as Example 1, except SNR $= -1$dB. Note the lower backgrounds and sharper peaks in the spectra.

Example 3. (Fig. 4.7). The data are of the form of (4.4.1b). $K = 1$, $(f_{1_1}, f_{1_2}) = (0.1000, 0.2150)$, $a_1^2 = 1.0$, and SNR $= -5.0$dB. Let $M = 4$. The MEM1, MEM2 and BT spectra are shown in Figs. 4.7a, b, c, respectively.

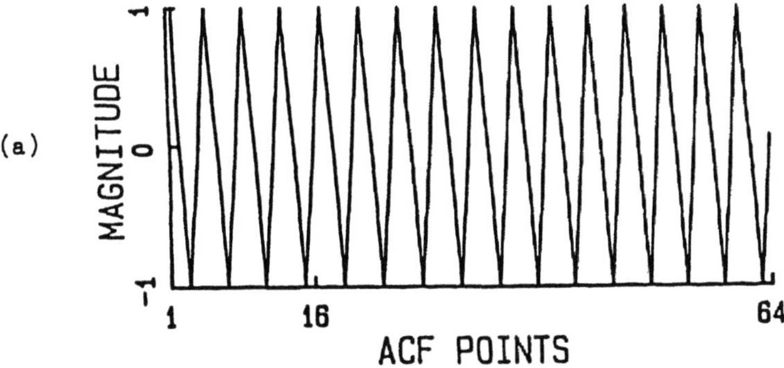

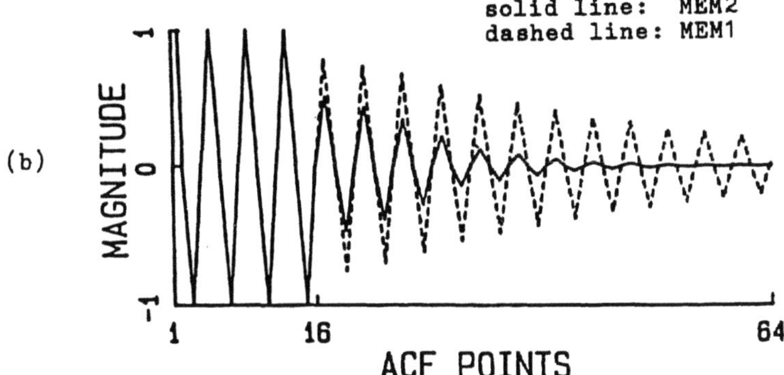

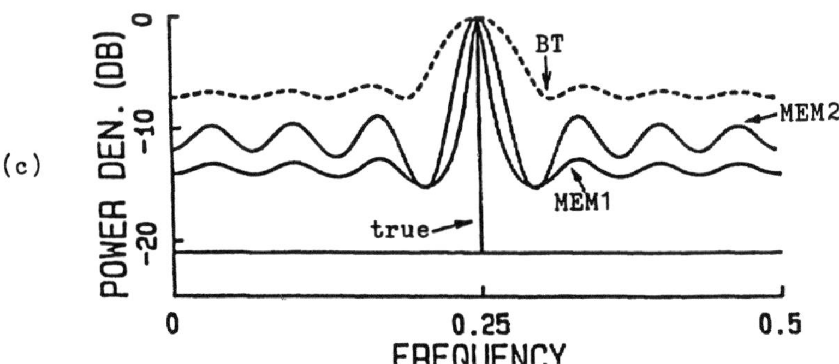

Fig. 4.5. The ACF, extrapolated ACF, and spectra in Example 1. $f = 0.25$ and $\text{SNR} = -3\text{dB}$. The peak values of spectra have been normalized to 0dB. (a) The partial ACF (ignoring the negative part). The zero-lag is actually 3. (b) The extrapolated ACF by MEM1 and MEM2 (ignoring the negative part). The zero-lag is actually 3. $M = 16$. (c) The spectra. The true spectrum is the isolated line plus the flat background

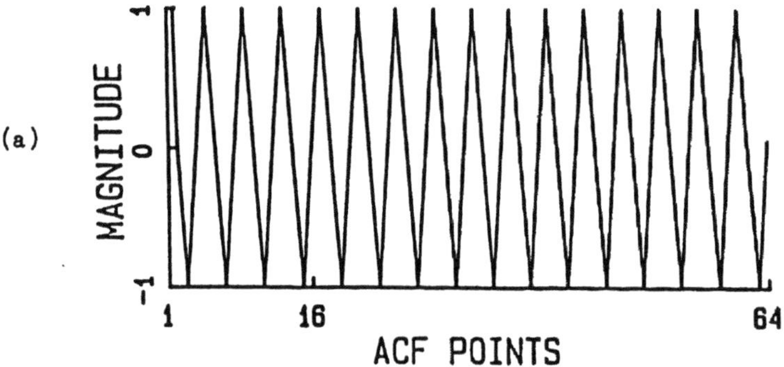

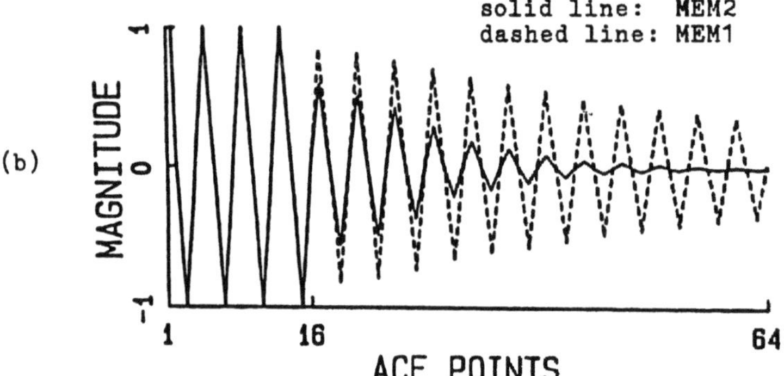

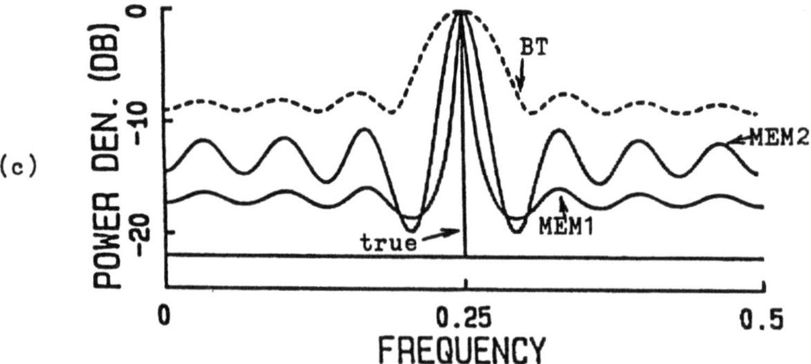

Fig. 4.6. The ACF, extrapolated ACF, and spectra in Example 2. $f = 0.25$ and SNR $= -1$dB. The peak values of spectra have been normalized to 0dB. (a) The zero-lag is actually 2.25, otherwise same as Fig. 4.5a. (b) The zero-lag is actually 2.25, otherwise same as Fig. 4.5b. (c) The spectra. Same as Fig. 4.5c except SNR $= -1$dB

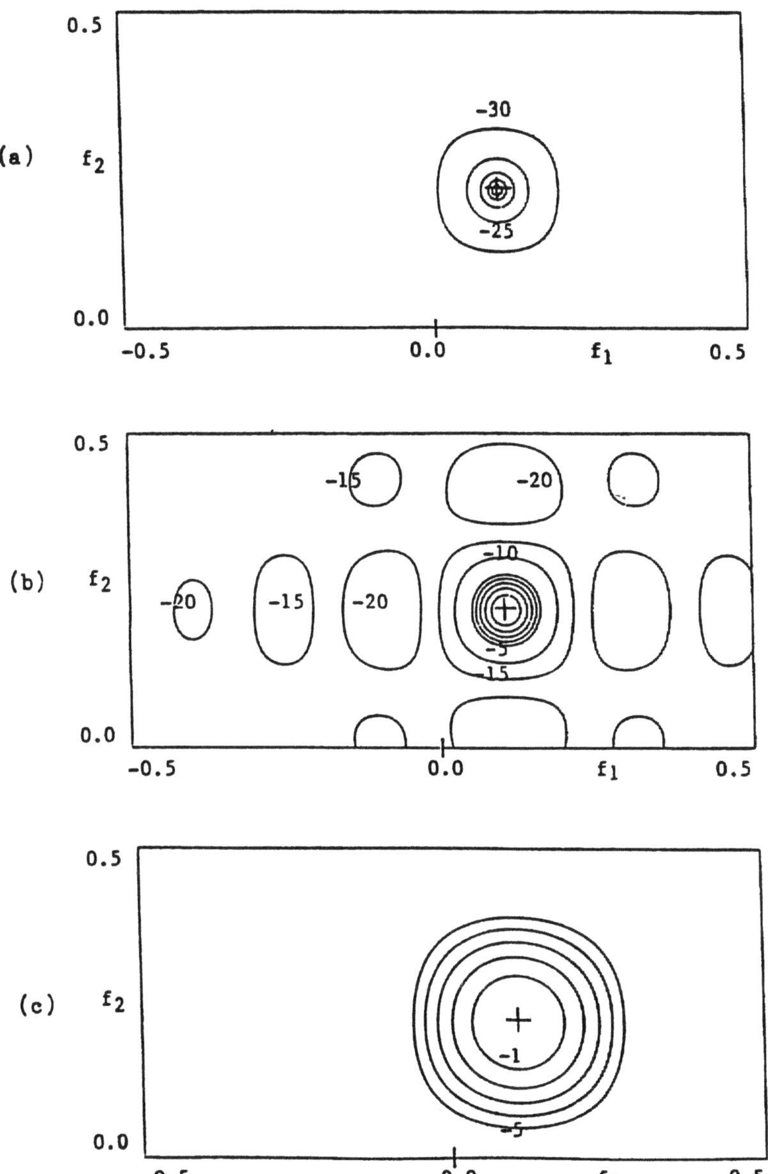

Fig. 4.7. Spectra in Example 3. $(f_{1_1}, f_{1_2}) = (0.1000, 0.2150)$ and SNR $= -5$dB. $M = 4$. The peak values of spectra have been normalized to 0dB. "+" represents the true peak locations. (a) MEM1, (b) MEM2, and (c) BT

Algorithms: the Lim-Malik algorithm of MEM1, the $R - \lambda$ procedure of MEM2, and the FT size is 128×128.

Example 4. (Fig. 4.8). Same as Example 3, except $K = 2$, $(f_{1_1}, f_{1_2}) = (0.1000, 0.1000)$, $(f_{2_1}, f_{2_2}) = (0.2000, 0.3500)$, and $a_1^2 = a_2^2 = 1.0$.

From the four examples above, we observe:

(1) MEM1, MEM2 and BT are in descending order as far as resolution is concerned.

(2) MEM1, MEM2 and BT are also in descending order in the effectiveness of data extension.

The consistency of resolution enhancement and the effectiveness of data extension is reasonable and expected.

(3) The resolution of both MEM1 and MEM2 is increasing as SNR is increasing (peaks are sharpening).

(4) The peak locations of MEM1, MEM2 and BT are all reasonably accurate.

4.4.2 Resolvability in 1-D Spectral Estimation

Now we turn to the quantitative studies of resolution. For this purpose, first of all we must precisely define the resolvability in spectral estimation [4.21].

The key issue is how to define two peaks to be "just resolved" when they are of finite width but not isolated lines. In Fig. 4.9a the two peaks of same shape and equal amplitude are obviously "well resolved". As the two peaks move towards each other, through the positions of "just resolved" shown in (b), and finally arrive at the positions of "not resolved" shown in (c). Note the changes in the peak amplitudes at f_1, f_2 and the central frequency

$$f_c = \frac{f_1 + f_2}{2} \ .$$

The difference

$$\Delta S = S(f_c) - \frac{S(f_1) + S(f_2)}{2}$$

is negative in (a) while positive in (c). Thus, a sensible definition for "just resolved" is $\Delta S = 0$. The corresponding frequency separation

$$\Delta f = |f_1 - f_2|$$

can be taken as the quantitative measure of the resolvability in the estimated spectra. A small Δf indicates high resolution.

The above definition can apply to the 2-D case. In the simplest case where each of the two peaks of same shape and equal amplitude has equal frequencies on the axes, that is, two peaks are, respectively, at (f_1, f_1) and (f_2, f_2), the above three equations can be used without modification.

After defining the "just resolved" frequency separation Δf, the normalized resolution is defined as

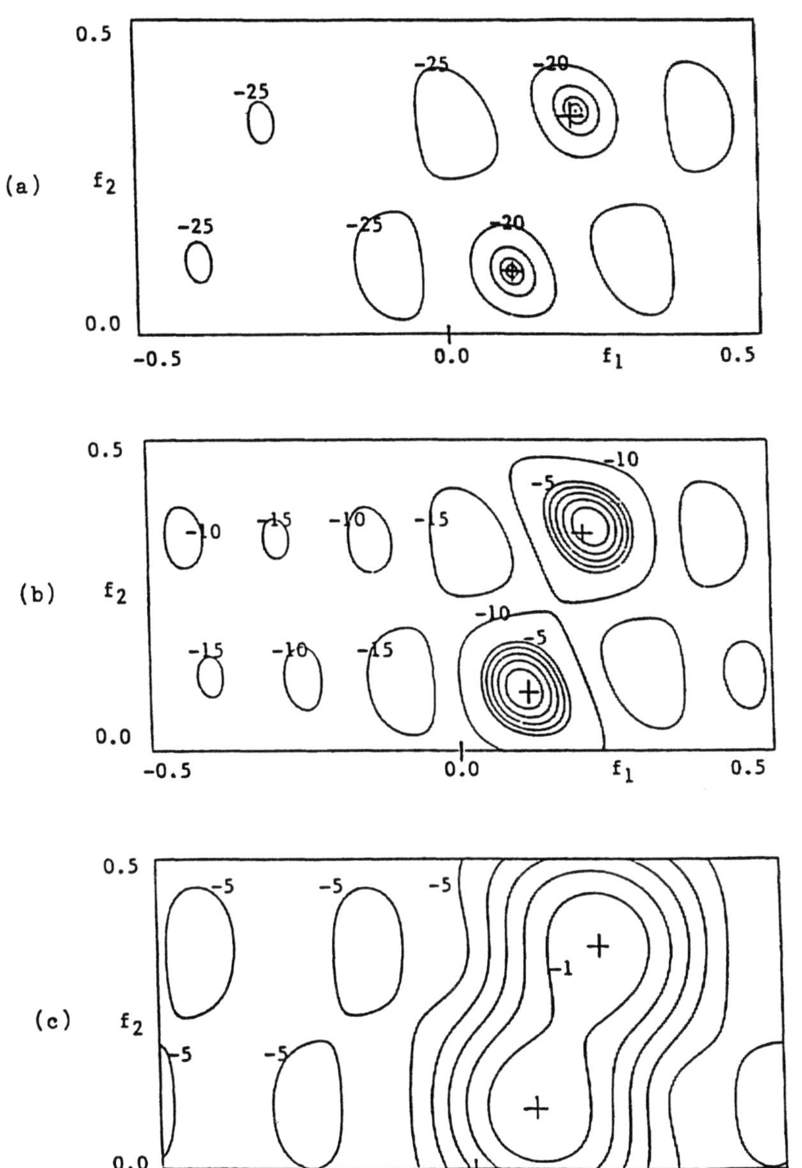

Fig. 4.8. Spectra in Example 4. $(f_{1_1}, f_{1_2}) = (0.1000, 0.1000)$, $(f_{2_1}, f_{2_2}) = (0.2000, 0.3500)$, otherwise same as Fig. 4.7. **(a)** MEM1, **(b)** MEM2, and **(c)** BT

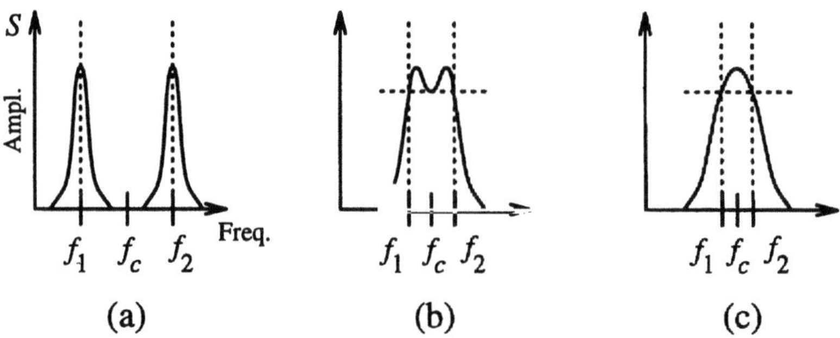

Fig. 4.9. Separation of two peaks. (a) Well resolved, (b) Just resolved, and (c) Not resolved

$$R = 2\pi M \Delta f \,, \tag{4.4.3}$$

where M is the ACF size on the nonnegative frequency axis.

In practical computation, there is little chance to make ΔS exactly zero by shifting f_1 and/or f_2 in an ad hoc manner. A simple bisection procedure can be used to effectively find sufficiently accurate Δf corresponding to sufficiently small $|\Delta S|$ [4.19].

Now it is time to investigate the experimental results.

Shown in Fig. 4.10 is the normalized resolution R as a function of the ACF size M and with SNR as a parameter in the spectra estimated by MEM1, MEM2 and BT. The results for MEM1 (i.e., AR method) and BT are adapted from [4.21]. SNR defined for real sinusoids buried in white Gaussian noise has been converted to its equivalent for complex exponentials by adding -3dB. The $R - \lambda$ procedure was used in the MEM2 spectral estimation.

Shown in Fig. 4.11 is the just resolved Δf as a function of SNR, $M = 20$ being fixed.

The main conclusions concerning the resolution in spectra estimation drawn from the two figures are as follows:

(1) The resolution of both MEM1 and MEM2 is increasing (R and Δf are decreasing) as SNR is increasing, M being fixed. For sufficiently low SNR (≤ -43dB for MEM1 and ≤ -23dB for MEM2) the results from both MEM1 and MEM2 approach that from BT. Note that the resolution of BT is independent of SNR.

(2) MEM1, MEM2 and BT are in descending order as far as resolution is concerned. This is concluded from the following observation: The only curve for BT is located above all the others (largest R and Δf); for a certain SNR, the curve for MEM2 is located above that for MEM1 (larger R and Δf). For a fairly high SNR, the difference in resolution between the three methods is quite significant.

(3) As the ACF size M increases, the resolution of MEM1 and MEM2 for high SNR improves rapidly at first. The rate of improvement then decreases. This is suggested by the rapid drop of the curves for small M and

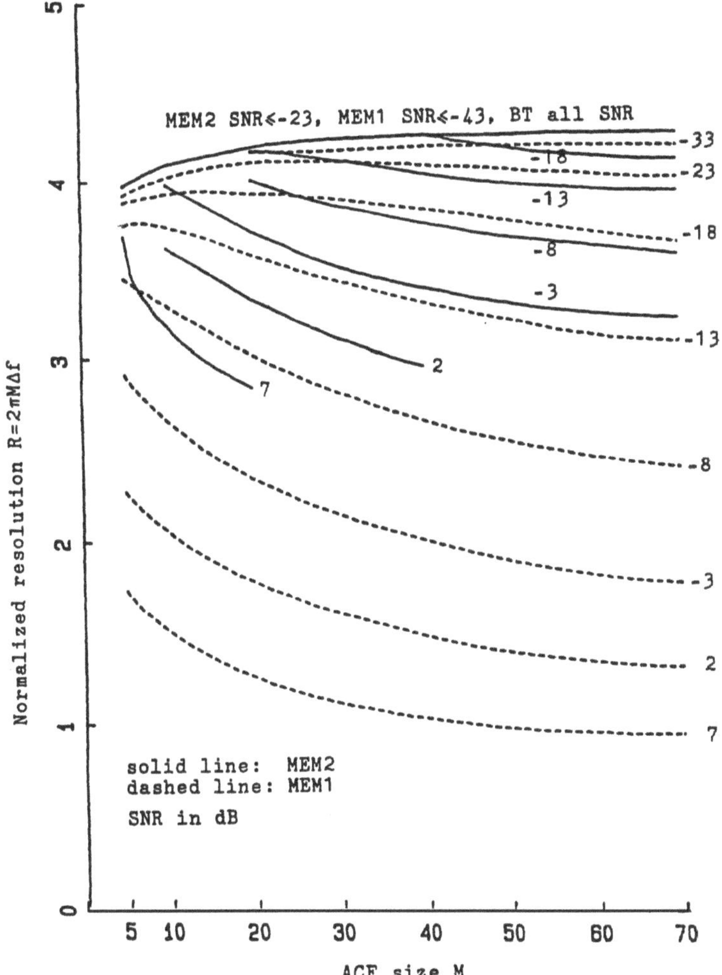

Fig. 4.10. 1-D spectral estimation. The normalized resolution R versus the ACF size M with SNR(dB) as a parameter

the somewhat flat part of the curves for large M. A flat straight line means R is constant and Δf is inversely proportional to M.

We give, by the way, the empirical formulae concerning resolution:

BT: $\quad\Delta f \approx 0.67/M, \quad M > 20 \,,$

MEM1: $\quad \Delta f \approx 1.03/\{M[(M+1)\cdot \mathrm{SNR}]^{0.31}\} \,, \quad M\cdot\mathrm{SNR} > 10 \,.$

4.4.3 Resolvability in 2-D Spectral Estimation

The experimental results concerning the resolution in the 2-D spectral estimation are shown in Figs. 4.12, 4.13. The Lim-Malik algorithm was used for

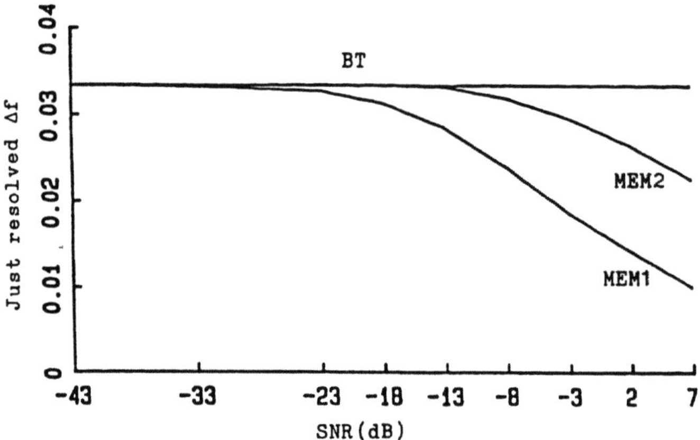

Fig. 4.11. 1-D spectral estimation. The just resolved Δf versus SNR(dB). $M = 20$

MEM1 while the $R - \lambda$ procedure was used for MEM2. Because the 2-D FFT of large size requires much space in computer memory and much computational time, the experimental datum points here are fewer than those in the 1-D case. Nevertheless, by comparing Figs. 4.10, 4.11 with Figs. 4.12, 4.13, it can be seen that the three conclusions drawn from the 1-D experimental results are applicable to the 2-D case.

4.4.4 Superresolution and Spectral Line Splitting

These two problems are particular to MEM1. In this regard, to the author's knowledge, nothing positive has been reported for MEM2.

1. Superresolution

The connotation of superresolution depends on the context. Here, roughly speaking, it means that the resolution in spectral estimation is "too good" so that false peaks arise. An experiment is described in the following [4.22].

The true spectrum $S(f)$ is composed of two narrow Gaussian peaks of equal amplitude and a flat background generated by white noise. These two peaks are so close to each other that they cannot be resolved (Fig. 4.14a). Inverse Fourier transforming the true spectrum yields the ACF sequence $R(n)$. If $R(n)$, $|n| = 0, 1, \ldots$ is Fourier transformed, then of course the true spectrum will be obtained, in which the two peaks are still unresolved. Now the ACF is truncated. Take $R(n)$, $|n| \leq 26$ ($M = 27$) and estimate the spectrum by MEM1, which results in $S_1(f)$. In the case where the background is sufficiently low (SNR $= 1.0$), something fantastic happens: The two peaks are resolved in $S_1(f)$, and their locations are more or less correct (Fig. 4.14b).

If we are given the ACF sequence $R(n)$ and nothing else, then the spectrum estimated from the truncated $R(n)$ by any method should not have higher resolution than the spectrum calculated from all $R(n)$, $|n| = 0, 1, \ldots$;

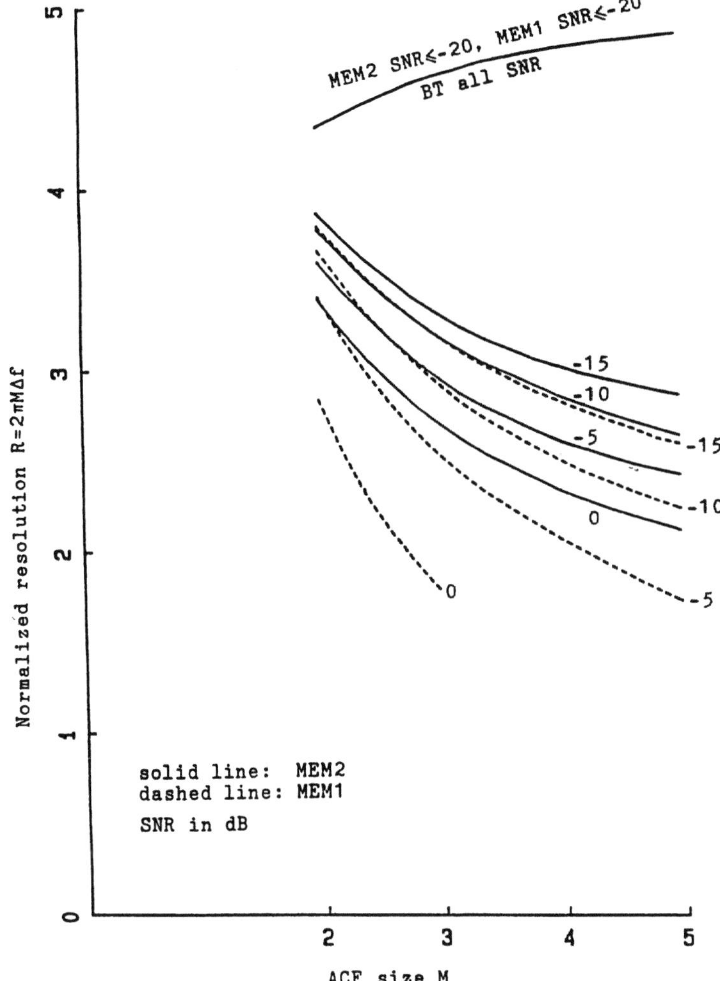

Fig. 4.12. 2-D spectral estimation. R versus M with SNR(dB) as a parameter

otherwise we say that "superresolution" takes place. The superresolution in the spectrum estimated by MEM1 described above should certainly be imputed to the method itself. The given ACF is innocent since it is exact. A simple explanation to the superresolution is that the ACF is wrongly extrapolated by MEM1 (extension decays not fast enough).

2. Line Splitting

Line splitting refers to a problem similar to superresolution. It means that when the signal is a harmonic process, its isolated spectral line splits to several (usually two) peaks of finite width and approximately equal amplitude in the

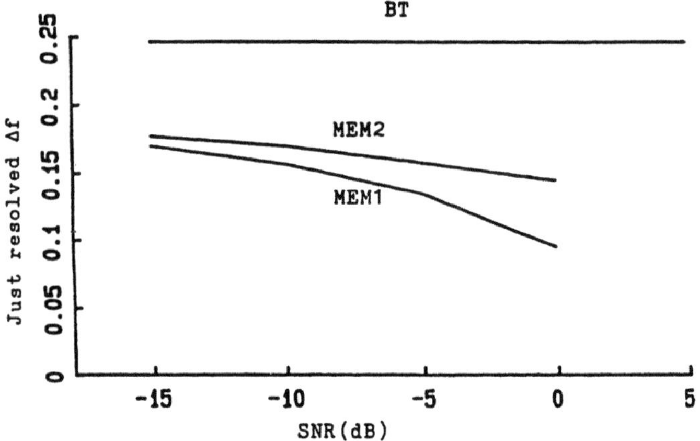

Fig. 4.13. 2-D spectral estimation. Δf versus SNR(dB). $M = 3$

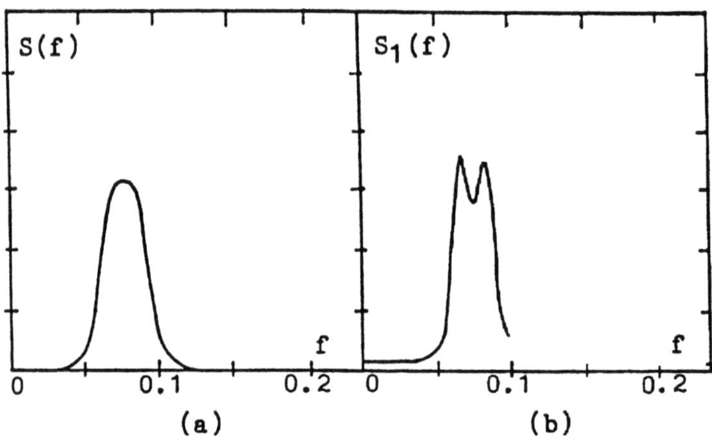

Fig. 4.14. Superresolution in the MEM1 spectral estimation [4.22]. (a) The true spectrum. (b) The MEM1 spectral estimate

estimated spectrum. Many examples have been reported in the literature, but only one is presented here [4.16].

The time series is composed of a sinusoid ($f = 0.05$) and white noise. The SNR (the effective value (i.e., *rms* value) ratio) is 20. The number of datum points is 24. The Burg algorithm is used. The AR order is chosen to be 20. The result is shown in Fig. 4.15. Two peaks of finite width located near $f = 0.05$ are observed.

Line splitting is related to many factors. Generally speaking, there is less chance for line splitting given a longer segment of a time series. When the length of the data segment is fixed, higher SNR and higher AR order will lead to a higher chance of line splitting. Line splitting is related to the initial phase of signal in a complex manner. Last but not least, line splitting depends on

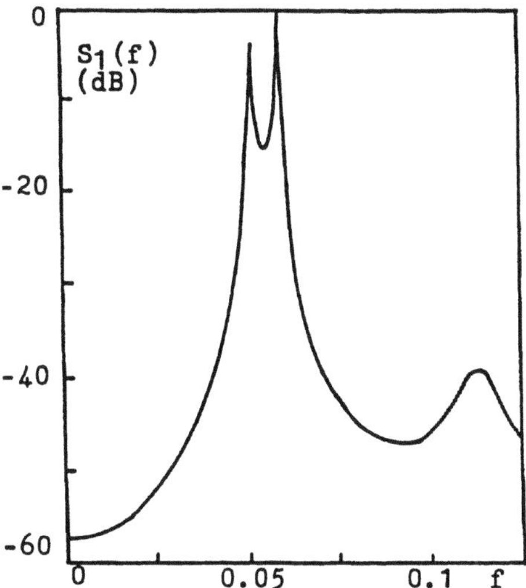

Fig. 4.15. Line splitting in the MEM1 spectral estimate [4.16]. The frequency range has been normalized to $[-1/2, 1/2]$ Hz

the algorithm. Line splitting in the MEM1 spectral estimation usually refers to the Burg algorithm.

The following conclusions are drawn from experiments under conditions that a segment of a single real sinusoid plus white noise is used as the data, SNR is very high, and the Burg algorithm is used [4.23]:

(1) If the data length is a multiple of 0.5 cycle of the sinusoid, there is no line splitting for any initial phase. This is the best case.

(2) If the data length is an odd multiple of 0.25 cycle and the initial phase of the sinusoid is 0 or $\pi/2$ (radians), there is no line splitting either.

(3) If the data length is an odd multiple of 0.25 cycle and the initial phase of the sinusoid is an odd multiple of $\pi/4$, then line splitting is the most severe. In this case, increasing the data length does not help much.

A single real sinusoid is equivalent to two complex exponentials having particular relation. For a signal containing two arbitrary complex exponentials, the results are far more complex than the above. Line splitting is related to many factors such as the data length, the phase difference between the two exponentials at the middle of the data record, and the frequency difference and the relative amplitudes of the two exponentials [4.24].

Line splitting was also observed in the case where a segment of the exact ACF of a harmonic process (three isolated spectral lines located closely) plus white noise was given. By noticing further many examples of line splitting in which the spectra were estimated directly from the time series (without the intermediate step of estimating ACF) having very high SNR, we may say that line splitting is a problem which should be imputed to MEM1 itself, but is not caused only by noise.

4.5 Resolution Enhancement and Data Extension (Theoretical Analysis)

This section is concerned with the theoretical analysis of resolution enhancement and data extension for the MEM1 and MEM2 spectral estimation. Our studies are concentrated on four respects: (1) Comparison of data extension in MEM1 and MEM2. (2) Comparison of resolution enhancement in MEM1 and MEM2. (3) Resolution of MEM1 and MEM2 for sufficiently low SNR. (4) Line splitting in MEM1. The analysis is carried out quantitatively or semi-quantitatively, or even qualitatively. This is done mainly in the case of 1-D noiseless data (ACF) [4.19, 4.25].

4.5.1 Data Extension in MEM1 and MEM2

Data extension takes place in both MEM1 and MEM2. In this regard, MEM1 is more effective than MEM2.

In MEM1 in defining the entropy $H1$ in the time domain, it is explicitly assumed that the segment of a time series is nonzero extrapolated; the extension is done in a manner of maximizing $H1$. The nonzero extension of the time series suggests a nonzero extension of the ACF. In the 1-D case a formula for the ACF extension is available.

In MEM2, on the other hand, the entropy $H2$ is defined in the frequency domain. In defining $H2$ the data extension in the time domain is not explicitly involved. In the 1-D case when the given data segment is part of a real, causal and minimum-phase time series, an explicit solution and data extension formula are available. Needless to say, a nonzero extension of a time series takes place; so the same thing happens to the ACF. If the given segment of the time series does not meet the above conditions, nonzero extension will not be so obvious.

In the case where the partial ACF is given, only an implicit solution is available in MEM2. Although it is possible to formally write out an extension formula for the ACF like (3.2.10a,b), yet λ_k cannot be solved for explicitly from the given ACF.

What we want to show now is that given the partial ACF $R(n)$, $|n| \leq M$, a (nonzero) ACF extension must take place.

A simplest way of proving the nonzero ACF extension is to utilize Property 3 of the complex cepstrum (Appendix A), which states that a sequence and its complex cepstrum both cannot be of finite duration. The Lagrange multipliers λ_n in MEM2 are the complex conjugate of the complex cepstrum of the ACF sequence $R(n)$ (given and extrapolated). Since λ_n has finite duration: $\lambda_n = 0$ for $|n| > M$, $R(n)$ must be of infinite duration. That is to say, $R(n)$ must be (nonzero) extrapolated in MEM2.

Alternatively, the proof can be done by reduction to absurdity utilizing the formula of implicit solution. Suppose that $R(n) \equiv 0$ for $|n| > M$. In formula (3.2.9)

$$R(n) = \sum_{k=-M}^{M} \left(\frac{k}{n}\right) \lambda_k^* R(n-k) \,, \quad n \neq 0 \,, \tag{4.5.1}$$

let $n = M+1, \ldots, 2M$; we obtain the following homogeneous linear equations in M unknowns, cf. (3.2.12):

$$\begin{bmatrix} 0 \\ 0 \\ \cdot \\ \cdot \\ \cdot \\ 0 \end{bmatrix} = \begin{bmatrix} \frac{1}{M+1}R(M) & \cdots & \cdots & \frac{M}{M+1}R(1) \\ & \frac{2}{M+2}R(M) & \cdots & \frac{M}{M+2}R(2) \\ & & \cdot & \cdot \\ & 0 & \cdot & \cdot \\ & & & \frac{M}{M+M}R(M) \end{bmatrix} \begin{bmatrix} \lambda_1^* \\ \lambda_2^* \\ \cdot \\ \cdot \\ \cdot \\ \lambda_M^* \end{bmatrix} \,.$$

For the solution $\lambda_1^*, \ldots, \lambda_M^*$ not to be identically zero, the determinant of the $M \times M$ coefficient matrix must be zero. Since the matrix is upper triangular, its determinant is the product of the diagonal elements, we have

$$\det[\text{coeff.}] = R^M(M) \prod_{j=1}^{M} \frac{j}{M+j} = 0 \,,$$

i.e., $R(M) = 0$. Utilizing this result and letting $n = M, \ldots, 2M-1$ in (4.5.1), we have

$$\begin{bmatrix} 0 \\ 0 \\ \cdot \\ \cdot \\ \cdot \\ 0 \end{bmatrix} = \begin{bmatrix} \frac{1}{M}R(M-1) & \cdots & \cdots & \frac{M}{M}R(0) \\ & \frac{2}{M+1}R(M-1) & \cdots & \frac{M}{M+1}R(1) \\ & & \cdot & \cdot \\ & 0 & \cdot & \cdot \\ & & & \frac{M}{2M-1}R(M-1) \end{bmatrix} \begin{bmatrix} \lambda_1^* \\ \lambda_2^* \\ \cdot \\ \cdot \\ \cdot \\ \lambda_M^* \end{bmatrix} \,.$$

Once again, for $\lambda_1^*, \ldots, \lambda_M^*$ not to be identically zero, we must have

$$\det[\text{coeff.}] = R^M(M-1) \prod_{j=1}^{M} \frac{j}{M-1+j} = 0 \,,$$

i.e., $R(M-1) = 0$. Proceeding likewise, we will come to the conclusion $R(M) = R(M-1) = \ldots = R(0) = 0$, which is contradictory to the given condition. Therefore, $R(n) \equiv 0$ for $|n| > M$ cannot be the solution of (4.5.1). That is, the ACF is, so is the time series, certainly (nonzero) extrapolated in MEM2.

The statement that MEM1 is more effective than MEM2 means that given the same data (ACF), the extension in MEM1 is decaying more slowly than in MEM2. Our theoretical analysis is restricted to the 1-D case since the extension formulae for $R(M+n)$ are available only in this case.

Now we investigate the $R(M+n)$ expressions. For MEM1, the extension formula is

$$R(M+n) = -\sum_{k=1}^{M} a_k R(M+n-k)\,, \quad n \geq 1\,, \tag{4.5.2}$$

which is (2.2.18). For MEM2, it is

$$R(M+n) = \frac{1}{M+n} \sum_{k=-M}^{M} (k\lambda_k^*) R(M+n-k)\,, \quad n \geq 1\,, \tag{4.5.3}$$

which is from (4.5.1). Note that (4.5.3) is not an extension formula in the ordinary sense since it represents a noncausal system. It is difficult to make quantitative comparison of the two formulae above. Intuitively, the coefficients a_k in (4.5.2) are shift-invariant while those in (4.5.3) are shift-variant. As n increases, the factor $1/(M+n)$ decreases rapidly so that $R(M+n)$ also decreases rapidly. This is true especially in the case of small M. The factor $1/n$ in the extension formula (3.3.12) for the time series in MEM2 plays the same role as does $1/(M+n)$ in the above. However, in MEM1 no formula corresponding to (3.3.12) is available for the purposes of comparison.

4.5.2 Resolution Enhancement of MEM1 and MEM2

MEM1 is more effective than MEM2 in resolution enhancement.

This conclusion can be understood in the following ways:

(1) From the point of view of data extension. MEM1 is more effective than MEM2 in data extension. Consequently, MEM1 is more effective than MEM2 in resolution enhancement. This consistency is reasonable and natural.

(2) From the point of view of GMEM. As mentioned in Sect. 4.1, the entropy $H1$ $(n=2)$ is "harder" than $H2$ $(n=1)$, so $H1$ is further away from the linear method and more capable of enhancing the resolution in spectral estimation. Moreover, the actual peak is fitted by a Lorentzian one in MEM1, while by a Gaussian one in MEM2. Generally speaking, a Gaussian peak is rising and falling more slowly.

(3) From the point of view of a signal model. The signal model of MEM1 is the reciprocal of a polynomial, and is an all-pole model. In contrast, the signal model of MEM2 is an exponential function (the exponent is a polynomial), and has no pole.

(4) From the point of view of a solution structure. The MEM2 solution (3.2.5) can be written in the form

$$\log S_2(f) = \sum_{k=-M}^{M} \lambda_k^* \exp(-2\pi i f k)\,. \tag{4.5.4}$$

If the ACF sequence to be processed has a rational Z-transform (ZT), then its complex cepstrum λ_k^* should decay at least as fast as $1/k$ according to

Property 4$'$ in Appendix A. We assume that in general the condition of having a rational ZT is approximately satisfied. λ_k^* hence decays fast. Thus, $\log S_2(f)$ has a finite number of Fourier components which decay fast. As a consequence, the resolution of $\log S_2(f)$ (the logarithmic spectrum of MEM2) cannot be high.

On the other hand, the MEM1 solution (2.2.14), i.e.,

$$S_1(f) = p_M \left/ \left| \sum_{k=0}^{M} a_k \exp(-2\pi i f k) \right|^2 \right. \tag{4.5.5}$$

can be written in the form

$$\log S_1(f) = \log p_M - 2 \log \left| \sum_{k=0}^{M} a_k \exp(-2\pi i f k) \right| . \tag{4.5.6}$$

Although the resolution of

$$S_a(f) = \left| \sum_{k=0}^{M} a_k \exp(-2\pi i f k) \right|$$

cannot be high because a_k, $(a_0 = 1)$, $k = 0, \ldots, M$ also decays rapidly (a minimum-phase sequence), yet the resolution of $\log S_1(f)$ may be high because of the logarithmic operation. Specifically, since

$$\mathrm{d} \log |x| = \frac{1}{|x|} \mathrm{d}|x| , \quad (x \neq 0) ,$$

small fluctuations in a low value area of $S_a(f)$ may well yield large fluctuations in a high value area of $\log S_1(f)$. This can also be seen from the reciprocal in (4.5.5). This reciprocal operation is a mixed blessing to MEM1: potential of high resolution, danger of superresolution and line splitting, and sensitivity to computational errors.

4.5.3 MEM1 and MEM2 Spectra at Low SNR

For sufficiently low SNR both MEM1 and MEM2 spectra approach the result from the Fourier method.

1. MEM1 [4.22]

Let $\varepsilon_k = R(k)/R(0)$. $|\varepsilon_k| \ll 1$ as SNR $\to -\infty$, i.e., as $R(0) \to \infty$. The given ACF matrix can be put in the form

$$R = R(0) \begin{bmatrix} 1 & \varepsilon_{-1} & \cdots & \varepsilon_{-M} \\ \varepsilon_1 & 1 & \cdots & \varepsilon_{-M+1} \\ \vdots & \vdots & \vdots & \vdots \\ \varepsilon_M & \varepsilon_{M-1} & \cdots & 1 \end{bmatrix} = R(0)(I + \varepsilon) , \tag{4.5.7}$$

where I is an identity matrix, and ε is a matrix whose elements are all very small in magnitude. Utilizing the (first-order) approximation

$$R^{-1} \approx 1/R(0) \cdot (I - \varepsilon) ,$$

the solution of the normal equations

$$R\boldsymbol{a} = p_M (1, 0, \ldots, 0)^{\mathrm{T}} ,$$

where $\boldsymbol{a} = (1, a_1, \ldots, a_M)^{\mathrm{T}}$, is

$$\boldsymbol{a} = p_M R^{-1} (1, 0, \ldots, 0)^{\mathrm{T}} \approx p_M/R(0) \cdot (1, -\varepsilon_1, \ldots, -\varepsilon_M)^{\mathrm{T}} .$$

$p_M/R(0) \approx 1$ since $a_0 = 1$, so that

$$\boldsymbol{a} \approx (1, -\varepsilon_1, \ldots, -\varepsilon_M)^{\mathrm{T}} .$$

Substituting this result in (4.5.5) yields the MEM1 spectral estimate

$$
\begin{aligned}
S_1(f) &\approx p_M \bigg/ \left| 1 - \sum_{k=1}^{M} \varepsilon_k \exp(-2\pi i f k) \right|^2 \\
&\approx R(0) \left| 1 + \sum_{k=1}^{M} \varepsilon_k \exp(-2\pi i f k) \right|^2 , \quad ((1-x)^{-1} \approx 1+x) , \\
&\approx R(0) \left[1 + 2\,\mathrm{Re}\left\{ \sum_{k=1}^{M} \varepsilon_k \exp(-2\pi i f k) \right\} \right] , \\
&\qquad\qquad\qquad\qquad\qquad\qquad (|1+x|^2 \approx 1 + 2\mathrm{Re}\{x\}) , \\
&= R(0) \left[1 + \sum_{\substack{k=-M \\ k\neq 0}}^{M} \varepsilon_k \exp(-2\pi i f k) \right] , \quad (\varepsilon_k \text{ is Hermitian}) ,
\end{aligned}
$$

i.e.,

$$S_1(f) \approx R(0) + \sum_{\substack{k=-M \\ k\neq 0}}^{M} R(k) \exp(-2\pi i f k) \tag{4.5.8}$$

$$= \sum_{k=-M}^{M} R(k) \exp(-2\pi i f k) . \tag{4.5.9}$$

From (4.5.9) we see that $S_1(f) \to S_{\mathrm{FT}}$ as $R(0) \to \infty$, where S_{FT} is the spectrum estimated by the Fourier method with a rectangular window, that is, the right-hand side of (4.5.9).

An approximate expression of $\log S_1(f)$ is derived from (4.5.8) for later use:

$$\log S_1(f) \approx \log \left\{ R(0) \left[1 + \frac{1}{R(0)} \sum_{\substack{k=-M \\ k\neq 0}}^{M} R(k) \exp(-2\pi i f k) \right] \right\}$$

$$\approx\ \log R(0) + \frac{1}{R(0)} \sum_{\substack{k=-M \\ k\neq 0}}^{M} R(k)\exp(-2\pi i f k)\,. \qquad (4.5.10)$$

2. MEM2

As $R(0) \to \infty$, from (4.5.1) or its expanded form (3.2.12) it follows (first-order approximation) that

$$[R(1),\ldots,R(M)]^{\mathrm{T}} \approx R(0)[I + O(\varepsilon)](\lambda_1^*,\ldots,\lambda_M^*)^{\mathrm{T}}\,.$$

Hence,

$$\begin{aligned}
(\lambda_1^*,\ldots,\lambda_M^*)^{\mathrm{T}} &\approx \frac{1}{R(0)}[I - O(\varepsilon)][R(1),\ldots,R(M)]^{\mathrm{T}} \\
&\approx \frac{1}{R(0)}[R(1),\ldots,R(M)]^{\mathrm{T}}\,.
\end{aligned}$$

Substituting it in (4.5.4) yields

$$\log S_2(f) \approx \lambda_0 + \frac{1}{R(0)} \sum_{\substack{k=-M \\ k\neq 0}}^{M} R(k)\exp(-2\pi i f k)\,. \qquad (4.5.11)$$

It can be shown that

$$\lambda_0 = \hat{R}(0) \to \log R(0) \quad \text{as} \quad R(0) \to \infty\,.$$

Comparing (4.5.10) with (4.5.11), we conclude

$$\begin{aligned}
\log S_1(f) &\approx \log S_2(f)\,, \\
S_1(f) &\approx S_2(f)\,.
\end{aligned}$$

Therefore, the conclusion concerning $S_1(f)$ applies to $S_2(f)$; that is,

$$S_2(f) \to S_{\mathrm{FT}} \quad \text{as} \quad R(0) \to \infty\,. \qquad (4.5.12)$$

Formulae (4.5.9, 4.5.12) can be used to explain the experimental result reported in the previous section: For sufficiently low SNR, the resolution of the spectral estimation by both MEM1 and MEM2 approaches that by the FT method (with triangular or rectangular window). As a matter of fact, this conclusion can be generalized: In problems involving a weak signal buried in very strong white noise, the difference between the spectra estimated by any linear or nonlinear method and by the simple FT method is insignificant.

4.5.4 Line Splitting of MEM1

Various explanations are available for the line splitting of MEM1. In the experiment a segment of a time series of a single real sinusoid or two complex exponentials plus white noise is given, the SNR is high, and the Burg algorithm is used. The experimental results can be explained mainly in two ways, both of which involve tedious mathematical manipulations, so they are only briefly introduced here. For details, see [4.26, 4.27].

(1) In the Burg algorithm, Levinson's recursion is upheld so as to ensure $1, a_{M,1}, \ldots, a_{M,M}$ to be a minimum-phase sequence and the modulus of the reflection coefficient $a_{M,M}$ to be less than unity, and to make the computational complexity to be of a magnitude of M^2. Each $a_{M,M}$ $(M = 1, \ldots, M_{\max})$ is determined by minimizing the prediction error power e_M with respect to $a_{M,M}$. $a_{M,M}$ obtained in such a way may have a modulus much less than one, and consequently the prediction error power may be rather large. (Note $p_M = (1 - |a_{M,M}|^2)p_{M-1}$.) This is not desirable. For example, for a noiseless harmonic process of single frequency, we should have $p_3 = p_4 = \ldots = 0$ (Sect. 4.7.2). This may be one of the causes of line splitting and peak shifting.

One way of improvement is to remove the constraint of Levinson's recursion, i.e., to minimize e_M with respect to all $a_{M,i}$, $i = 1, \ldots, M$, e.g., the Marple algorithm. As shown by experiment, this is indeed effective for eliminating line splitting and reducing peak shift.

Another way of improvement is to force the modulus of $a_{m,m}$ to be close to unity. Let

$$a_{m,m} = U \sin \theta_m , \quad 1 \leq m \leq M ,$$

where U is slightly less than 1, e.g., $U = 0.999999$ (its value depends on the computational accuracy). Beginning with $M = 2$, for every M adjust all θ_m to minimize e_M, so as to determine $a_{m,m}$ $(1 \leq m \leq M)$. The parameters $a_{m,i}$ are still calculated using Levinson's recursion, which ensures $1, a_{m,1}, \ldots, a_{m,m}$ to be a minimum-phase sequence.

Indeed, experiment showed that by this method line splitting could be eliminated and peak shift could be reduced. However, this method is too complicated to be put into practical use.

(2) Line splitting is caused by the error of $\hat{a}_{2,1}$ with respect to $a_{2,1}$, where $\hat{a}_{2,1}$ is the estimate of the true second-order AR parameter $a_{2,1}$ of a harmonic process of single frequency. This statement is based on the following observation in experiment: Given a segment of a time series of cosinusoid (or sinusoid), the peak shift caused by the error of $\hat{a}_{2,1}$ (with respect to $a_{2,1}$) estimated by the Burg algorithm is related to the data length and the initial phase of the signal in the same way as is line splitting (described in Sect. 4.4.4)

The above explanations are suggestive but not conclusive. Recalling that superresolution and line splitting also arise for given exact ACF, we may say that it is unfair if one imputes line splitting to any single factor.

(3) Improvement concerning the peak shift and line splitting with the Burg algorithm can be achieved by tapering or windowing [4.28, 4.29]. Specifically, include the optimal taper $W_{M,k}$ in the expression of the prediction error energy in (2.5.3), which now becomes

$$e_M = \sum_{k=1}^{N-M} W_{M,k}[|f_{M,k}|^2 + |b_{M,k}|^2] \, .$$

Then the formula for calculating the reflection coefficient $a_{M,M}$, (2.5.7), becomes

$$a_{M,M} = \frac{-2 \sum_{k=1}^{N-M} f_{M-1,k+1} b_{M-1,k} W_{M,k}}{\sum_{k=1}^{N-M} [|f_{M-1,k+1}|^2 + |b_{M-1,k}|^2]} \, , \qquad M = 1, 2, \dots \, .$$

The taper $W_{M,k}$ is chosen in such a way that the average (over all possible phases) variance of the estimated frequency of a real sinusoid is minimum. (All the quantities are real now and therefore the conjugation is dropped out in the above expression.)

The optimal taper $W_{M,k}$ is worked out to be

$$W_{M,k}^0 = \frac{6(k+1)(N-M-k+1)}{(N-M+1)(N-M+2)(N-M+3)} \, .$$

Levinson's recursion starts with (1) $a_{1,1}$ and $a_{2,2}$ obtained by a procedure described in (1) above ($a_{1,1} = \sin\theta_1$, $a_{2,2} = \sin\theta_2$); or (2) $a_{1,1}$ obtained by minimizing (a) the weighted prediction error energy of order 2 with respect to $a_{2,1}$ and $a_{2,2}$ (if the magnitude of $a_{1,1}$ so obtained is less than 1), or (b) that of order 1 with respect to $a_{1,1}$ ($a_{1,1}$ so obtained has magnitude always less than 1).

4.6 Peak Location and Relative Power Estimation (Experimental Results)

In this section we continue to report the experimental results concerning the properties of the MEM1, MEM2 and BT spectral estimation, including the accuracy of peak location and relative power (i.e., power ratio) estimation. The next section presents theoretical analysis.

4.6.1 Peak Location (Given ACF)

The accuracy of peak location is represented by the error defined as follows:

$$1\text{-D} : \text{error} = \sum_{k=1}^{K} |f_{ke} - f_{kt}| \, , \tag{4.6.1a}$$

$$2\text{-D} : \text{error} = \sum_{k=1}^{K} [(f_{k_1 e} - f_{k_1 t})^2 + (f_{k_2 e} - f_{k_2 t})^2]^{1/2} \, , \tag{4.6.1b}$$

where $K = 1$ or 2 is the number of complex exponential(s), f_{ke} (f_{k_1e}, f_{k_2e}) represents the estimated frequency location of the kth peak, and f_{kt} (f_{k_1t}, f_{k_2t}) the true peak (line) location. Obviously, a small value of error indicates high accuracy.

The experiment shows that for $K = 1$ (single peak), the error in the MEM1, MEM2 and BT spectral estimates are all zero in both the 1-D and 2-D cases provided that the FT size is sufficiently large. To save space, therefore, the statistics are not given here. For $K = 2$ (two peaks), only the results in the "well resolved" case are presented.

1. 1-D Case

The given data are the exact ACF of complex exponentials buried in white noise as shown in (4.4.1a), $|n| \leq m$, $M = m+1$. The explicit solution (solution of the normal equations) is used for MEM1 while the $R - \lambda$ procedure for MEM2.

The experiment was carried out for many different true peak locations and combinations of SNR and M. But only a set of results are shown in Table 4.2. The main conclusions are as follows:

(1) Errors for BT are independent of SNR but reduced with increasing M. This a well-known fact.
(2) Errors for MEM1 and MEM2 are reduced with increasing SNR and M.
(3) None of the three methods shows clear superiority. MEM1 shows marginally better results perhaps because the explicit solution was used.

Table 4.2. Peak locations and errors for 1-D MEM1, MEM2 and BT. SNR = 0dB. $M = 5$

True	MEM1		MEM2		BT	
locations	Estim.	Error	Estim.	Error	Estim.	Error
$-.4000$	$-.4100$		$-.4150$		$-.3878$	
$-.2135$	$-.2035$	0.0200	$-.1985$	0.0301	$-.2257$	0.0244
$-.1231$	$-.1263$		$-.1282$		$-.1293$	
0.1235	0.1267	0.0065	0.1285	0.0101	0.1296	0.0123
0.1100	0.1073		0.1056		0.1041	
0.3600	0.3628	0.0055	0.3664	0.0088	0.3658	0.0117
$-.1010$	$-.1010$		$-.1060$		$-.1022$	
0.3990	0.3990	0.0001	0.3940	0.0099	0.3978	0.0023

2. 2-D Case

The given data are as shown in (4.4.1b), $|n_1|$, $|n_2| \leq m$, $M = m + 1$. A set from among many experimental results are shown in Table 4.3. The data concerning MEM1 and BT are adopted from [4.17]. For MEM2 the $R - \lambda$ procedure was used.

From the table it can be seen that the conclusions drawn in the 1-D case also apply to the 2-D case. Particularly, the three methods all show some finite errors; none of them shows clear superiority.

Table 4.3. Peak locations and errors (in the parentheses) for 2-D MEM1 (SNR = 5dB), MEM2 (SNR = 0dB), and BT (SNR = 0dB). $M = 3$

True locs.	Estimated locs.		
(f_1, f_2)	MEM1	MEM2	BT
$-.4000, 0.0000$	$-.4010, 0.0040$	$-.4023, 0.0039$	$-.4043, 0.0000$
$0.0745, -.4456$	$0.0755, -.4496$	$0.0762, -.4492$	$0.0781, -.4453$
	(0.0082)	(0.0085)	(0.0079)
$0.3000, 0.4120$	$0.3050, 0.4060$	$0.3047, 0.4082$	$0.2969, 0.4004$
$-.0500, -.0500$	$-.0550, -.0440$	$-.0547, -.0469$	$-.0469, -.0391$
	(0.0156)	(0.0117)	(0.0234)
$0.2000, 0.3125$	$0.1950, 0.3135$	$0.1937, 0.3125$	$0.1836, 0.3027$
$-.1125, 0.0330$	$-.1075, 0.0320$	$-.1094, 0.0332$	$-.0957, 0.0430$
	(0.0102)	(0.0059)	(0.0386)
$-.3000, -.2000$	$-.2900, -.2040$	$-.2949, -.2012$	$-.2910, -.2090$
$0.1000, 0.4430$	$0.0900, 0.4470$	$0.0938, 0.4453$	$0.0898, 0.4512$
	(0.0215)	(0.0119)	(0.0257)

4.6.2 Peak Location (Given Time Series)

1. 1-D Case

The given data are a segment of a time series of complex exponential(s) or real sinusoid(s) buried in white noise. Only experimental results concerning MEM1 and BT are available.

The accuracy of peak location in the spectra estimated by MEM1 depends on the data length, SNR, the initial phase of the signal, the algorithm used to estimate AR parameters, and the order chosen; the situation is very complex. This was mentioned in Sect. 4.3.4. General conclusions are as follows:

(1) Unconstrained least square methods (e.g., the Marple and LUD algorithms) have a smaller peak shift in the estimated spectrum than constrained ones (e.g., the Burg algorithm).

(2) For the Burg algorithm, single peak. The peak shift decreases in an oscillating manner as the data length increases. The (quasi-)frequency of oscillation is twice the frequency of the sinusoid (Fig. 4.16). For short data segments, the peak shifts are usually large and greatly affected by the initial phase of the signal (Fig. 4.17). For a data segment of fixed length, low AR

orders lead to broad peaks; on the contrary, too high AR orders cause line splitting easily. If line splitting takes place in the extreme case, it makes no sense to talk about peak location. Finally, strong noise results in large peak shift.

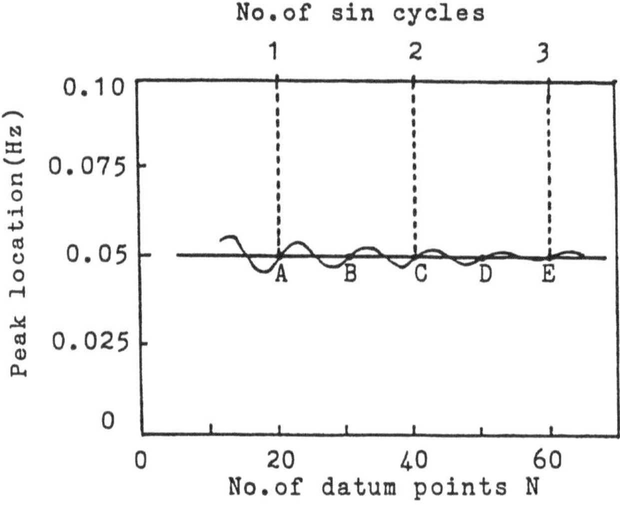

Fig. 4.16. Peak location in the MEM1 spectrum varies with N. The exact location is 0.05 Hz. SNR(rms) = 20. The AR order is 8 [4.16]. The frequency has been normalized

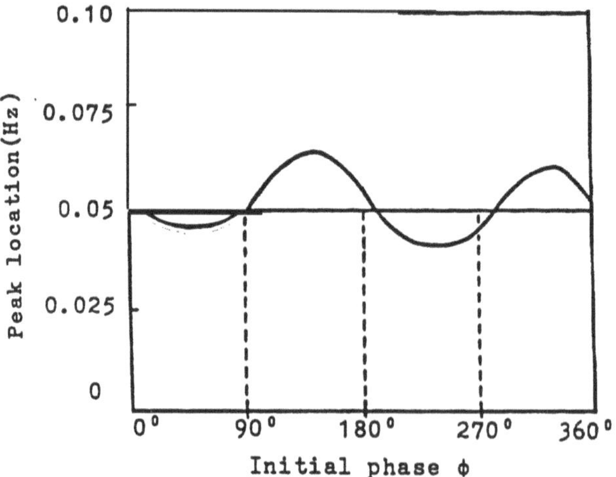

Fig. 4.17. Peak location in the MEM1 spectrum varies with ϕ. $N = 15$, and the other three parameters are the same as in Fig 4.16 [4.16]

(3) The performance of MEM1 and BT is more or less the same as far as peak shift is concerned.

In order to get a feeling of how the estimated peak location varies with the number of datum points N and the initial phase ϕ of signal, two experimental results by the Burg algorithm are presented graphically in Figs. 4.16, 4.17, respectively [4.16]. From the figures it can be seen that the peak location varies approximately at "double frequency" as N or ϕ changes.

In the experiment the time series is a real sinusoid plus white noise of unit power, i.e.,

$$x(n) = \sqrt{2}a \sin(2\pi f n + \phi) + w_n , \quad n = 0,\ldots,N-1 . \tag{4.6.2}$$

The SNR is defined by the effective (rms) value ratio as

$$\mathrm{SNR}(rms) = a . \tag{4.6.3}$$

2. 2-D Case

Only experimental results concerning MEM1 are available [4.17]. In the experiment the biased ACF estimate (triangular window) is obtained from the given segment of a time series of complex exponential plus white noise. The Lim-Malik algorithm is then used to estimate the spectrum. The experimental results show that the peak location in the MEM1 spectrum is oscillating in the vicinity of the true location; the amplitude of oscillation decreases as the number of datum points and SNR increase.

4.6.3 Relative Power Estimation (Given ACF)

It is well known that the FT method is a linear one. Given two complex exponentials or real sinusoids, their true power ratio is equal to the peak value (amplitude) ratio in the FT spectrum. This linear relationship does not hold for spectra estimated by MEM1 and MEM2 since they are nonlinear methods. The situation is complex especially for MEM2.

1. 1-D Case

(1) MEM1. Given the partial ACF

$$R(n) = a_1^2 \cos(2\pi f_1 n) + a_2^2 \cos(2\pi f_2 n) + \delta_n , \quad |n| \leq 10 ,$$

where $f_1 = 0.1$, $f_2 = 0.2$, $a_2^2/a_1^2 = 2.0$, using the explicit solution, the experimental results are shown in Table 4.4. The "area" refers to the area under the spectral curve in the interval $(0.08, 0.12)$ or $(0.18, 0.22)$ around the peak locations. The SNR has been converted to its equivalent of a complex exponential, i.e.,

$$\mathrm{SNR} = 10\log_{10}(a_1^2 + a_2^2) - 3 \quad (\mathrm{dB}) .$$

The conclusion from Table 4.4 is that in the 1-D MEM1 spectral estimates for high SNR, the ratio between the amplitudes of two peaks tends to be equal

Table 4.4. Relative power in the 1-D MEM1 spectral estimates. The true power ratio is $a_2^2/a_1^2 = 2.0$. (Adapted from [4.14])

SNR	Estimated ratio	
(dB)	Ampl.	Area
−3	2.74	1.83
0	3.30	1.87
3	3.68	1.94
6	4.02	1.98
9	4.06	1.98

to the squared true power ratio; the ratio between the areas under the peak tends to be equal to the true power ratio.

(2) MEM2. The given data are the ACF of two complex exponentials plus white noise as shown in (4.4.1a), $f_1 = 0.1$, $f_2 = 0.3$, $|n| \leq 7$. The $R - \lambda$ procedure is used. A set of experimental results are shown in Table 4.5. The true power ratio is defined by a_1^2/a_2^2. The intervals for calculating the "area" are $(0.05, 0.15)$ and $(0.25, 0.35)$.

Table 4.5. Relative power in the 1-D MEM2 spectral estimates

True ratio	1.0		2.0		4.0	
SNR	Estimated ratio					
(dB)	Ampl.	Area	Ampl.	Area	Ampl.	Area
−5	1.000	0.995	2.062	1.912	3.984	3.623
−3	1.000	0.996	2.105	1.900	4.184	3.610
0	1.001	0.999	2.165	1.916	4.505	3.636
1	1.000	0.999	2.188	1.928	4.566	3.650
3	0.999	1.000	2.197	1.934	4.808	3.831

From Table 4.5 we see that in the 1-D MEM2 spectral estimates, the ratio between the amplitudes of two peaks deviates from the squared true power ratio; the ratio between the areas under the peaks deviates from the true power ratio (except for the trivial case where the ratio is one). In other words, the conclusion drawn from the MEM1 results is no longer valid in MEM2. No certain relationship exists between the true power ratio and the estimated amplitude ratio or area ratio.

2. 2-D Case

(1) MEM1. The given data are the ACF of two complex exponentials plus white noise as shown in (4.4.1b), $f_{1_1} = f_{1_2} = 0.20$, $f_{2_1} = -0.30$, $f_{2_2} = -0.40$, $|n_1|$, $|n_2| \leq 5$. The Lim-Malik algorithm is used. A set of experimental results are shown in Table 4.6. The "volume" refers to the sum of the spectral values around a peak and above a certain threshold.

Table 4.6. Relative power in the 2-D MEM1 spectral estimates. (Adapted from [4.17])

True ratio	1.5		2.0	
SNR	Estimated ratio			
(dB)	Ampl.	Vol.	Ampl.	Vol.
−5	1.765	1.511	2.663	2.038
−3	2.297	1.513	4.168	2.072
0	5.815	1.502	4.020	2.097

From Table 4.6 it may be concluded that in the 2-D MEM1 spectral estimates for high SNR, the ratio between the volumes under two peaks tends to be equal to the true power ratio. This is similar to the conclusion in the 1-D case (area ratio). But the ratio between the amplitudes of two peaks deviates from the squared true power ratio, which is different from the conclusion in the 1-D case.

(2) MEM2. The given data are also the ACF as shown in (4.4.1b), $f_{1_1} = f_{1_2} = -0.10$, $f_{2_1} = f_{2_2} = 0.30$, $|n|, |n_2| \leq 2$. The $R - \lambda$ procedure is used. A set of experimental results are shown in Table 4.7. The integral region for calculating the "volume" under the peak is a disk centered at the true location with a radius of 0.16.

Table 4.7. Relative power in the 2-D MEM2 spectral estimates

True ratio	1.0		2.0		4.0	
SNR	Estimated ratio					
(dB)	Ampl.	Vol.	Ampl.	Vol.	Ampl.	Vol.
−5	1.001	1.000	2.646	2.252	7.246	5.952
−3	0.999	1.000	2.667	2.169	7.353	5.464
0	1.001	1.000	2.740	2.108	7.813	5.208
3			2.865	2.146	8.130	5.128
5			2.857	2.141	8.333	4.926

From Table 4.7 we can see that compared with 1-D MEM2, in the 2-D spectral estimates the amplitude ratio deviated further from the true ratio; so does the volume (under the peak) ratio; the situation here is even worse (the relationships are more uncertain). (The trivial case of one-to-one ratio is an exception.)

4.6.4 Summary and Comments

To summarize, in order to find the true power ratio between two complex exponentials or two real sinusoids, we need to calculate the square root of the peak amplitude ratio or the area (under the peak) ratio for the 1-D MEM1 spectral estimation, and the volume (under the peak) ratio for 2-D MEM1. On the other hand, for MEM2 no rules can be followed in either the 1-D or 2-D case (except that two peaks have approximately equal amplitudes). In citing these conclusions we should keep the following points in mind:

(1) The above conclusion concerning MEM1 is drawn from the case of two isolated spectral lines far apart from each other. It may not be valid in other cases. For instance, in the example of Sect. 2.4, given the sum of Gaussian functions, the peaks in the MEM1 spectral estimate are almost identical with the true ones; hence the estimated amplitude ratio is just equal to the true amplitude ratio. As another example, if two isolated spectral lines are located so closely that they cannot be well resolved in the MEM1 spectral estimate, the "area" or "volume" under the peak may not be well defined.

(2) What if the number of complex exponentials or real sinusoids is greater than two? A reasonable conjecture is that the above conclusion concerning MEM1 holds provided that the spectral lines are sufficiently far apart from each other so that they can be well resolved in the MEM1 spectral estimate.

(3) The above conclusion concerning 1-D MEM1 can be explained theoretically. In contrast, the above conclusion concerning 2-D MEM1 lacks theoretical explanation. It needs to be checked by more experiments.

(4) The above conclusion concerning MEM2 is also drawn from the case of two isolated spectral lines which are well separated. It may not be valid in other cases. For instance, in the example of Sect. 3.5, in which the true spectrum is the sum of Gaussian functions, the MEM2 and true spectra are similar, especially around the peaks.

4.7 Peak Location and Relative Power Estimation (Theoretical Analysis)

This section is devoted to the theoretical analysis of peak location and relative power estimation for the MEM1 and MEM2 spectral estimation as to explain the experimental results reported in the previous section. Most of the conclusions drawn from the experimental results are qualitative or approximate. Similarly, approximate calculation is inevitable in the theoretical analysis, even in simple cases. For these reasons, in the so-called theoretical analysis we will be necessarily conjecturing sometimes and the argument may not be unassailable. At the end of this section we will summarize the properties of the MEM1 and MEM2 spectral estimation.

Now we begin to analyze theoretically peak shifting and relative power estimation.

4.7.1 Interference Between Peaks Causes Peak Shifting

Interference between peaks is one of the causes for peak shifting. (Noise may be another cause.) In the MEM1, MEM2 and BT spectral estimates, the isolated lines in the true spectrum become gradually rising and falling peaks of finite width, and hence the interference takes place. The above argument is supported by the following experimental results: (1) Given the partial exact ACF of a complex exponential of single frequency, the peak shift is zero in all the spectra estimated by the three methods (without the interference between the peaks). (2) For a signal composed of two complex exponentials of two frequencies, peak shift is observed (with the interference). Note that a sinusoid of single frequency can be viewed as a special case of this. (3) Given a segment of a time series of a real sinusoid of single frequency plus white noise, the peak shift in the MEM1 spectral estimates (Burg algorithm) can be reduced by reducing or eliminating the interference between the positive and negative frequency components. Interference reduction is usually done by increasing the data length, raising SNR, choosing higher AR order and so on. Interference elimination is carried out by substituting the real sinusoid with its analytic signal (i.e., a complex exponential of the same frequency) to eliminate the negative frequency component, and halving the sampling rate (to keep the noise to be white). Indeed, in this way the peak shift can be greatly reduced as reported in [4.30].

4.7.2 Explanation of the Peak Shifting in MEM1 Spectra

In this subsection we use the second-order model of harmonic process to explain the peak shifting in the MEM1 spectral estimates [4.27, 4.24].

Given a segment of a time series of a single real sinusoid of frequency f plus white noise, for high SNR it can be approximated by

$$x(k) = \cos[(k-1)\omega + \phi]\,, \quad k = 1,\ldots,N\,, \tag{4.7.1}$$

($\omega = 2\pi f$). The second-order model of this harmonic process refers to a second-order AR process whose spectrum is the same as that of this harmonic process. The exact parameters of this model are

$$(1, a_{2,1}, a_{2,2}; \, p_2)^{\mathrm{T}} = (1, -2\cos\omega, 1; \, 0)^{\mathrm{T}}\,. \tag{4.7.2}$$

The exact ACF of $x(k)$ is $R(n) = \frac{1}{2}\cos n\omega$. The normal equations

$$R_2(1, a_{2,1}, a_{2,2})^{\mathrm{T}} = p_2(1, 0, 0)^{\mathrm{T}}$$

is

$$\frac{1}{2}\begin{bmatrix} 1 & \cos\omega & \cos 2\omega \\ \cos\omega & 1 & \cos\omega \\ \cos 2\omega & \cos\omega & 1 \end{bmatrix} \begin{bmatrix} 1 \\ a_{2,1} \\ a_{2,2} \end{bmatrix} = \begin{bmatrix} p_2 \\ 0 \\ 0 \end{bmatrix}.$$

Solving the last two equations

$$\begin{cases} \cos\omega + a_{2,1} + a_{2,2}\cos\omega &= 0\,, \\ \cos 2\omega + a_{2,1}\cos\omega + a_{2,2} &= 0 \end{cases}$$

yields $a_{2,2} = 1$, $a_{2,1} = -2\cos\omega$. Substituting these in the first equation yields $p_2 = 0$ Thus, the derivation of (4.7.2) has been completed.

It is easy to verify that the spectrum of this second-order AR process is indeed composed of two isolated lines at $\pm\omega$.

Why is only the second-order model under consideration? Simple calculation shows that the exact parameters of the first-order model are

$$(1, a_{1,1};\ p_1)^{\mathrm{T}} = \left(1, -\cos\omega;\ \frac{1}{2}\sin^2\omega\right)^{\mathrm{T}}.$$

Its peaks cannot be isolated lines. For AR processes of higher order than two, since $p_2 = 0$, the exact parameters obtained from the normal equations are in the form

$$(1, a_{M,1}, a_{M,2}, 0, \ldots;\ 0)^{\mathrm{T}}\,, \tag{4.7.3}$$

which in fact still represents a second-order process.

Now we estimate the second-order AR parameters $(1, a_{2,1}, a_{2,2};\ p_2)^{\mathrm{T}}$ by the Burg algorithm. But before doing this we should estimate the first-order ones $(1, a_{1,1};\ p_1)^{\mathrm{T}}$ (Sect. 2.5).

Let $M = 1$ in (2.5.7), $f_{0,k+1} = x(k+1) = \cos(k\omega + \phi)$, $b_{0,k} = x(k) = \cos[(k-1)\omega + \phi]$, and

$$a_{1,1} = \frac{-2\displaystyle\sum_{k=1}^{N-1}\cos(k\omega + \phi)\cos[(k-1)\omega + \phi]}{\displaystyle\sum_{k=1}^{N-1}\left\{\cos^2(k\omega + \phi) + \cos^2[(k-1)\omega + \phi]\right\}}\,. \tag{4.7.4}$$

Utilizing formulae for the product of triangular functions, double angle formulae, and the summation formula

$$\sum_{n=0}^{N-p}\cos(2n\omega + \alpha) = \cos[(N+1-p)\omega - \omega + \alpha]\frac{\sin(N+1-p)\omega}{\sin\omega}\,, \tag{4.7.5}$$

it follows from (4.7.4), after some mathematical manipulation, that

$$a_{1,1} = \frac{\cos\omega + \cos[(N-1)\omega + 2\phi]B_{N-1}}{1 + \cos\omega\cos[(N-1)\omega + 2\phi]B_{N-1}}\,, \tag{4.7.6}$$

where

$$B_{N-1} = \frac{\sin(N-1)\omega}{(N-1)\sin\omega}\,.$$

(In the calculation, let $n = k - 1$ in (4.7.4) and $p = 2$ in (4.7.5).)

$|B_{N-1}| = 1$ for $\omega = 0, \pm\pi$. When $N - 1 \gg 1$, even a small deviation of ω from $0, \pm\pi$ will result in a value of $|B_{N-1}|$ much less than 1. Consequently, (4.7.6) can be (first-order) approximated by

$$\begin{aligned}
a_{1,1} &\approx -\{\cos\omega + \cos[(N-1)\omega + 2\phi]B_{N-1}\} \\
&\quad \times \{1 - \cos\omega\cos[(N-1)\omega + 2\phi]B_{N-1}\} \\
&\approx -\{\cos\omega + \sin^2\omega\cos[(N-1)\omega + 2\phi]B_{N-1}\} \,.
\end{aligned} \tag{4.7.7}$$

Put $a_{1,1}$ in the form

$$\begin{aligned}
a_{1,1} &= -\cos(\omega - \delta) \\
&\approx -\cos\omega - \sin\omega\sin\delta \,,
\end{aligned} \tag{4.7.8}$$

where $|\delta| \ll 1$. Comparing the last expression with (4.7.7), we get

$$\begin{aligned}
\delta &\approx \sin\omega\cos[(N-1)\omega + 2\phi]B_{N-1} \\
&= \frac{1}{N-1}\cos[(N-1)\omega + 2\phi]\sin(N-1)\omega \,.
\end{aligned} \tag{4.7.9}$$

(For our purposes, the expression of p_1 is not needed.)

In order to estimate the second-order parameters, let $M = 2$ in (2.5.7); we have

$$a_{2,2} = \frac{-2\sum\limits_{k=1}^{N-2} f_{1,k+1}b_{1,k}}{\sum\limits_{k=1}^{N-1}[(f_{1,k+1})^2 + (b_{1,k})^2]} \,. \tag{4.7.10}$$

From (2.5.5, 2.5.6), it follows that

$$\begin{aligned}
f_{1,k+1} &= f_{0,k+2} + a_{1,1}b_{0,k+1} = \cos[(k+1)\omega + \phi] + a_{1,1}\cos(k\omega + \phi) \,, \\
b_{1,k} &= b_{0,k} + a_{1,1}f_{0,k+1} = \cos[(k-1)\omega + \phi] + a_{1,1}\cos(k\omega + \phi) \,.
\end{aligned}$$

Substituting these in (4.7.10), utilizing formulae for triangular functions and (4.7.5), yields

$$a_{2,2} = -\frac{\frac{X}{Y}\{1 + (\frac{Y}{X})\cos[(N-1)\omega + 2\phi]B_{N-2}\}}{1 + (\frac{X}{Y})\cos[(N-1)\omega + 2\phi]B_{N-2}} \,, \tag{4.7.11}$$

where

$$\begin{aligned}
B_{N-2} &= \frac{\sin(N-2)\omega}{(N-2)\sin\omega} \,, \\
X &= a_{1,1}^2 + 2a_{1,1}\cos\omega + \cos 2\omega \,, \\
Y &= a_{1,1}^2 + 2a_{1,1}\cos\omega + 1 \,.
\end{aligned}$$

Utilizing (4.7.8), we have

$$\begin{aligned}
X &\approx -(1 - \delta^2)\sin^2\omega \,, \\
Y &\approx (1 + \delta^2)\sin^2\omega \,.
\end{aligned}$$

Therefore,

$$\frac{X}{Y} \approx -\frac{(1-\delta^2)}{(1+\delta^2)} \approx -1 \approx \frac{Y}{X} \ .$$

From this result and (4.7.11), it follows that

$$a_{2,2} \approx 1 \ , \quad (|\delta| \ll 1 \ , \text{ first-order approximation}).$$

Utilizing this result and (2.5.4, 4.7.8), the other two second-order AR parameters are found to be

$$\begin{aligned} a_{2,1} &= a_{1,1} + a_{2,2}a_{1,1} \approx 2a_{1,1} \approx -2\cos(\omega - \delta) \ , \\ p_2 &= (1 - a_{2,2}^2)p_1 \approx 0 \ . \end{aligned}$$

To sum up, the estimated second-order parameters are

$$(1, a_{2,1}, a_{2,2}; \ p_2)^{\mathrm{T}} \approx (1, -2\cos(\omega - \delta), 1; 0)^{\mathrm{T}} \ , \tag{4.7.12}$$

where δ is determined in (4.7.9), $|\delta| \ll 1$. Equation (4.7.12) indicates that the spectral lines are located at $\pm(\omega - \delta)$.

By comparison of (4.7.2) and (4.7.12), we see that the peaks (being approximately isolated lines) in the estimated spectrum are shifted by

$$\Delta f = \mp \delta / 2\pi = \mp \frac{1}{2\pi(N-1)} \cos[2\pi(N-1)f + 2\phi] \sin 2\pi(N-1)f \ ,$$

$$(N - 1 \gg 1) \ , \tag{4.7.13}$$

where N is the number of datum points, i.e., data length as usually called, $(N-1)$ is the "temporal length" or duration of data segment. Normally $N-1$ can be replaced by N since $N - 1 \gg 1$.

From the above argument we see that to the extent of first-order approximation, the peak shifting for a single real sinusoid is due to the deviation of the estimated second-order AR parameter $a_{2,1}$ from its exact value. Since $a_{2,1} \approx 2a_{1,1}$, the deviation of $a_{2,1}$ stems from that of $a_{1,1}$, it may also be said that the peak shifting (and its extreme case, line splitting) is due to the deviation of the estimated reflection coefficient (i.e., partial correlation function) $a_{1,1}$. However, considering that the order of AR model for a harmonic process is at least 2, it is more appropriate to relate peak shifting (and line splitting) to the deviation of $a_{2,1}$. If a higher order approximation is taken, then the estimated $a_{2,2} \not\approx 1$, $p_2 \not\approx 0$, and hence the estimates of the higher order parameters $a_{M,3}, \ldots \ldots$ in (4.7.3) will not be zero. The estimated spectrum is no longer composed of isolated lines.

(1) Formula (4.7.13) can be used to explain the experimental results concerning peak shifting in Sect. 4.6.2 (1. 1-D case, (2) For the Burg algorithm, single peak). By comparison of (4.6.2) and (4.7.1), we see that in the experiment given there for high SNR the formula for peak shift should be

$$\Delta f = \mp \frac{1}{2\pi(N-1)} \cos[2\pi(N-1)f + 2\phi - \pi] \sin 2\pi(N-1)f \ . \tag{4.7.14}$$

From (4.7.14) it can be seen that Δf decreases in an oscillating manner as N increases. If the temporal length of data segment $(N-1)$ is a multiple of 0.5 cycle, then $\sin 2\pi(N-1)f = 0$ and $\Delta f = 0$, which corresponds to points A, B, C, D, and E in Fig. 4.16. If $(N-1)$ is an odd multiple of 0.25 cycle, then for $\phi = 0, \pi/2$, $\cos[2\pi(N-1)f + 2\phi - \pi] = 0$ and $\Delta f = 0$; for ϕ equal to an odd multiple of $\pi/4$, $|\cos[2\pi(N-1)f + 2\phi - \pi]\sin 2(N-1)f| = 1$ and $|\Delta f|$ attains its maximal value which decreases with $1/[2\pi(N-1)]$ and cannot be zero.

Let us turn our attention to the variation of Δf with ϕ. Substituting the parameters of Fig. 4.17 in (4.7.14) yields

$$\Delta f = \mp \frac{1}{28\pi} \cos(2\phi + 0.4\pi)\sin(1.4\pi) . \tag{4.7.15}$$

The zeros, (local) maximal and minimal points calculated from the above equation are more or less the same as shown in Fig. 4.17 except that the calculated maximal and minimal values are ± 0.011, which is quite different from the experimental result shown in the figure that Δf is oscillating between $+0.015$ and -0.0085. This difference is due to the fact that here $|B_{N-1}| = |\sin 1.4\pi/(14\sin 0.1\pi)| \approx 0.22$, and hence the condition $|B_{N-1}| \ll 1$ is not satisfied.

Finally, putting (4.7.13) into the form

$$\Delta f = \mp \frac{1}{4\pi(N-1)} \{\sin[2\pi(N-1)\cdot 2f + 2\phi] - \sin 2\phi\} , \tag{4.7.16}$$

it can be seen that as N varies while ϕ is fixed, or ϕ varies while N is fixed, the frequency of Δf variation is twice that of the signal (f) or of ϕ variation. This agrees more or less with what is shown in Figs. 4.16, 4.17.

(2) Now we revert to conclusions (1–3) from the experiment concerning line splitting in Sect. 4.4.4. If line splitting is viewed as an extreme case of peak shifting, then the severity of line splitting and the amount of peak shift $|\Delta f|$ should depend on N and ϕ in the same manner. The dependence of $|\Delta f|$ on N and ϕ described in (1) above indeed agrees with aforementioned conclusions (1–3) concerning line splitting. Therefore, the second-order AR model can also be used to explain line splitting.

4.7.3 Relative Power Estimation for MEM1

An explicit solution is necessary for calculating the power ratio, so the theoretical analysis can be carried out only in the 1-D case of given partial ACF. Our goal is to derive the expressions of peak value and peak width in the MEM1 spectrum of a single complex exponential plus white noise, and then to explain the experimental results concerning the relative power estimation of MEM1 [4.14].

1. For a single complex exponential plus white noise, the given partial ACF can be expressed by

$$R(n) = A\exp(\mathrm{i}\omega_1 n) + \delta_n , \quad |n| \le M , \tag{4.7.17}$$

where $\omega_1 = 2\pi f_1$, f_1 is the signal frequency, and A is the signal power.

Define vectors

$$\begin{aligned}
\boldsymbol{v} &= (1, \mathrm{e}^{\mathrm{i}\omega}, \ldots, \mathrm{e}^{\mathrm{i}\omega M})^{\mathrm{T}} , \\
\boldsymbol{v}_1 &= (1, \mathrm{e}^{\mathrm{i}\omega_1}, \ldots, \mathrm{e}^{\mathrm{i}\omega_1 M})^{\mathrm{T}} .
\end{aligned}$$

It is easy to verify that the ACF matrix is

$$R = I + A\boldsymbol{v}_1\boldsymbol{v}_1^{\mathrm{H}} , \tag{4.7.18}$$

where I is the $(M+1) \times (M+1)$ identity matrix, and H is the Hermitian operator.

The AR parameter vector obtained by solving the normal equations is, cf. (2.2.17),

$$\boldsymbol{a} = R^{-1}\boldsymbol{p} = p_M R^{-1}\boldsymbol{u} , \tag{4.7.19}$$

where

$$\begin{aligned}
\boldsymbol{a} &= (1, a_1, \ldots, a_M)^{\mathrm{T}} , \\
\boldsymbol{p} &= p_M \boldsymbol{u} , \\
\boldsymbol{u} &= (1, 0, \ldots, 0)^{\mathrm{T}} .
\end{aligned}$$

The formula for the MEM1 spectral estimate is

$$S(\omega) = \frac{p_M}{|\boldsymbol{v}^{\mathrm{H}}\boldsymbol{a}|^2} . \tag{4.7.20}$$

Utilizing the formula of matrix inversion, from (4.7.18) we have

$$R^{-1} = I - \frac{A\boldsymbol{v}_1\boldsymbol{v}_1^{\mathrm{H}}}{1 + A(M+1)} .$$

(This can be verified by noticing $\boldsymbol{v}_1^{\mathrm{H}}\boldsymbol{v}_1 = M+1$.) Substituting this result in (4.7.19) yields

$$\begin{aligned}
\boldsymbol{a} &= p_M \left[I - \frac{A\boldsymbol{v}_1\boldsymbol{v}_1^{\mathrm{H}}}{1 + A(M+1)} \right] \boldsymbol{u} \\
&= p_M \left[1 - \frac{A}{1 + A(M+1)}, \frac{-A\mathrm{e}^{\mathrm{i}\omega_1}}{1 + A(M+1)}, \cdots, \frac{-A\mathrm{e}^{\mathrm{i}\omega_1 M}}{1 + A(M+1)} \right]^{\mathrm{T}} .
\end{aligned}$$

Since $a_0 = 1$, by comparison of the first components on the two sides, we get

$$p_M^{-1} = 1 - \frac{A}{1 + A(M+1)} . \tag{4.7.21}$$

Besides, we have, by direct calculation,

$$\boldsymbol{v}^{\mathrm{H}}\boldsymbol{a} = p_M \left[1 - \frac{A(M+1)}{1 + A(M+1)} B_{M+1}(\omega - \omega_1) \right] , \tag{4.7.22}$$

where

$$B_{M+1}(\omega) = \frac{1}{M+1} \sum_{n=0}^{M} e^{-i\omega n} . \tag{4.7.23}$$

Substituting (4.7.21, 4.7.22) in (4.7.20) yields

$$S(\omega) = \left[1 - \frac{A}{1 + A(M+1)}\right] \left|1 - \frac{A(M+1)}{1 + A(M+1)} B_{M+1}(\omega - \omega_1)\right|^{-2} . \tag{4.7.24}$$

This is the expression of the MEM1 spectral estimate.

From (4.7.24) we can conclude that the peak location in the MEM1 estimate is exactly ω_1. Noticing $B_{M+1}(0) = 1$, it is easy to show the peak value of $S(\omega)$ to be

$$\begin{aligned}
S(\omega_1) &= [1 + A(M+1)](1 + AM) \\
&\approx A^2(M+1)^2 , \qquad (A(M+1) \gg 1) .
\end{aligned} \tag{4.7.25}$$

The half-power point of $S(\omega)$, $\omega_3 = 2\pi f_3$, is defined by

$$S(\omega_3) = 0.5 \, S(\omega_1) ,$$

i.e.,

$$\left|1 - \frac{A(M+1)}{1 + A(M+1)} B_{M+1}(\omega_3 - \omega_1)\right|^2 = 2 \left|1 - \frac{A(M+1)}{1 + A(M+1)}\right|^2 . \tag{4.7.26}$$

Assuming that $M + 1 \gg 1$, the peak is sharp, i.e., $|\Delta\omega| = |\omega_3 - \omega_1| \ll 1$, we have a good approximation:

$$B_{M+1}(\Delta\omega) \approx B_{M+1}(0) + B'_{M+1}(0)\Delta\omega .$$

From (4.7.23) we have

$$\begin{aligned}
B_{M+1}(0) &= 1 , \\
B'_{M+1}(0) &= -\frac{i}{M+1} \sum_{n=0}^{M} n = -i\frac{M}{2} .
\end{aligned}$$

Utilizing the three expressions above, (4.7.26) becomes approximately

$$\left|1 - \frac{A(M+1)}{1 + A(M+1)} \left(1 - i\frac{M\Delta\omega}{2}\right)\right|^2 = 2 \left|\frac{1}{1 + A(M+1)}\right|^2 .$$

By a straightforward mathematical manipulation, we obtain

$$\left|1 - i \cdot \frac{1}{2} A(M+1) M\Delta\omega\right|^2 = 1 ,$$

$$\Delta\omega = \pm\frac{2}{A(M+1)M} .$$

Hence, the full half-power peak width (i.e., 3dB bandwidth of the peak) in the spectral estimate is

$$\mathrm{BW} = 2|\Delta\omega|/2\pi \approx \frac{2}{A(M+1)^2\pi} \, , \quad (M+1 \gg 1) \, . \tag{4.7.27}$$

For the purposes of comparison, we cite, without derivation, the following formula for the BT spectral estimate [4.14]:

$$S_{\mathrm{BT}}(\omega) = \frac{1}{M+1}[1 + A(M+1)|B_{M+1}(\omega - \omega_1)|^2] \, . \tag{4.7.28}$$

So we see that the peak in the BT spectral estimate is also located exactly at ω_1. The peak value and peak width calculated from (4.7.28) are, respectively,

$$S_{\mathrm{BT}}(\omega_1) \approx \frac{1}{M+1}[1 + A(M+1)] \approx A \, , \quad (A(M+1) \gg 1) \, , \tag{4.7.29}$$

$$\mathrm{BW}_{\mathrm{BT}} \approx \frac{\sqrt{6}}{\pi(M+1)} \, , \quad (M+1 \gg 1) \, . \tag{4.7.30}$$

By comparison of (4.7.27) and (4.7.30), we conclude that $\mathrm{BW} < \mathrm{BW}_{\mathrm{BT}}$, $(A(M+1) > 2/\sqrt{6})$; that is, the MEM1 spectral estimate has higher resolution than BT.

We see from (4.7.29, 4.7.25) that the peak values of the BT and MEM1 spectral estimates are proportional to A and A^2 ($A(M+1) \gg 1$), respectively. The product of peak-value and peak-width (or amplitude and bandwidth) of the MEM1 spectral estimate is approximately

$$A^2(M+1)^2 \cdot \frac{2}{A(M+1)^2\pi} = \frac{2}{\pi}A \, , \quad ((M+1) \gg 1 \, , \ A(M+1) \gg 1) \, . \tag{4.7.31}$$

This indicates that for MEM1 the area under the peak is approximately proportional to the signal power A.

2. In the case of two complex exponentials plus white noise, it is also possible to write out the ACF matrix and then carry out calculation. But the calculation now would be much more complex than that in the previous case 1. This calculation can be avoided by the following argument in drawing conclusions: Provided that the two peaks are well separated and hence the interference between them is negligible, formulae (4.7.25, 4.7.27, 4.7.31) hold approximately for each component exponential. Consequently, the true power ratio is approximately equal to the square root of the peak-value ratio or to the area (under the peak) ratio in the MEM1 spectral estimate. The above argument and conclusion apply to the case of multiple exponentials plus white noise, as long as the peaks are well separated and hence the interference between them is sufficiently small. The above argument and conclusion may apply to the case of real sinusoid(s) plus white noise subject to an additional condition that the signal frequencies are reasonably far away from $0, \pm\frac{1}{2}$ and hence the interference between the positive and negative frequency components is insignificant. Thus, we have explained theoretically the experimental results concerning the relative power estimation for 1-D MEM1 reported in the previous section.

4.7.4 Summary for Sects. 4.4–4.7

By the experimental studies and theoretical analysis, we have obtained a general understanding of the properties of the MEM1 and MEM2 spectral estimation.

(1) Both MEM1 and MEM2 make nonzero extension of data. They have higher resolution in the estimated spectrum than does the conventional method (windowing and then Fourier transforming) which makes zero extension of data. In this respect the effect is particularly evident when the number of data is small and/or SNR is high. MEM1 is more effective than MEM2 in data extension and resolution enhancement. However, for very low SNR, either MEM1 or MEM2 shows no superiority over the conventional method.

(2) The peak location for the MEM1 and MEM2 spectral estimation is quite accurate and comparable with that for the conventional method. The peak shift is quite small for all three methods unless the given data contain very strong noise or the peaks are located closely so that the interference between them is serious.

(3) Both MEM1 and MEM2 tend to suppress a lower peak relatively to a higher one. As a result, the ratio between the high and low peak values in the estimated spectrum is greater than their true ratio. For MEM1, this effect, which is unfavorable to the low peak, can be compensated for by the broadening of the peak width in such a way that the integral power (area or volume under the peak) ratio keeps somewhat constant. In contrast, for MEM2 this compensation does not obey a rule, so that there is no way to estimate the true power ratio. This problem may be alleviated by using a good prior estimate. As a matter of fact, the quantitative estimation of relative power for MEM2 has been a challenging issue. Up to now no satisfactory solution has been found. The general advice at present is to never trust any ratio in the MEM2 estimated spectrum with respect to the estimation of the true power ratio except when very close to unity.

4.8 Comments on the Three Schools of Thought on MEM

We have investigated the three schools of thought on MEM, namely, MEM1, MEM2 and GMEM, pointing out their relationships, similarities and differences. In this section, we make brief comments on the three schools. Some main points presented previously are to be mentioned again.

1. Basic Ideas

MEM1 and MEM2 involve the concept of information-theoretic entropy. They choose, among all the feasible solutions, the one that has the maximum entropy. In the language of spectral analysis, since there exist the time domain and the frequency domain, the entropy can be defined in either domain and hence MEM1 and MEM2. Specifically, MEM1 chooses the solution having the maximum entropy from the point of view of nonzero data extension in the time domain, while MEM2 does it from the point of view of configurational entropy in the frequency domain.

GMEM bears no relation to information theory but borrows the term "entropy". GMEM views the maximization of "entropy" as a remedy to the incompleteness of condition for the solution. Conditions that the function $f(B)$ in the "entropy" expression must meet are determined from the point of view of nonlinear transformation. On the basis of these conditions, a number of possible "entropy" expressions, including $H1$ and $H2$, are derived. GMEM is a noninformation-theoretic and pragmatic school of thought.

The essential requirement to spectral estimation is the nonnegativity of solution. All the aforementioned maximum "entropy" methods can be viewed as a spectral estimation method making nonzero data extension subject to the nonnegativity constraint. Among them MEM1 has the "maximal flexibility".

The MEM1 and MEM2 solutions satisfy the consistency requirement, which is the consequence of the composition law of entropy. On the other hand, the GMEM solution does not satisfy the consistency requirement unless taking $H_G = H1$ or $H_G = H2$. This is the most important difference between GMEM and MEM1, MEM2. This is, therefore, the focus of the argument between the "fundamentalists" and the "pragmatists".

2. Entropy Expressions and Formulation

The entropy expressions for the three schools of thought are listed in Table 4.1. Among them the commonly used ones are (1-D continuous form)

$$H1 = \int \log S(f) \mathrm{d}f \,,$$

$$H2 = -\int S(f) \log S(f) \mathrm{d}f \,,$$

$$H_G = \int S^{1/2}(f) \mathrm{d}f \,, \quad \ldots\ldots \,.$$

In mathematics, all the problems to be solved are constrained maximization. In general, an explicit solution is available only for 1-D MEM1 given partial exact ACF.

3. Present Situation

Among the three schools, MEM1 and MEM2 are dominating the research while GMEM is not widely accepted. So we are concerned only with MEM1 and MEM2 in the rest of this section.

Generally speaking, MEM1 is mature in theory, but MEM2 is not. The main reason for this is that the mathematical manipulation in MEM2 is much more difficult.

In practical application MEM1 is used mainly for (1-D) spectral analysis while MEM2 mainly for (2-D) image restoration. The main reasons for this are as follows:

(1) *Historical Reason.* In 1967 MEM1 was introduced into spectral analysis by Burg; its firm theoretical and experimental foundation has been established. On the other hand, in 1972 MEM2 was introduced into image restoration by Frieden; a wide variety of algorithms have been developed ever since. This state of affairs has been continuing up to now.

(2) *Basic Ideas.* MEM1 approaches ill-posed problems from the point of view of (nonzero) data extension; its extension has "maximal flexibility". On the other hand, MEM2 considers the configuration of the whole image, which is of the greatest concern in image restoration.

(3) *Solution Properties.* The MEM1 and MEM2 solutions are same or similar in quite a number of respects. They are both positive and unique. They have higher resolution, lower sidelobes and lower level noise compared with the solution from the conventional Fourier method. Their accuracy of estimated peak location is more or less the same. They both show nonlinearity of the estimated peak amplitude. However, MEM1 is more effective that MEM2 in resolution enhancement as well as in data extension. In the MEM1 spectrum sometimes false peaks arise while in the MEM2 spectrum such "superreso-lution" has not been observed. In spectral estimation of a time series, the resolution can be altered by adjusting the filter (or AR) order in MEM1. In contrast, in image restoration the measured data are usually the ACF (visibility), and the filter order cannot be changed. MEM2, which has lower resolution than MEM1 but is safer to use, is preferable in image restoration.

(4) *Available Algorithms.* The existing algorithms of MEM1 are all for spectral analysis while those of MEM2 are mostly for image restoration. This situation may be changed as MEM is developing.

5. Applications of the Maximum Entropy Method in Mathematics and Physics

This chapter is concerned with the MEM applications in mathematics and physics. First of all, we would like to say a few words about the terminology and notation to be used.

(1) In this chapter only MEM2 is involved, that is, only the expression of the information-theoretic entropy $H2$ is used. Therefore, the designator 2 is omitted. The entropy will be denoted by S instead of H.

(2) Sometimes the entropy just mentioned in (1) is called explicitly the information-theoretic entropy and denoted by S_I so as to distinguish it from the entropies in thermodynamics and statistical mechanics.

(3) According to the convention in mathematics and physics, the natural logarithm is denoted by ln instead of log used in the previous chapters. Thus, expression of the information-theoretic entropy is

$$S_I = \sum_i p_i \ln(p_i/m_i)$$

or

$$S_I = - \int p(x) \ln[p(x)/m(x)]\mathrm{d}x \ .$$

(4) According to the convention in physics, the ensemble average or mathematical expectation or mean is denoted by $< \cdot >$ instead of $\mathrm{E}[\ \cdot \ ]$.

(5) In physics, the method is usually called the *principle of Maximum Entropy (ME)* instead of the Maximum Entropy Method (MEM).

Mathematics and physics are closely related. Usually the solution of a physical problem can be divided into three steps. (1) Establish a mathematical model. (2) Solve the mathematical problem. (3) Analyze and interpret the result. Step (2) will be emphasized in our studies. Our main goal is to show how to use MEM to solve pure mathematical problems and mathematical problems in physics. In choosing examples, both simplicity and typicalness are considered with the goal of demonstrating the procedure of applying MEM and the accuracy of results.

This chapter is organized as follows: The first three sections are devoted to the MEM applications in mathematics, namely, the solution of moment problems in Sect. 5.1, of integral equations in Sect. 5.2, and of partial differential equations in Sect. 5.3. In each section, the problem is stated and formulated, a solution is worked out, and numerical algorithm and examples are presented. The remaining four sections are concerned with the MEM applications in physics. Section 5.4 gives the general idea and formulae in Predictive Statistical Mechanics. Section 5.5 derives the distribution of particles among energy levels by MEM for both classical and quantum systems. Sections 5.6, 5.7 derive the system distributions in classical and quantum statistical ensembles, respectively. In all the cases the conventional methods and their results are cited for the purposes of comparison.

5.1 Solution of Moment Problems

Suppose that $f(x)$ is a density function defined on $[a, b]$, $f(x) \geq 0$. The moment of a function $g(x)$ with respect to this density function $f(x)$ is defined by

$$< g(x) >= \int_a^b g(x)f(x)\mathrm{d}x \ . \tag{5.1.1}$$

So in fact the moment is an expectation or average.

In this section $f(x)$ is taken as a probability density function (p.d.f.) $p(x)$, $g_n(x)$ as a power function x^n, $n = 0, 1, \ldots$. Then (5.1.1) becomes

$$\mu_n =< x^n >= \int_a^b x^n p(x)\mathrm{d}x \ . \tag{5.1.2}$$

μ_0 is the zeroth moment. The normalization of the density function is equivalent to taking $\mu_0 = 1$. μ_1 is the first moment, i.e., the mean of x. Generally μ_n is the nth moment. In this section we deal only with the case of finite a and b, i.e., the Hausdorff moment problem. The following four subsections are concerned, respectively, with the general theory, numerical methods, noisy moment problems, and examples in mathematical and the physical applications [5.1].

5.1.1 General Theory

If the density $p(x)$ is known, then all the moments μ_n, $n = 0, 1, \ldots \ldots$ can be calculated by (5.1.2). Conversely, if all μ_n, $n = 0, 1, \ldots \ldots$ are known, then the density $p(x)$ can be uniquely determined providing the Maclaurin series of the characteristic function $\phi(t) = \int_a^b \mathrm{e}^{\mathrm{i}tx} p(x)\mathrm{d}x$ of $p(x)$, $\sum_{n=0}^{\infty} \frac{1}{n!} \phi^{(n)}(0)t^n$, converges in $(-\infty, +\infty)$. This is because if all μ_n are known, then all the derivatives of the characteristic function at the origin, $\phi^{(n)}(0)$, are known: $\phi^{(n)}(0) = \mathrm{i}^n \mu_n$. Hence, $\phi(t)$ is uniquely determined. So is $p(x)$ since $p(x)$ and $\phi(t)$ are in a one-to-one relation.

A problem in which the density $p(x)$ is sought from moments of up to order N, μ_n, $n = 0, 1, \ldots, N$, is referred to as a *moment problem*. This is an ill-posed problem (the solution is not unique). Assigning different values to the unknown higher $(n > N)$ moments results in different $p(x)$'s. The number of solutions is infinite. To solve this moment problem, various methods are available such as series truncation, orthogonal polynomial expansions, and Padé approximates. Now another one, MEM, is added to the list. Virtually, the problem is solved by a nonzero extension of the higher moments. The formal statement is as follows:

$$\text{Maximization}: \quad S = -\int_a^b p(x)\ln p(x)\mathrm{d}x , \tag{5.1.3}$$

$$\text{Constraints}: \quad \mu_n = \int_a^b x^n p(x)\mathrm{d}x , \quad (\mu_0 = 1) , \quad n = 0, 1, \dots, N ; \tag{5.1.4}$$

Determine the MEM solution $p_N(x)$.

As usual, this constrained maximization problem is solved by the Lagrange multiplier method. Specifically, form the objective functional

$$Q = -\int_a^b p(x)\ln p(x)\mathrm{d}x - \sum_{n=0}^N \lambda_n \left[\int_a^b x^n p(x)\mathrm{d}x - \mu_n \right] ,$$

where λ_n, $n = 0, 1, \dots, N$ are the Lagrange multipliers. Let the variation of Q with respect to $p(x)$ be zero, i.e.,

$$\begin{aligned}
0 &= \delta Q \\
&= -\int_a^b [1 + \ln p(x)]\delta p(x)\mathrm{d}x - \sum_{n=0}^N \lambda_n \int_a^b x^n \delta p(x)\mathrm{d}x \\
&= -\int_a^b \left[1 + \ln p(x) + \sum_{n=0}^N \lambda_n x^n \right] \delta p(x)\mathrm{d}x .
\end{aligned}$$

Since $\delta p(x)$ is arbitrary, the quantity in the square brackets must be zero; we get

$$p_N(x) = \exp\left(-\lambda_0 - \sum_{n=1}^N \lambda_n x^n \right) . \tag{5.1.5}$$

Note that in this expression $\lambda_0 + 1$ has been replaced by λ_0. The Lagrange multipliers λ_0, λ_1, ..., λ_N are determined by the constraints

$$\mu_n = \int_a^b x^n p_N(x)\mathrm{d}x , \quad (\mu_0 = 1) , \quad n = 0, 1, \dots, N . \tag{5.1.6}$$

Substituting (5.1.5) in (5.1.6), letting $n = 0$, yields

$$\int_a^b \exp\left(-\lambda_0 - \sum_{n=1}^N \lambda_n x^n \right) \mathrm{d}x = \mu_0 = 1 .$$

Define the partition function by

$$Z(\lambda_1,\ldots,\lambda_N) = e^{\lambda_0} = \int_a^b \exp\left(-\sum_{n=1}^N \lambda_n x^n\right) dx \ . \tag{5.1.7}$$

Then, the MEM density can be put in the form

$$p_N(x) = \frac{1}{Z} \exp\left(-\sum_{n=1}^N \lambda_n x^n\right) \ . \tag{5.1.8}$$

The nth moment with respect to $p_N(x)$ is

$$
\begin{aligned}
< x^n >_N &= \int_a^b x^n p_N(x) dx \\[2mm]
&= \int_a^b x^n \exp\left(-\lambda_0 - \sum_{n=1}^N \lambda_n x^n\right) dx \\[2mm]
&= \frac{\displaystyle\int_a^b x^n \exp\left(-\sum_{n=1}^N \lambda_n x^n\right) dx}{\displaystyle\int_a^b \exp\left(-\sum_{n=1}^N \lambda_n x^n\right) dx} \\[2mm]
&= \frac{1}{Z} \int_a^b x^n \exp\left(-\sum_{n=1}^N \lambda_n x^n\right) dx \ , \quad n=0,1,\ldots \ . \tag{5.1.9}
\end{aligned}
$$

The constraints (5.1.6) can then be written as

$$< x^n >_N = \mu_n \ , \quad n=1,\ldots,N \ . \tag{5.1.10}$$

It should be pointed out that e^{λ_0} can be moved out of the integration and the partition function Z can be defined because $g_0(x)$ corresponding to λ_0 is a constant, otherwise the above manipulation cannot be done and λ_0 must be treated the same as other λ_n.

We have found the MEM solution (5.1.5) or (5.1.8) of the density function and (5.1.10) for determining the Lagrange multipliers $\lambda_1,\ldots,\lambda_N$. Now the following two issues should be discussed: (1) The existence and uniqueness of solution. The value of μ_n cannot be arbitrary since it is the nth moment. Equations (5.1.10) may have no solution or multiple solutions. (2) The convergence of solution, i.e., the behavior of $p_N(x)$ as $N \to \infty$. We present two relevant theorems without proof [5.1, 5.2].

Theorem 1. *A necessary and sufficient condition for the existence and uniqueness of the MEM solution is that the given moments μ_n, $n = 0, 1, \ldots, N$ should be a part of a completely monotonic sequence. For $[a,b] = [0,1]$, a completely monotonic sequence μ_n, $n = 0, 1, \ldots\ldots$ satisfies*

$$\Delta^k \mu_n \stackrel{\triangle}{=} \sum_{m=0}^{k} (-1)^m \begin{pmatrix} k \\ m \end{pmatrix} \mu_{n+m} > 0 , \quad n,k = 0,1,\dots .$$

(For $[a,b] \neq [0,1]$, the expression is more complex.)

Under the above condition, the set of Lagrange multipliers in the MEM solution $p_N(x)$ is a unique (global) minimum point of the potential function

$$\Gamma(\lambda_1,\dots,\lambda_N) \stackrel{\triangle}{=} \ln Z + \sum_{n=1}^{N} \mu_n \lambda_n , \tag{5.1.11}$$

and hence satisfies

$$\partial \Gamma / \partial \lambda_n = 0 , \quad n = 1,\dots,N . \tag{5.1.12}$$

(It is easy to show the equivalence between (5.1.12) and (5.1.10).)

As far as convergence is concerned, the pointwise convergence, i.e.,

$$\lim_{N \to \infty} p_N(x) = p(x) , \quad x \in [a,b]$$

has not been proved yet. What has been proved is the convergence of the average with respect to $p_N(x)$ (sometimes called MEM average for short).

Theorem 2. *The average with respect to the MEM solution $p_N(x)$ converges to the true average, i.e., if $F(x)$ is a continuous function on $[a,b]$, then the sequence of the moments of $F(x)$ with respect to $p_N(x)$ converges to the moment with respect to $p(x)$:*

$$\lim_{N \to \infty} < F(x) >_N = < F(x) > ,$$

i.e.,

$$\lim_{N \to \infty} \int_a^b F(x) p_N(x) \mathrm{d}x = \int_a^b F(x) p(x) \mathrm{d}x . \tag{5.1.13}$$

Although the convergence of the average with respect to $p_N(x)$ is weaker than the pointwise convergence, it suffices in usual physical applications.

5.1.2 Numerical Methods

From Theorem 1 we see that the key to determining the MEM solution $p_N(x)$ is to find the minimum point of the potential function $\Gamma(\lambda_1,\dots,\lambda_N)$. Numerical methods are necessary for $N > 1$. In principle, various optimization methods (finding the minimum point of an objective functional) can be used. Among them the Newton method is a successful one.

In the Newton method, starting with initial values $\lambda_n^{(0)}$, $n = 1,\dots,N$ (zero or small positive numbers), the formula for iteration is

$$\lambda_n^{(k+1)} = \lambda_n^{(k)} - a_n , \quad n = 1,\dots,N ,$$

where the changes a_m are the solution of the simultaneous linear equations

$$\sum_{m=1}^{N} H_{nm} a_m = \mu_n - < x^n >_N , \quad n = 1, \ldots, N . \tag{5.1.14}$$

The coefficient matrix (H_{nm}) is the Hessian of Γ, whose elements are

$$H_{nm} = \partial^2 \Gamma / \partial \lambda_n \partial \lambda_m .$$

H_{nm} can be expressed in terms of the moments in the iteration. From the definitions of Γ and Z, (5.1.11, 5.1.7), respectively, and the rule of differentiation of a product, it is easy to show that

$$H_{nm} = < x^{n+m} >_N - < x^n >_N < x^m >_N , \quad n, m = 1, \ldots, N . \tag{5.1.15}$$

The moments $< x^n >_N$ are calculated from the kth iterates $\lambda_n^{(k)}$ by (5.1.9). The iteration proceeds until the (relative or absolute) error of each calculated moment $< x^n >_N$ with respect to the given μ_n or the (absolute or percentage) difference between $\lambda_n^{(k+1)}$ and $\lambda_n^{(k)}$ for each n is less than a prescribed small positive number ε.

In each iteration, it is needed to calculate $2N$ moments by numerical integration to form a positive definite (symmetrical) matrix (H_{nm}), and then to solve a set of N linear equations. This requires much computational time due mainly to the numerical integration. However, the Newton method is efficient. (Some improvements are possible.) Experience has shown that five to six (depending on the prescribed ε) iterations is typically sufficient for convergence. There is no difficulty in dealing with the cases of up to $N = 10 \sim 12$. For numerical integration and the solution of linear equations, suitable algorithms are easily found. For example, relevant FORTRAN subroutines can be simply called from the NAG or IMSL library.

In a numerical method slightly different from the preceding one, all λ_n, $n = 0, 1, \ldots, N$ are treated equally. For details, see the next section, where $g_n(x)$ is usually not the power function of x, x^n.

5.1.3 Noisy Moment Problems

Before proceeding to present numerical examples, we digress to the noisy moment problems in this subsection [5.3].

Suppose that the moment data, except μ_0, are contaminated by noise in measurement. Then we use the χ^2-statistic, instead of (5.1.4), for the constraint. Specifically, the measured moments (data)

$$\mu_n = \int_a^b x^n p(x) \mathrm{d}x + \varepsilon_n , \quad n = 1, \ldots, N ,$$

where ε_n represents zero-mean white Gaussian noise with standard deviation σ_n. The problem is formulated as follows:

$$\text{Maximization}: \qquad S = - \int_a^b p(x)\ln p(x)\mathrm{d}x \qquad (\,(5.1.3)\,)\,,$$

$$\text{Constraints}: \qquad 1 = \mu_0 = \int_a^b p(x)\mathrm{d}x\,, \tag{5.1.16}$$

$$\chi^2 = \sum_{n=1}^N \frac{1}{\sigma_n^2}\left[\int_a^b x^n p(x)\mathrm{d}x - \mu_n\right]^2 = N\,; \tag{5.1.17}$$

Determine the MEM solution $p_N(x)$.

Form the objective functional

$$\begin{aligned} Q \;=\; & -\int_a^b p(x)\ln p(x)\mathrm{d}x - \lambda_0\left[\int_a^b p(x)\mathrm{d}x - 1\right]\\ & -\lambda_1\left\{\sum_{n=1}^N \frac{1}{\sigma_n^2}\left[\int_a^b x^n p(x)\mathrm{d}x - \mu_n\right]^2 - N\right\}\,, \end{aligned}$$

where λ_0 and λ_1 are the Lagrange multipliers. By the variational method, from $\delta Q = 0$, it is not difficult to find the solution to be

$$p_N(x) = \exp\left(\sum_{n=0}^N a_n x^n\right)\,, \tag{5.1.18}$$

where the coefficients of the polynomial of degree N in x satisfy the equations

$$\begin{aligned} a_0 \;&=\; -\lambda_0 - 1\,,\\ a_n \;&=\; -\frac{2\lambda_1}{\sigma_n^2}\left[-\mu + \mathrm{e}^{a_0}\int_a^b y^n \exp\left(\sum_{m=1}^N a_m y^m\right)\mathrm{d}y\right]\,,\\ & \hspace{6cm} n = 1,\dots,N\,. \end{aligned} \tag{5.1.19}$$

Substituting (5.1.18) in (5.1.16) yields

$$\mathrm{e}^{a_0} = \left[\int_a^b \exp\left(\sum_{m=1}^N a_m y^m\right)\mathrm{d}y\right]^{-1}\,. \tag{5.1.20}$$

Thus, we can eliminate e^{a_0} in (5.1.19), and the N equations therein become

$$\begin{aligned} a_n \;=\; -\frac{2\lambda_1}{\sigma_n^2}\Bigg\{ & -\mu + \int_a^b y^n \exp\left(\sum_{m=1}^N a_m y^m\right)\mathrm{d}y\\ & \times \left[\int_a^b \exp\left(\sum_{m=1}^N a_m y^m\right)\mathrm{d}y\right]^{-1}\Bigg\}\,,\\ & \hspace{5cm} n = 1,\dots,N\,. \end{aligned} \tag{5.1.21}$$

The constraint in (5.1.17) gives one more equation:

$$\sum_{n=1}^{N} \frac{1}{\sigma_n^2} \left\{ \int_a^b y^n \exp\left(\sum_{m=1}^{N} a_m y^m\right) \mathrm{d}y \right.$$

$$\left. \times \left[\int_a^b \exp\left(\sum_{m=1}^{N} a_m y^m\right) \mathrm{d}y\right]^{-1} - \mu_n \right\}^2 = N \ . \qquad (5.1.22)$$

There are $N+1$ unknowns $a_1, \ldots, a_N, \lambda_1$ in $N+1$ equations (5.1.21, 5.1.22). They can be solved iteratively. Then the MEM solution $p_N(x)$ can be evaluated using (5.1.18, 5.1.20).

A practical procedure to solve the above $N+1$ equations is as follows: Solve (5.1.21) iteratively for a fixed λ_1. Then evaluate the χ^2-statistic on the left-hand side of (5.1.22) and compare it with N. Adjust the value of λ_1 and repeat the iteration until the constraint (5.1.22) is satisfied.

Since the data are noisy, it is not guaranteed that a solution to (5.1.21, 5.1.22) can be found. Nevertheless, provided that the data are precise enough and σ_n are appropriately estimated, the monotonicity conditions may be satisfied, and therefore the solution $p_N(x)$ exists and is unique.

In the above we treated the noise moment problems in the usual way, that is, using the global χ^2 constraint in place of the exact data fit. Another more sophisticated way is to solve these problems by the Bayesian method [5.4].

5.1.4 Numerical Examples

1. Elementary Example

Consider the density function on $[0, 1]$

$$p(x) = \alpha + 2(1 - \alpha)x \ . \qquad (5.1.23)$$

$p(x) \geq 0$ for $0 \leq \alpha \leq 1$. The moments are

$$\mu_n = \int_0^1 x^n p(x)\mathrm{d}x = \frac{\alpha}{n+1} + \frac{2(1-\alpha)}{n+2} \ , \quad (\mu_0 = 1), \quad n = 0, 1, \ldots \ .$$

$$(5.1.24)$$

Note that here the true density $p(x)$ is a polynomial of degree 1 in x while the MEM (approximate) solution $p_N(x)$ is an exponential function whose exponent is a polynomial in x. Now we investigate the accuracy of $p_N(x)$.

(1) *Pointwise Accuracy of $p_N(x)$.* Shown in Table 5.1 are the true densities $p(x) = x+1/2$ $(\alpha = 1/2)$ and $p(x) = 2x$ $(\alpha = 0)$, and the MEM solution $p_8(x)$ for various x on $[0, 1]$. Also shown in the table are the absolute errors $\Delta(x) = p_8(x) - p(x)$ and the root-mean-square error $rms = [1/11 \sum_{i=1}^{11} \Delta^2(x_i)]^{1/2}$. $p(x)$ is calculated by (5.1.23). $p_8(x)$ is the MEM solution with $\mu_0, \ldots, \mu_8$ calculated by (5.1.24) as given data.

The pointwise convergence of $p_N(x)$ has not been proved in theory, so it is not ensured theoretically that $p_N(x)$ is a good pointwise approximation

Table 5.1. Comparison of the MEM solution $p_8(x)$ and the true density $p(x)$ in Example 1. ($p_8(x)$ and $p(x)$ from [5.1])

x	$\alpha = 1/2,\ p(x) = x + 1/2$			$\alpha = 0,\ p(x) = 2x$		
	$p_8(x)$	$p(x)$	$\Delta(x),\times 10^{-7}$	$p_8(x)$	$p(x)$	$\Delta(x),\times 10^{-4}$
0.0	0.5000035	0.5	35	0.0172	0.0	172
0.1	0.6000006	0.6	6	0.2032	0.2	32
0.2	0.6999999	0.7	-1	0.3998	0.4	-2
0.3	0.7999993	0.8	-7	0.5946	0.6	-54
0.4	0.9000009	0.9	9	0.8062	0.8	62
0.5	1.0000000	1.0	0	1.0005	1.0	5
0.6	1.0999991	1.1	-9	1.1934	1.2	-66
0.7	1.2000007	1.2	7	1.4047	1.4	47
0.8	1.3000002	1.3	2	1.6015	1.6	15
0.9	1.3999993	1.4	-7	1.7949	1.8	51
1.0	1.4999963	1.5	-37	1.9736	2.0	-264
rms	16×10^{-7}			32×10^{-4}		

to $p(x)$. Nevertheless, from Table 5.1 it can be seen that $p_8(x)$ is a fairly accurate approximation to $p(x)$ especially when $\alpha = 1/2$, $p(x)$ has no zero.

(2) *Accuracy of Average.* Let

$$F(x) = \frac{1}{2}\sqrt{x}\,, \tag{5.1.25}$$

and calculate $< F(x) >_N$ and $< F(x) >$ according to the definition (5.1.13). Shown in Table 5.2 are $< F(x) >_N$ for $N = 0, 1, \ldots, 10$, $< F(x) >$, and the errors $\Delta_N =< F(x) >_N - < F(x) >$ when $\alpha = 0$, $p(x) = 2x$.

Table 5.2. The MEM average $< F(x) >_N$, true average $< F(x) >$, and the errors. $\alpha = 0$, $p(x) = 2x$. $F(x) = \frac{1}{2}\sqrt{x}$ in Example 1. ($< f(x) >_N$ and $< F(x) >$ from [5.1])

N	$< F(x) >_N$	$\Delta_N,\times 10^{-7}$
0	0.3333333	-666667
1	0.3969133	-30867
2	0.3995009	-4991
3	0.3998648	-1352
4	0.3999518	-482
5	0.3999794	-206
6	0.3999900	-100
7	0.3999947	-53
8	0.3999970	-30
9	0.3999982	-18
10	0.3999988	-12
$< F(x) >$	0.4000000	

From Table 5.2 we see that the MEM average $< F(x) >_N$ approaches the true average $< F(x) >$ rapidly as N increases. The estimated $< F(x) >_N$ is quite accurate even when a relatively small number of moments are given. The situation here is much better than that for pointwise accuracy. These are the general characteristics of the MEM solution. Since the pointwise accuracy when $\alpha = \frac{1}{2}$, $p(x) = x + \frac{1}{2}$ is higher than that when $\alpha = 0$, $p(x) = 2x$, we can predict that in the former case $< F(x)_N >$ approaches $< F(x) >$ even more rapidly and the accuracy of the average is even higher.

2. Density of Angular Frequency of Face-Centered Cubic Crystal Lattice

(1) *The Problem.* The total vibrational energy of a cubic crystal lattice consisting of N interactive atoms can be expressed in terms of the sum of energies of $3N$ independent one-dimensional harmonic oscillators. In the thermodynamic limit ($N \to \infty$ and the number density of atoms remains constant), the internal energy U and specific heat C are given by

$$\frac{U}{3N\hbar\omega_M} = \tau \int_0^1 \left[\left(\frac{\omega}{2\tau}\right) \coth\left(\frac{\omega}{2\tau}\right) \right] g(\omega)\mathrm{d}\omega , \qquad (5.1.26)$$

$$\frac{C}{3Nk} = \int_0^1 \left[\frac{(\omega/2\tau)}{\sinh(\omega/2\tau)} \right]^2 g(\omega)\mathrm{d}\omega , \qquad (5.1.27)$$

where $\hbar = h/2\pi$ and h is the Plank constant, k the Boltzmann constant, ω_M the maximum angular frequency of vibration, ω the normalized angular frequency (actual angular frequency times $1/\omega_M$), τ the normalized temperature (actual temperature times $k/(\hbar\omega_M)$), and $g(\omega)$ the normalized angular frequency density (actual density times ω_M).

From (5.1.26, 5.1.27) we see that the key to the calculation of the internal energy and specific heat (including other thermodynamic quantities in fact) is the density $g(\omega)$. In the Einstein model, it is assumed that all the atoms of the crystal vibrate with the same angular frequency. On the other hand, in the Debye model, the crystal is considered as a continuous elastic body. In these two models no specific manner in which the atoms interact is involved in determining $g(\omega)$. Only in very few simple cases is it possible to calculate $g(\omega)$ for a three-dimensional crystal lattice according to the interactions between atoms. Usually, approximate methods are needed. For a face-centered cubic (fcc) crystal lattice, if only the nearest-neighbor interactions between atoms are considered, then the first 34 even moments (excluding $\mu_0 = 1$) can be obtained by numerical method. Hence, various methods for solving moment problems can be used to estimate the density $g(\omega)$.

The even moments

$$\mu_n = \int_0^1 \omega^{2n} g(\omega)\mathrm{d}\omega \qquad (5.1.28)$$

may be converted to consecutive ones by variable substitution. Specifically, let

$$x = \omega^2, \quad g(\omega)\mathrm{d}\omega = p(x)\mathrm{d}x \ ,$$

then

$$\mathrm{d}x = 2\omega\mathrm{d}\omega \ , \quad g(\omega) = 2\omega p(x) \ ,$$

and (5.1.28) becomes

$$\mu_n = \int_0^1 x^n p(x)\mathrm{d}x \ . \tag{5.1.29}$$

Normalizing the internal energy and specific heat by $3N\hbar\omega_M$ and $3Nk$, respectively, (5.1.26, 5.1.27) become

$$U \;=\; \tau \int_0^1 \left[\left(\frac{\sqrt{x}}{2\tau}\right)\coth\left(\frac{\sqrt{x}}{2\tau}\right)\right] p(x)\mathrm{d}x \ , \tag{5.1.30}$$

$$C \;=\; \int_0^1 \left[\frac{(\sqrt{x}/2\tau)}{\sinh(\sqrt{x}/2\tau)}\right]^2 p(x)\mathrm{d}x \ . \tag{5.1.31}$$

The internal energy at zero temperature is

$$U_0 = \lim_{\tau\to 0} U = \frac{1}{2}\int_0^1 \sqrt{x}\,p(x)\mathrm{d}x \ . \tag{5.1.32}$$

The three expressions above have the form of $<F(x)>$, where $F(x)$ is continuous on $[0,1]$ (defining $F(0) = F(+0)$ if necessary).

(2) *MEM Solution.* The first 11 moments (including $\mu_0 = 1$) used to determine the MEM solution $p_{10}(x)$ are listed in Table 5.3. Listed in the table are also the values of the Lagrange multipliers $\lambda_0,\ldots,\lambda_{10}$ by numerical calculation.

Table 5.3. The moments μ_n for an fcc crystal lattice and the Lagrange multipliers λ_n in the MEM solution for Example 2. (μ_n from [5.5] and λ_n from [5.1])

n	μ_n	λ_n
0	1.00000000000000000000000000000 D+00	2.6887514 D+00
1	5.00000000000000000000000000000 D−01	−6.4706499 D+01
2	3.12500000000000000000000000000 D−01	9.9456353 D+02
3	2.22656250000000000000000000000 D−01	−8.6422019 D+03
4	1.71630859375000000000000000000 D−01	4.2355688 D+04
5	1.38824462890625000000000000000 D−01	−1.2478595 D+05
6	1.15787506103515625000000000000 D−01	2.2876071 D+05
7	9.85897779464721679687500000000 D−02	−2.6063951 D+05
8	8.51962082087993621826171875000 D−02	1.7726248 D+05
9	7.44508523494005203247070312500 D−02	−6.4746924 D+04
10	6.56434122356586158275604248050 D−02	9.5046571 D+03

(i) *Pointwise accuracy of $p_{10}(x)$.* The curve of the MEM density calculated by

$$p_N(x) = \frac{1}{Z} \exp\left(-\sum_{n=1}^{N} \lambda_n x^n\right) , \quad N = 10$$

is plotted in Fig. 5.1. Also plotted in the figure is the curve of the Debye density, whose equation is

$$p_{DB}(x) = 3x^2 .$$

The pointwise accuracy of $p_{10}(x)$ cannot be determined since no exact solution $p(x)$ is available. Nevertheless, the double-peak shape of the $p_{10}(x)$ curve is reasonable while the parabolic shape of $p_{DB}(x)$ is far from the truth.

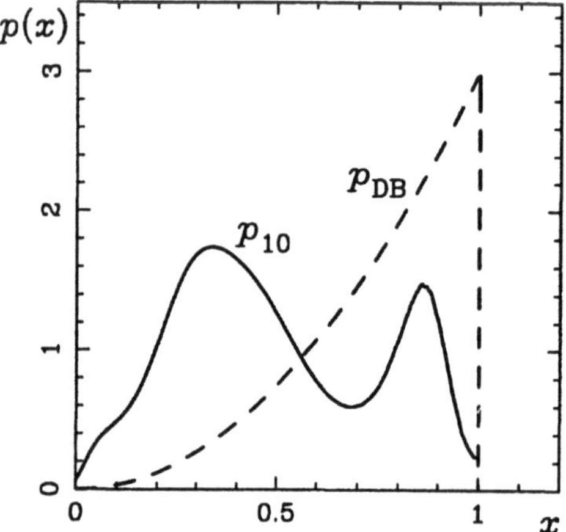

Fig. 5.1. The MEM density $p_{10}(x)$ and the Debye density $p_{DB}(x)$ for Example 2

(ii) *Accuracy of average.* The MEM solutions of the specific heat C and the internal energy at zero temperature U_0 can be calculated by substituting $p_N(x)$ for $p(x)$ in (5.1.31, 5.1.32), respectively. The results, along with the upper and lower bounds of the exact U_0 determined using the first 30 moments (including $\mu_0 = 1$), are listed in Table 5.4. It can be seen from the table that the MEM solution falls within the bounds for $N \geq 9$. Thus, we see, once again, that the accuracy of the MEM average $< F(x) >_N$ is fairly high even when a relatively small number of moments are given. As far as the specific heat is concerned, we see desirably $C \to 0$ as $\tau \to 0$, but it is not clear whether $C \propto \tau^3$.

Table 5.4. The internal energy at zero temperature U_0 and the specific heat C. (Adapted from [5.1])

N	U_0	$C\ (\tau = 0.05)$		$N = 10$	
				τ	C
0	0.3333333				
1	0.3333333			1.0	0.9595995696
2	0.3391339			0.5	0.8520405755
3	0.3406157	0.01189947		0.1	0.09866593
4	0.3410812	0.00768781		0.05	0.01001197
5	0.3407759	0.01175413			
6	0.3408676	0.01028976			
7	0.3408796	0.01008638			
8	0.3408653	0.01031767			
9	0.3408845	0.01003605			
10	0.3408864	0.01001197			

U_0 bounds		
upper	0.3408883	
lower	0.3408807	

5.2 Solution of Integral Equations

The solution of an integral equation by MEM consists of two steps: the conversion of the integral equation to a moment problem, and the solution of the resultant moment problem. In this section these two steps are described first. Then numerical examples are demonstrated. Finally, some issues are discussed [5.6].

5.2.1 Conversion of Integral Equations to Moment Problems

Consider the following Fredholm (Fr) integral equation of the second kind:

$$f(x) = \Phi(x) - \int_a^b K(x,y)f(y)\mathrm{d}y , \tag{5.2.1}$$

where $f(x)$ is the unknown function to be solved for, the free term $\Phi(x)$ and the kernel $K(x,y)$ are given functions, and $[a,b]$ is a finite interval. The corresponding Fr integral equation of the first kind is

$$\Phi(x) = \int_a^b K(x,y)f(y)\mathrm{d}y . \tag{5.2.2}$$

Suppose that $M_n(x)$, $n = 0,1,\ldots$ are a set of functions linearly independent on $[a,b]$. Multiply the two sides of (5.2.1) by $M_n(x)$ and then integrate them with respect to x over $[a,b]$, it follows that

$$\int_a^b M_n(x)f(x)\mathrm{d}x$$

$$= \int_a^b M_n(x)\Phi(x)\mathrm{d}x - \int_a^b M_n(x)\left[\int_a^b K(x,y)f(y)\mathrm{d}y\right]\mathrm{d}x .$$

Assuming that $M_n(x)$ and $K(x,y)$ are sufficiently well-behaved (e.g., continuous), interchanging the order of two integrations in the last term yields, after moving terms,

$$\int_a^b M_n(x)\Phi(x)\mathrm{d}x$$
$$= \int_a^b M_n(x)f(x)\mathrm{d}x + \int_a^b \left[\int_a^b M_n(x)K(x,y)\mathrm{d}x\right] f(y)\mathrm{d}y \,.$$

In the double integration substituting t for x and then x for y yields

$$\int_a^b M_n(x)\Phi(x)\mathrm{d}x = \int_a^b \left[M_n(x) + \int_a^b M_n(t)K(t,x)\mathrm{d}t\right] f(x)\mathrm{d}x \,. \tag{5.2.3}$$

Define the calculable quantity

$$\mu_n = \int_a^b M_n(x)\Phi(x)\mathrm{d}x \tag{5.2.4}$$

and the calculable function

$$G_n(x) = M_n(x) + \int_a^b M_n(t)K(t,x)\mathrm{d}t \,. \tag{5.2.5}$$

Equation (5.2.3) then becomes

$$\mu_n = \int_a^b G_n(x)f(x)\mathrm{d}x \,. \tag{5.2.6}$$

If $f(x)$ is considered as a density function, then μ_n is the moment of $G_n(x)$ with respect to $f(x)$. μ_n and $G_n(x)$ can be calculated from the known functions by (5.2.4, 5.2.5), respectively. Let $n = 0, 1, \ldots, N$, then to solve (5.2.6) for $f(x)$ is a moment problem. MEM can be used to determine its (approximate) solution.

Thus, we have formally converted the integral equation (5.2.1) to the moment problem (5.2.6), $n = 0, 1, \ldots, N$. In the same way the Fr integral equation of the first kind, (5.2.2), can be converted to a moment problem of the same form except that $G_n(x)$ is calculated by

$$G_n(x) = \int_a^b M_n(t)K(t,x)\mathrm{d}t \,. \tag{5.2.7}$$

5.2.2 Solution of Moment Problems by MEM

The problem and solution here are similar to those in Sect. 5.1. Specifically, the statement of the problem is

$$\text{Maximization}: \quad S = -\int_a^b f(x)\ln f(x)\mathrm{d}x \,,$$

$$\text{Constraints}: \quad \mu_n = \int_a^b G_n(x)f(x)\mathrm{d}x \,, \quad (\mu_0 = 1), \quad n = 0, 1, \ldots, N,$$

Determine the MEM solution $f_N(x)$;

which is different from (5.1.4) in that usually $G_n(x)$ is not x^n and μ_n is not the nth moment. By a proper choice of $M_n(x)$, $\mu_0 = 1$ is always possible, though this may not be done in practice.

The MEM solution is

$$f_N(x) = \exp\left[-\sum_{n=0}^{N} \lambda_n G_n(x)\right] , \tag{5.2.8}$$

where the Lagrange multipliers λ_n are determined by the constraints

$$\mu_n = \int_a^b G_n(x) f_N(x) \mathrm{d}x , \quad n = 0, 1, \ldots, N . \tag{5.2.9}$$

μ_n and $G_n(x)$ are calculated by (5.2.4, 5.2.5), respectively. When $G_n(x) \neq x^n$, we also use the the Newton method to solve (5.2.9). But the formulae are somewhat different from those in Sect. 5.1. Let the initial $\lambda_n^{(0)}$, $n = 0, 1, \ldots, N$ be zero or small positive numbers. The formula for iteration is

$$\lambda_n^{(k+1)} = \lambda_n^{(k)} - a_n , \quad n = 0, 1, \ldots, N .$$

The changes a_m are the solution of the following simultaneous linear equations:

$$\sum_{m=0}^{N} H_{nm} a_m = \mu_n - <G_n(x)>_N , \quad n = 0, 1, \ldots, N , \tag{5.2.10}$$

where

$$\left\{ \begin{aligned} <G_n(x)>_N &= \int_a^b G_n(x) f_N^{(k)}(x) \mathrm{d}x , \\ H_{nm} &= <G_n(x) G_m(x)>_N , \quad n, m = 0, 1, \ldots, N . \end{aligned} \right. \tag{5.2.11}$$

In calculation the symmetry of the matrix (H_{nm}) should be utilized to reduce the number of numerical integrations for calculating the moments. This is the key to reducing the computational time for each iteration. To generate the upper triangular part of (H_{nm}), $(N+1)(N+2)/2$ moments need to be calculated. The $(N+1)$ moments $<G_n(x)>_N$ on the right-hand side of (5.2.10) also need to be calculated. The $(N+1)$ numerical integrations for this calculation can be eliminated when $G_0(x)$ is a constant, in which case H_{0m} and $<G_n(x)>_N$ are different only by a constant factor.

In general, (H_{nm}) is not a positive definite matrix, and therefore a general method must be used to solve the $(N+1)$ linear equations in (5.2.10). This has little effect on the total computational time because in each iteration, (5.2.10) needs to be solved only once, while the calculation of the moments needs to be carried out many times.

The numerical procedure described above, of course, applies to the case where $G_n(x) = x^n$ and hence $H_{nm} = <x^{n+m}>_N$, $n, m = 0, 1, \ldots, N$. In each iteration, totally $(2N + 1)$ moments need to be calculated. (H_{nm}) is a positive definite matrix, and hence the equations can be solved faster.

Some points should be discussed before demonstrating numerical examples.

(1) We have converted the integral equation to the moment problem formally. It can be proved [5.7, 5.8] that for the Fr integral equation of the second kind having a parameter λ,

$$f(x) = \Phi(x) - \lambda \int_a^b K(x,y)f(y)\mathrm{d}y \, ,$$

its corresponding moment problem (with the parameter λ before the integration in (5.2.5)) possesses a unique solution $f_N(x)$, as shown in (5.2.8), for every N, and $f_N(x)$ weakly converges to the true density function $f(x)$ (i.e., the exact solution of the above integral equation) as $N \to \infty$, provided that λ (equal to 1 in our case) is not an eigenvalue of the integral equation and its exact solution is nonnegative.

In the general case, this nonnegativity condition may not be satisfied. Then $f_N(x)$ doesn't converge to $f(x)$. This is obvious because $f_N(x) \geq 0, \forall x$.

In practical applications, however, we may know beforehand that the function, say a probability density or a mass density, $f(x) \geq 0$, so that it is likely that no difficulty would be encountered in applying MEM. If we can determine the range of the function $f(x)$, then it is possible to convert the unknown function to a positive-valued one by variable substitution. In the case where nothing is known about $f(x)$ beforehand, a trial-and-error procedure is inevitable. (Cf. Example 2 in the next subsection.)

(2) Generally speaking, $M_n(x)$ should be so chosen that the calculation of μ_n and $G_n(x)$ is as easy as possible, and the resultant moment problem is relatively easy to solve. However, these requirements often cannot be met simultaneously. In general, μ_n and $G_n(x)$ can be calculated by an analytic method while the resultant problem must be solved by a numerical method. Whenever it is possible, we will choose $M_n(x) = \alpha x^n$ (α being a constant). Sometimes some tricks are also used, cf. Example 3 in the next subsection. But we do not have specific rules to follow in choosing $M_n(x)$.

(3) For the Fr integral equations of the second kind, the accuracy of solution can be remarkably improved by one iteration starting with the MEM solution $f_N(x)$. Specifically, substituting $f_N(x)$ in the right-hand side of (5.2.1) yields

$$f_N^M(x) \stackrel{\triangle}{=} \Phi(x) - \int_a^b K(x,y)f_N(y)\mathrm{d}y \, , \tag{5.2.12}$$

which is taken as the solution of the integral equation. According to Theorem 2 in Sect. 5.1, for $K(x,y)$ sufficiently well-behaved, we have

$$\lim_{N \to \infty} \int_a^b K(x,y)f_N(y)\mathrm{d}y = \int_a^b K(x,y)f(y)\mathrm{d}y \, .$$

Hence,

$$\lim_{N \to \infty} f_N^M(x) = \Phi(x) - \int_a^b K(x,y)f(y)dy$$
$$= f(x) .$$

So we see that the solution after one iteration, $f_N^M(x)$, also converges to the exact solution $f(x)$ of the integral equation. The accuracy improvement suggests that $f_N^M(x)$ converges to $f(x)$ more rapidly than $f_N(x)$ does. The iteration also makes it possible for $f_N^M(x)$ to assume zero or negative values.

5.2.3 Numerical Examples

Example 1. The Fr integral equation of the second kind

$$f(x) = -1 + \frac{3}{2}\int_0^1 e^{|x-y|}f(y)dy . \tag{5.2.13}$$

Here $\Phi(x) = -1$ and $K(x,y) = -\frac{3}{2}e^{|x-y|}$. It is easy to verify the exact solution

$$f(x) = -\frac{1}{4} + Ae^{2x} + Be^{-2x} , \quad (f(x) > 0) ,$$

$$A = \frac{3}{4(e^2 - 3)} , \quad B = e^2 A .$$

Choose

$$M_n(x) = -x^n , \quad n = 0, 1, \dots .$$

Calculation by (5.2.4, 5.2.5) yields

$$\begin{cases} \mu_n = \dfrac{1}{n+1} , & (\mu_0 = 1) , \quad n = 0, 1, \dots , \\ G_n(x) = Q_n(x) + C_n e^x + D_n e^{-x} , & n = 0, 1, \dots , \end{cases}$$

where the polynomial

$$Q_n(x) = -4x^n - 3\left\{ n(n-1)x^{n-2} + n(n-1)(n-2)(n-3)x^{n-4} + \cdots \right.$$

$$\left. + \frac{1}{2}[1 + (-1)^{n-1}]n!x + \frac{1}{2}[1 + (-1)^n]n! \right\} ,$$

and the constants are

$$C_n = \frac{3}{2}n! ,$$

$$D_n = \frac{3}{2}e[1 - n + n(n-1) - \cdots + (-1)^{n-2}n(n-1)\cdots 4 \cdot 3] .$$

The MEM solution determined from the first 10 moments ($N = 9$) is

$$f_9(x) = \exp\left[-\sum_{n=0}^{9} \lambda_n G_n(x)\right] .$$

$\lambda_n, n = 0, \ldots, 9$ are listed in Table 5.5.

Table 5.5. λ_n in Example 1. $N = 9$

n	λ_n
0	0.52615927
1	$-.21806538$ D+01
2	0.13640400 D+01
3	0.21100743 D+01
4	$-.61335158$ D+01
5	0.16044239 D+02
6	$-.19848681$ D+02
7	0.93359993 D+01
8	0.40946008
9	$-.11006575$ D+01

Table 5.6. $f_9(x)$, $f_9^M(x)$ and $f(x)$ in Example 1

x	$f_9(x)$	$f_9^M(x)$	$f(x)$
0.0	1.1830436	1.1835184	1.1835183
0.1	0.9925615	0.9924733	0.9924739
0.2	0.8512271	0.8512950	0.8512944
0.3	0.7543053	0.7543141	0.7543137
0.4	0.6976715	0.6976393	0.6976396
0.5	0.6790006	0.6789976	0.6789976
0.6	0.6976204	0.6976398	0.6976396
0.7	0.7543236	0.7543137	0.7543137
0.8	0.8512989	0.8512944	0.8512944
0.9	0.9924580	0.9924741	0.9924739
1.0	1.1834027	1.1835184	1.1835183

Listed in Table 5.6 are the MEM solution $f_9(x)$, the solution after one iteration $f_9^M(x)$, and the exact solution $f(x)$ for various x on $[0, 1]$. From the table it can be seen that the accuracy is indeed improved by one iteration: The absolute errors of $f_9(x)$ are of a magnitude of 10^{-4} or 10^{-5} while those of $f_9^M(x)$ are 10^{-7}.

Example 2. The Fr integral equation of the second kind

$$f(x) = \left(\frac{3}{2} - x\right) + 2\int_0^1 (x + 6y)f(y)\mathrm{d}y . \tag{5.2.14}$$

Here $\Phi(x) = \frac{3}{2} - x$ and $K(x,y) = -2(x + 6y)$. It is easy to verify that the exact solution is

$$f(x) = \frac{1}{2} - x \ . \tag{5.2.15}$$

$f(x)$ is negative in $(\frac{1}{2}, 1]$. So it is safe to say that the direct use of MEM will fail to yield success. To see this more clearly, let us try once by choosing

$$M_n(x) = x^n \ .$$

Calculations yield

$$\left\{ \begin{aligned} \mu_n &= \frac{n+4}{2(n+1)(n+2)} > 0 \ , \quad (\mu_0 = 1) \ , \quad n = 0, 1, \dots, \\ G_n(x) &= x^n - \frac{12}{n+1}x - \frac{2}{n+2} \ , \quad n = 0, 1, \dots \ . \end{aligned} \right.$$

The first 13 $G_n(x)$,

$$G_0(x) = -12x \ , \quad G_1(x) = -(\frac{2}{3} + 5x) \ , \quad \dots \ , \quad G_{12}(x)$$

assume negative values in $(0, 1]$. The other $G_n(x)$ assume negative values in the vicinity of 0 in $(0, 1]$. In the equations

$$\mu_n = \int_0^1 G_n(x) f_N(x) \mathrm{d}x \ , \quad n = 0, 1, \dots, 12 \ ,$$

on the left-hand side $\mu_n > 0$; but on the right-hand side $G_n(x) < 0$ and $f_N(x) > 0$, and hence the integral takes a negative value. So apparently no solution exists for the above equations.

By variable substitution, we may make the unknown function assume only nonnegative values. This is done in a simple way in this particular example. It suffices to let

$$g(x) = f(x) + \frac{1}{2} \ . \tag{5.2.16}$$

Substituting $f(x) = g(x) - \frac{1}{2}$ in (5.2.14) yields a new integral equation

$$g(x) = -(1 + 2x) + 2 \int_0^1 (x + 6y) g(y) \mathrm{d}y \ , \tag{5.2.17}$$

with $\Phi(x) = -(1 + 2x)$ and $K(x,y) = -2(x + 6y)$.

Choose

$$M_n(x) = -\frac{1}{2}x^n \ , \quad n = 0, 1, \dots \ .$$

Calculations yield

$$\left\{ \begin{aligned} \mu_n &= \frac{3n+4}{2(n+1)(n+2)} > 0 \ , \quad (\mu_0 = 1) \ , \quad n = 0, 1, \dots, \\ G_n(x) &= -\frac{1}{2}x^n + \frac{6x}{n+1} + \frac{1}{n+2} \ , \quad n = 0, 1, \dots \ . \end{aligned} \right.$$

Now $G_n(x) > 0$ for $n \le 12$, $x \in (0, 1]$. For $n > 12$, $G_n(x)$ assumes negative values in the vicinity of 1 in $(0, 1]$. Let $N = 7$. The MEM solution $g_7(x)$ is

$$g_7(x) = \exp\left[-\sum_{n=0}^{7} \lambda_n G_n(x) \right] .$$

The corresponding $f_7(x) = g_7(x) - \frac{1}{2}$.

The multipliers λ_n, $n = 0, 1, \ldots, 7$ are listed in Table 5.7. Listed in Table 5.8 are the MEM solution $g_7(x)$, the solution after one iteration $g_7^M(x)$, and the exact solution $g(x) = 1 - x$ for various x on $[0, 1]$. From the table it can be determined easily that the absolute errors of $g_7(x)$ are of a magnitude of 10^{-3} (except 10^{-4} and 10^{-2} for $x = 0.2$ and 1.0, respectively) while those of $g_7^M(x)$ are 10^{-8} (except 10^{-7} for $x = 0.9$, 1.0).

Table 5.7. λ_n in Example 2. $N = 7$

n	λ_n
0	0.140944545
1	0.139093302 D+01
2	0.204164566 D+02
3	$-$.166994280 D+03
4	0.607711388 D+03
5	$-$.112796689 D+04
6	0.102568236 D+04
7	$-$.364782390 D+03

Table 5.8. $g_7(x)$, $g_7^M(x)$ and $g(x)$ in Example 2

x	$g_7(x)$	$g_7^M(x)$	$g(x)$
0.0	1.00509941	0.99999998	1.0
0.1	0.90198115	0.89999997	0.9
0.2	0.80049333	0.79999996	0.8
0.3	0.69621766	0.69999995	0.7
0.4	0.60144192	0.59999994	0.6
0.5	0.50375941	0.49999993	0.5
0.6	0.39768384	0.39999992	0.4
0.7	0.29661849	0.29999991	0.3
0.8	0.20442739	0.19999991	0.2
0.9	0.09792993	0.09999990	0.1
1.0	0.01130707	$-$.11419072 D$-$06	0.0

Example 3. The Fr integral equation of the first kind

$$\frac{1 + e^{-\pi x}}{1 + x^2} = \int_0^\infty e^{-xy} f(y)\,dy .$$

$$(5.2.18)$$

Here $\Phi(x) = (1+\mathrm{e}^{-\pi x})/(1+x^2)$, $K(x,y) = \mathrm{e}^{-xy}$, and the integration interval is $[0, \infty)$. By the Laplace transform relationship or direct calculation, it is easy to verify that the exact solution is

$$f(x) = \begin{cases} \sin x , & 0 \le x \le \pi , \\ 0 , & \pi < x . \end{cases}$$

Instead of choosing $M_n(x)$ and then calculating μ_n and $G_n(x)$ by (5.2.4, 5.2.5), another approach is taken now. Expanding $\Phi(x)$ and $K(x,y)$ of (5.2.18) in Maclaurin series in x yields

$$
\begin{aligned}
\sum_{n=0}^{\infty} \frac{1}{n!} \Phi^{(n)}(0) x^n &= \int_0^{\infty} \left[\sum_{n=0}^{\infty} \frac{1}{n!} (-1)^n y^n x^n \right] f(y)\mathrm{d}y \\
&= \sum_{n=0}^{\infty} \frac{1}{n!} (-1)^n \cdot \int_0^{\infty} y^n f(y)\mathrm{d}y \cdot x^n \\
&= \sum_{n=0}^{\infty} \frac{1}{n!} (-1)^n \mu_n x^n ,
\end{aligned}
\tag{5.2.19}
$$

where

$$\mu_n = \int_0^{\infty} y^n f(y)\mathrm{d}y ,$$

which is compared with (5.2.6) and we see that

$$G_n(x) = x^n .$$

By comparison of the coefficients of x^n on the two sides of (5.2.19), we get

$$\mu_n = (-1)^n \Phi^{(n)}(0) .$$

For large values of n and this particular function $\Phi(n)$, the calculation of $\Phi^{(n)}(0)$ is tedious. Nevertheless, this can be done anyway. The calculated first 10 moments are

$$
\begin{aligned}
\mu_0 &= 2 , \\
\mu_1 &= \pi , \\
\mu_2 &= \pi^2 - 4 , \\
\mu_3 &= \pi^3 - 6\pi , \\
\mu_4 &= \pi^4 - 12\pi^2 + 48 , \\
\mu_5 &= \pi^5 - 20\pi^3 + 120\pi , \\
\mu_6 &= \pi^6 - 30\pi^4 + 360\pi^2 - 1440 , \\
\mu_7 &= \pi^7 - 42\pi^5 + 840\pi^3 - 5040\pi , \\
\mu_8 &= \pi^8 - 56\pi^6 + 1680\pi^4 - 20160\pi^2 + 80640 , \\
\mu_9 &= \pi^9 - 72\pi^7 + 3024\pi^5 - 60480\pi^3 + 362880\pi .
\end{aligned}
$$

Using these 10 moments $(N = 9)$, the MEM solution is

$$f_9(x) = \exp\left[-\sum_{n=0}^{9} \lambda_n G_n(x)\right].$$

λ_n, $n = 0, 1, \ldots, 9$ are listed in Table 5.9. Listed in Table 5.10 are the MEM solution $f_9(x)$ and the exact solution $f(x)$ for various x. From the table it can be seen that $f_9(x)$ is not accurate; its absolute errors are of a magnitude of 10^{-2} (except 10^{-3} for $x = 3\pi/4$). As expected, $f_9(x)$ drops to 0 rapidly when x exceeds π. Unlike for the Fr integral equation of the second kind, for the Fr integral equation of the first kind the accuracy of solution cannot be improved simply by one iteration resulting in $f_9^M(x)$.

Table 5.9. λ_n in Example 3. $N = 9$

n	λ_n
0	0.370185879 D+01
1	−.161277315 D+02
2	0.382530313 D+02
3	−.508501939 D+02
4	0.345046446 D+02
5	−.720265160 D+01
6	−.499214265 D+01
7	0.374660972 D+01
8	−.955788906
9	0.887647046 D−01

Table 5.10. $f_9(x)$ and $f(x)$ in Example 3

x	$f_9(x)$	$f(x)$
0.0	0.247 D−01	0.000
$\pi/8$	0.395	0.383
$\pi/4$	0.695	0.707
$3\pi/8$	0.937	0.924
$\pi/2$	0.987	1.000
$5\pi/8$	0.935	0.924
$3\pi/4$	0.699	0.707
$7\pi/8$	0.396	0.383
π	0.143 D−01	0.000
3.5	0.133 D−13	0.000
4.0	0.000	0.000
4.5	0.000	0.000

5.2.4 Discussion

(1) From the introduced theory and demonstrated numerical examples, we see that some questions are still open in the theory and practice of the solution of integral equations by MEM, e.g., the choice of $M_n(x)$, and the variable substitution by which the (unknown) solution becomes nonnegative. MEM is not always feasible. On the other hand, it may be the case that MEM is usable while other methods are of no use. On this basis the introduction of MEM is justified. However, what is more important is perhaps that in addition to the already existing approximate solutions, now one more, the MEM solution, is available for the purposes of comparison and reference. This is particularly valuable in the case where the exact solution cannot be determined. We admit that the three numerical examples demonstrated above seem to be a bit too simple. Nevertheless, we see that the MEM solutions are well-behaved in these cases of known exact solutions. These examples also serve the purpose of demonstrating specifically the steps in the solution of integral equations by MEM.

(2) MEM is a method having nonlinearity of high degree. This renders the quantitative error analysis of the MEM solution very difficult. The error bounds can be determined for the MEM solutions of integral equations through complex calculation [5.8].

5.3 Solution of Partial Differential Equations

Partial differential equations are often used to represent processes related to time and space. If the function to be solved for is a density, then it is positive, and its spatial distribution tends to be uniform spontaneously with time, which means that the entropy of the distribution increases with time and tends towards the maximum. Diffusion is a typical example. The physical mechanism of these kinds of processes reminds us of using MEM to solve the corresponding partial differential equations.

5.3.1 Theory

The solution of a partial differential equation consists of three steps. Steps 1 and 2 are similar to those in the solution of an integral equation. They are the conversion of the partial differential equation to a moment problem, and the solution of the resultant moment problem by MEM. The very first step, called Step 0, is to confirm that the entropy of the process represented by the partial differential equation does increase monotonically with time. Sometimes the conclusion that the entropy increases monotonically with time can be drawn based on the physical mechanism of the process. But this conclusion will be proved rigorously by mathematical means whenever it is possible.

In the following, the three steps are described first. Then a numerical example is demonstrated. Finally, a brief discussion on some issues is presented. For simplicity, we are concerned only with the (spatially) one-dimensional case [5.9].

Step 0. Confirmation of the monotonic increase of entropy with time. Let the solution sought be $f(x,t)$. The *time-varying entropy* is defined by

$$S(t) = - \int_a^b f(x,t)\ln f(x,t)\mathrm{d}x \ . \tag{5.3.1}$$

What is to be proved is

$$\frac{\mathrm{d}S(t)}{\mathrm{d}t} > 0 \ . \tag{5.3.2}$$

Usually, the proof is not straightforward. Since $f(x,t)$ is unknown, the proof must be based on the given partial differential equation and utilize the boundary conditions.

Step 1. Conversion of the partial differential equation to a moment problem. Choose a set of linearly independent functions $g_n(x,t)$, $n = 0, 1, \ldots, N$ such that the moments

$$\mu_n(t) = <g_n> = \int_a^b g_n(x,t)f(x,t)\mathrm{d}x \ , \quad n = 0,1,\ldots,N \tag{5.3.3}$$

are calculable. Usually, what is obtained from the given partial differential equation and the boundary conditions is an ordinary differential equation of $\mu_n(t)$. To solve for $\mu_n(t)$, the given initial conditions must be used. The situation here is more complex than that in solving the Fredholm integral equation. The choice of g_n depends on the given specific partial differential equation. g_n may not contain the argument t.

Step 2. Solution of the moment problem by MEM. The problem and solution here are similar to those in the previous sections with the exception that the functions contain the temporal variable t. Specifically, the problem is stated as

$$\text{Maximization}: \quad S(t) = - \int_a^b f(x,t)\ln f(x,t)\mathrm{d}x \ ,$$

$$\text{Constraints}: \quad \mu_n(t) = <g_n(x,t)> = \int_a^b g_n(x,t)f(x,t)\mathrm{d}x \ ,$$

$$n = 0,1,\ldots,N \ ,$$

Determine the MEM solution $f_N(x,t)$;

where $g_n(x,t)$ have been chosen, and $\mu_n(t)$ have been calculated in Step 1. It is worthwhile to point out that it is not a logical consequence of the monotonic increase of $S(t)$ with time, but a reasonable requirement only, that $S(t)$ at each instant t attains its maximum.

The MEM solution is

$$f_N(x,t) = \exp\left[-\sum_{n=0}^{N} \lambda_n(t)g_n(x,t)\right] , \qquad (5.3.4)$$

where the Lagrange multipliers $\lambda_n(t)$ are determined from the constraints

$$\mu_n(t) = \int_a^b g_n(x,t)f_N(x,t)\mathrm{d}x , \quad n = 0,1,\ldots,N . \qquad (5.3.5)$$

μ_n is a function of t. So is λ_n. In solving for λ_n by numerical method, the numerical procedure must be carried out for each chosen value of t. Therefore, the total computational time is usually very large.

5.3.2 Numerical Example

Consider the following initial value problem of partial differential equation:

$$\begin{cases} \dfrac{\partial f(\phi,t)}{\partial t} = \dfrac{\partial^2 f(\phi,t)}{\partial \phi^2} , \\ f(\phi,0) = \delta(\phi) , \quad \phi \in [-\pi,\pi] . \end{cases} \qquad (5.3.6)$$

Step 0. Equation (5.3.6) represents a diffusion process. Suppose that the linear density of mass along a circle is initially zero everywhere. At time $t = 0$, a substance of unit mass is introduced onto the position $\phi = 0$. Then the density f along the circle varies as a function of the angular coordinate ϕ and the time t. This equation can also represent a heat conduction process, f being the temperature. Based on the physical interpretation of the equation, it is easy to understand that the entropy in the process must increase monotonically with time. This conclusion can also be proved rigorously based on (5.3.6) as follows:

By the definition of entropy, we have

$$\begin{aligned} \frac{\mathrm{d}S(t)}{\mathrm{d}t} &= \frac{\mathrm{d}}{\mathrm{d}t}\left[-\int_{-\pi}^{\pi} f(\phi,t)\ln f(\phi,t)\mathrm{d}\phi\right] \\ &= -\int_{-\pi}^{\pi} \frac{\partial}{\partial t}(f\ln f)\mathrm{d}\phi \\ &= -\int_{-\pi}^{\pi} \frac{\partial f}{\partial t}(\ln f + 1)\mathrm{d}\phi . \end{aligned}$$

Utilizing (5.3.6), the above expression becomes

$$\frac{\mathrm{d}S(t)}{\mathrm{d}t} = -\int_{-\pi}^{\pi} \frac{\partial^2 f}{\partial \phi^2}(\ln f + 1)\mathrm{d}\phi . \qquad (5.3.7)$$

On the one hand,

$$\int_{-\pi}^{\pi} \frac{\partial^2}{\partial \phi^2}(f \ln f)\mathrm{d}\phi = \frac{\partial}{\partial \phi}(f \ln f)\Big|_{-\pi}^{\pi} = 0$$

due to the periodicity of f and ϕ, and on the other,

$$\int_{-\pi}^{\pi} \frac{\partial^2}{\partial \phi^2}(f \ln f)\mathrm{d}\phi$$

$$= \int_{-\pi}^{\pi} \frac{\partial}{\partial \phi}\left[\frac{\partial f}{\partial \phi}\ln f + \frac{\partial f}{\partial \phi}\right]\mathrm{d}\phi$$

$$= \int_{-\pi}^{\pi} \left[\frac{\partial^2 f}{\partial \phi^2}\ln f + \frac{1}{f}\left(\frac{\partial f}{\partial \phi}\right)^2 + \frac{\partial^2 f}{\partial \phi^2}\right]\mathrm{d}\phi$$

$$= \int_{-\pi}^{\pi} \frac{\partial^2 f}{\partial \phi^2}(\ln f + 1)\mathrm{d}\phi + \int_{-\pi}^{\pi} \frac{1}{f}\left(\frac{\partial f}{\partial \phi}\right)^2 \mathrm{d}\phi .$$

Therefore, we have

$$-\int_{-\pi}^{\pi} \frac{\partial^2 f}{\partial \phi^2}(\ln f + 1)\mathrm{d}\phi = \int_{-\pi}^{\pi} \frac{1}{f}\left(\frac{\partial f}{\partial \phi}\right)^2 \mathrm{d}\phi ,$$

which is substituted in (5.3.7) and we obtain

$$\frac{\mathrm{d}S(t)}{\mathrm{d}t} = \int_{-\pi}^{\pi} \frac{1}{f}\left(\frac{\partial f}{\partial \phi}\right)^2 \mathrm{d}\phi > 0 , \quad (t > 0,\ f > 0) .$$

Step 1. Choose $g_n = \cos n\phi$, $n = 0, 1, \ldots, N$, derive the ordinary differential equations of $\mu_n(t) = <\cos n\phi>$ which are solved for $\mu_n(t)$. Note that $\sin n\phi$, $n = 0, 1, \ldots, N$ are not chosen as g_n since the solution of (5.3.6) must be an even function in ϕ as seen from that equation.

$$\begin{aligned}
\frac{\mathrm{d}}{\mathrm{d}t}\mu_n(t) &= \frac{\mathrm{d}}{\mathrm{d}t}\int_{-\pi}^{\pi} f(\phi, t)\cos n\phi \mathrm{d}\phi \\
&= \int_{-\pi}^{\pi} \frac{\partial^2 f}{\partial \phi^2}\cos n\phi \mathrm{d}\phi \\
&= \frac{\partial}{\partial \phi}\cos n\phi\Big|_{-\pi}^{\pi} + n\int_{-\pi}^{\pi} \frac{\partial f}{\partial \phi}\sin n\phi \mathrm{d}\phi \\
&= nf\sin n\phi\Big|_{-\pi}^{\pi} - n^2\int_{-\pi}^{\pi} f\cos n\phi \mathrm{d}\phi \\
&= -n^2\mu_n(t) ,
\end{aligned}$$

hence

$$\mu_n(t) = Ce^{-n^2 t} ,$$

where C is a constant. From the initial condition, we have

$$\begin{aligned} C &= \mu_n(0) \\ &= \int_{-\pi}^{\pi} f(\phi,0)\cos n\phi \, d\phi \\ &= \int_{-\pi}^{\pi} \delta(\phi)\cos n\phi \, d\phi \\ &= 1 \, . \end{aligned}$$

The particular solution is

$$\mu_n(t) = e^{-n^2 t} \, . \tag{5.3.8}$$

For $n = 0$,

$$\mu_0(t) = \int_{-\pi}^{\pi} f(\phi,t)\mathrm{d}\phi = 1$$

represents the conservation law.

Step 2. The MEM solution is

$$f_N(\phi,t) = \exp\left[-\sum_{n=0}^{N}\lambda_n(t)\cos n\phi\right] \, . \tag{5.3.9}$$

The equations for determining the Lagrange multipliers are

$$\int_{-\pi}^{\pi}\exp\left[-\sum_{n=0}^{N}\lambda_n(t)\cos n\phi\right]\cos n\phi \, d\phi = e^{-n^2 t} \, , \quad n = 0, 1, \ldots, N \, . \tag{5.3.10}$$

For a particular t, $\lambda_n(t)$ can be solved for by the Newton method introduced in Sect. 5.2. For $N = 1$, the following faster method can also be used, whereby the computational time may be reduced to about $1/8$.

For $N = 1$, (5.3.10) become

$$\begin{cases} \displaystyle\int_{-\pi}^{\pi} e^{-\lambda_0(t)-\lambda_1(t)\cos\phi}\mathrm{d}\phi = 1 \, , \\[2mm] \displaystyle\int_{-\pi}^{\pi} e^{-\lambda_0(t)-\lambda_1(t)\cos\phi}\cos\phi \, \mathrm{d}\phi = e^{-t} \, . \end{cases}$$

Utilizing the integral representation for modified Bessel functions,

$$I_0(x) = \frac{1}{2\pi}\int_{-\pi}^{\pi} e^{x\cos\theta}\mathrm{d}\theta \, , \quad I_1(x) = \frac{1}{2\pi}\int_{-\pi}^{\pi} e^{x\cos\theta}\cos\theta \, \mathrm{d}\theta \, ,$$

the above equations in turn become

$$\begin{cases} I_1(-\lambda_1(t))/I_0(-\lambda_1(t)) = e^{-t} \, , & \tag{5.3.11a} \\[2mm] e^{-\lambda_0(t)} = [2\pi I_0(-\lambda_1(t))]^{-1} \, . & \tag{5.3.11b} \end{cases}$$

They can only be solved by numerical method but without numerical integration. Specifically, for a particular $t_1 > 0$, solve (5.3.11a) for $-\lambda_1(t_1)$ by a trial-and-error procedure or other methods. Then it is easy to find $-\lambda_1(t)$ at $t_1 < t_2 < \ldots < t_m$ by taking advantage of the monotonicity of $I_1(x)/I_0(x)$ and e^{-t}. Finally, $e^{-\lambda_0(t)}$ or $-\lambda_0(t)$, $t = t_1, \ldots, t_m$ are calculated by (5.3.11b).

The exact solution of (5.3.6) is

$$f(\phi, t) = \frac{1}{2\pi} + \frac{1}{\pi} \sum_{n=1}^{\infty} e^{-n^2 t} \cos n\phi \ . \tag{5.3.12}$$

Its asymptotic value is

$$f(\infty) = f(\phi, \infty) = \frac{1}{2\pi} \ .$$

$e^{-t} \to 0$ as $t \to \infty$. $I_1(-\lambda_1(\infty)) = 0$ from (5.3.11a), hence $\lambda_1(\infty) = 0$. It is substituted in (5.3.11b) to get $e^{-\lambda_0(\infty)} = [2\pi I_0(0)]^{-1} = (2\pi)^{-1}$. Consequently, the MEM solution in (5.3.9) is asymptotically

$$f_1(\phi, \infty) = \frac{1}{2\pi} \ .$$

So we see $f(\phi, \infty) = f_1(\phi, \infty)$ as expected. In fact, it is easy to infer $f_N(\phi, \infty) = (2\pi)^{-1}$ by the conservation law.

The Fourier cosine series of $f(\phi, t)$ in ϕ is

$$f(\phi, t) = \frac{1}{2} a_0(t) + \sum_{n=1}^{\infty} a_n(t) \cos n\phi \ ,$$

where the Fourier coefficients are

$$a_n(t) = \frac{1}{\pi} \int_{-\pi}^{\pi} f(\phi, t) \cos n\phi \, d\phi \ .$$

The moment expressions are

$$\mu_n(t) = \int_{-\pi}^{\pi} f(\phi, t) \cos n\phi \, d\phi \ .$$

Comparing the last two expressions we see that

$$a_n(t) = \frac{1}{\pi} \mu_n(t) = \frac{1}{\pi} e^{-n^2 t} \ .$$

Given $\mu_n(t)$, $n = 0, 1, \ldots, N$, the sum of the truncated cosine series of $f(\phi, t)$, i.e., the partial sum of the series in the exact solution (5.3.12), can be calculated. From the above relationship, $a_n(t) = \frac{1}{\pi} \mu_n(t)$, we see that in fact, MEM makes a nonzero extension of the Fourier coefficients by recalling that the moments are extrapolated in MEM.

Shown graphically in Figs. 5.2a–e are the exact solution $f(\phi, t)$, the MEM solution $f_1(\phi, t)$, and the sum of the truncated cosine series being normalized ($f/f(\infty) = 2\pi f$) as functions of t, respectively, for $\phi = 0, \pi/4, \pi/2, 3\pi/4$ and

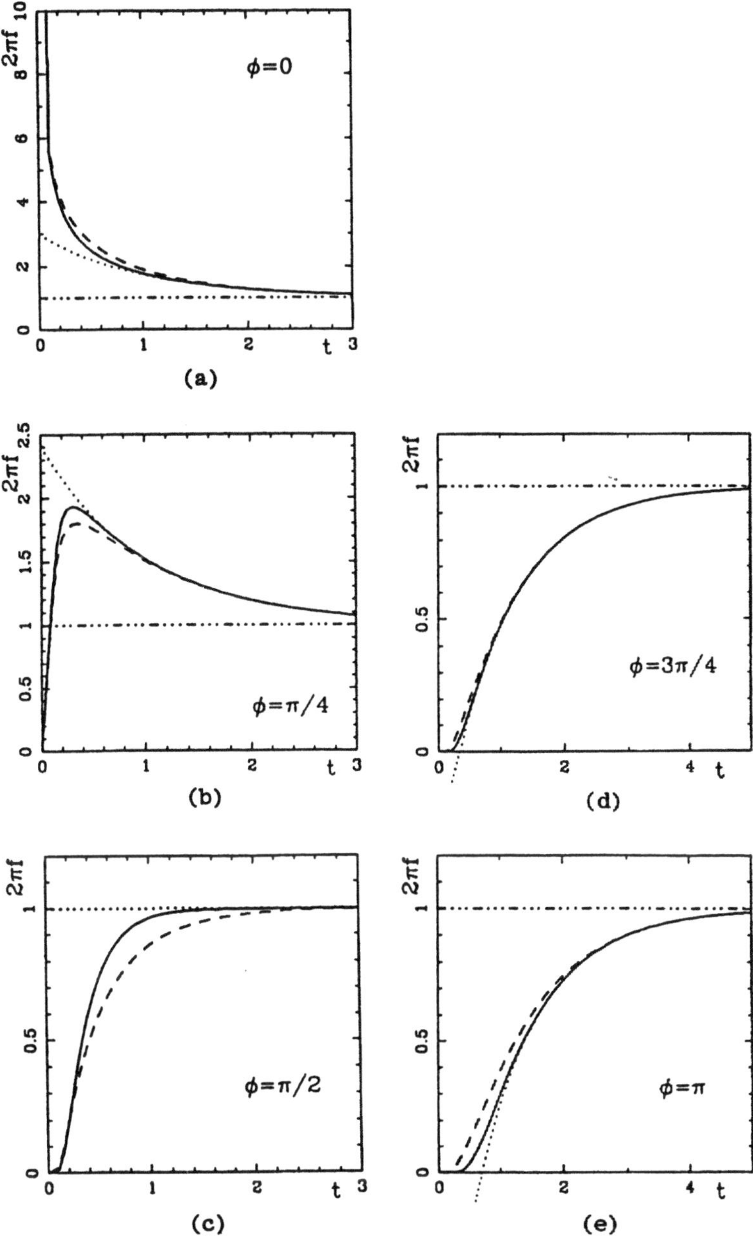

Fig. 5.2. The normalized $(2\pi f)$ exact solution $f(\phi, t)$ (*solid line*), the MEM solution $f_1(\phi, t)$ (*dashed line*), and the sum of the truncated cosine series (*dotted line*). (a) $\phi = 0$. (b) $\phi = \pi/4$. (c) $\phi = \pi/2$. (d) $\phi = 3\pi/4$. The dotted line extends at $t = 0.35$ downwards until intersecting the vertical axis at -0.414. (e) $\phi = \pi$. The dotted line extends at $t = 0.70$ downwards until intersecting the vertical axis at -1.00

π. From the figures we see that on the whole f_1 and f are similar in shape, but in some regions they are noticeably discrepant and f_1 is not accurate. The sum of the truncated cosine series is poorly-behaved especially in the vicinity of $t = 0$. All these are because only two moments are used, one of which just comes from the normalization condition for the density function.

The accuracy of the MEM solution can be improved effectively by increasing the number of moments used. But a great deal of computational time is consequently needed for calculating $f_N(\phi, t)$ for reasonably many values of t to plot the curves. Shown graphically in Figs. 5.3a–e are the normalized $f(\phi, t)$, $f_3(\phi, t)$ and the sum of the truncated cosine series (the case $N = 3$). From the figures we see that the MEM solution f_3 is very close to the exact solution f. They are indistinguishable in large regions. However, the sum of the truncated cosine is still poorly-behaved in the vicinity of $t = 0$. For $N = 4$, the accuracy of $f_4(\phi, t)$ is further improved, and f_4 coincides with f almost everywhere. (Figures are not presented here.) Listed in Table 5.11 are the maximum absolute errors $\Delta = |f(\phi, t) - f_N(\phi, t)|$ and the corresponding t for various angular coordinates ϕ in the cases of $N = 1, 2, 3, 4$.

Table 5.11. Errors of the MEM solution $f_N(\phi, t)$

N	1		2	
ϕ	Δ	t	Δ	t
0	4.57 E$-$02	0.35	1.32 E$-$02	0.40
$\pi/4$	2.35 E$-$02	0.25	1.10 E$-$02	0.40
$\pi/2$	2.32 E$-$02	0.60	5.26 E$-$03	0.35
$3\pi/4$	7.19 E$-$03	0.40	2.94 E$-$03	0.70
π	1.73 E$-$02	0.70	5.14 E$-$03	0.65

N	3		4	
ϕ	Δ	t	Δ	t
0	5.58 E$-$03	0.40	2.70 E$-$03	0.35
$\pi/4$	4.57 E$-$03	0.45	1.04 E$-$03	0.50
$\pi/2$	3.21 E$-$03	0.50	1.07 E$-$03	0.35
$3\pi/4$	2.27 E$-$03	0.60	1.06 E$-$03	0.50
π	1.99 E$-$03	0.60	8.86 E$-$04	0.55

5.3.3 Discussion

(1) We have introduced the three steps in solving partial differential equations by MEM, and used the diffusion (parabolic) equation as an example. A question arises immediately: Is it possible to use MEM to solve wave (hyperbolic) equations or Laplace (elliptic) equations? For wave equations, in principle, the answer is *No* because the entropy in a (nondamping) wave process does

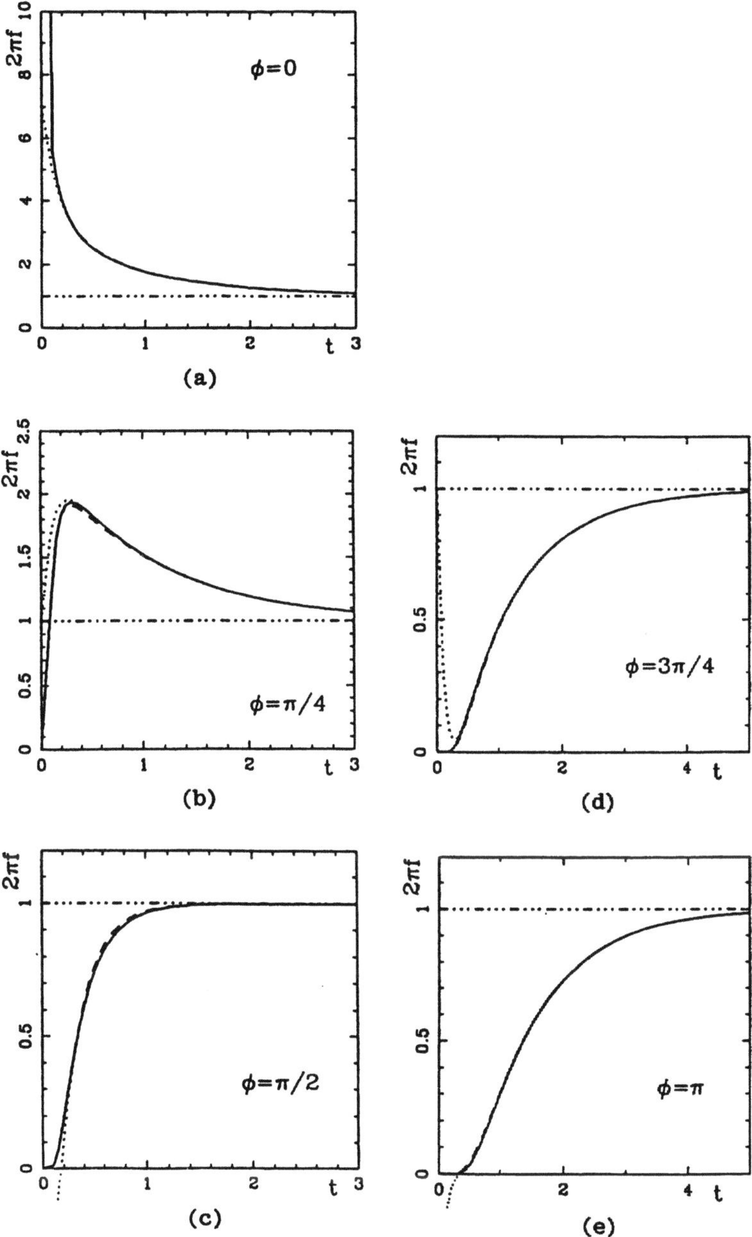

Fig. 5.3. The normalized $(2\pi f)$ exact solution $f(\phi, t)$ (*solid line*), the MEM solution $f_3(\phi, t)$ (*dashed line*), and the sum of the truncated cosine series (*dotted line*). (a) $\phi = 0$. (b) $\phi = \pi/4$. (c) $\phi = \pi/2$. The dotted line extends at $t = 0.15$ downwards until intersecting with the vertical axis at -1.00. (d) $\phi = 3\pi/4$. (e) $\phi = \pi$. The dotted line extends at $t = 0.35$ downwards until intersecting with the vertical axis at -1.00

not tend towards the maximum. From the point of view of mathematical operation, the unknown function (displacement) may assume positive and negative values. Even if this function has been converted to a positive-valued one, we cannot derive reasonably simple differential equations of the moments. For Laplace equations, in principle, the answer is *Yes* because they are the steady-state limits of diffusion equations. However, in mathematics we cannot derive moments as the constraints in a simple way. Hopefully, the situation may be changed as the research advances.

(2) From the example demonstrated, we see that what is derived from the chosen $g_n(x,t)$ (t may be left out) are ordinary differential equations of $\mu_n(t)$. The situation here is much more difficult than that in solving an integral equation. The proper choice of g_n is important, however, we still have no specific rules to follow.

(3) We cannot determine the error bounds of MEM solutions to partial differential equations. In general, the MEM solutions are fairly accurate even if only a relatively small number of moments are given or calculated. But it will be very difficult if one wants to improve the accuracy of the solution of a partial differential equation by increasing the number of moments. The difficulties are both in solving for the moments and in determining the Lagrange multipliers. Here, the greatest value of MEM is still to provide one more approximate solution when the exact solution of the given equation cannot be determined.

5.4 Predictive Statistical Mechanics

Statistical mechanics is a microscopic theory corresponding to the macroscopic one, thermodynamics. In statistical mechanics, we deal with various microscopic quantities such as particle velocity, collision frequency, and energy level. These quantities cannot be directly measured. What can be directly measured are macroscopic quantities such as temperature, pressure, and volume. The relationship between these two kinds of quantities is: Macroscopic quantities are averages (means or expectations as usually called) of microscopic ones. Thus, the central task in statistical mechanics is to determine the probability distributions of particles (or systems in ensemble theory) among microstates.

5.4.1 Formulation and Solution

In conventional statistical mechanics, the theory is established on the basis of motion equations of particles. In order to apply the statistical method, some assumptions (stipulations), such as ergodicity and equiprobabilities, are needed. Entropy and the Principle of the increase of entropy arise as result. But as suggested by *Jaynes* [5.10, 5.11], we can go in the opposite direction, that is, use entropy as a starting concept. In so-called *Predictive Statistical Mechanics*, statistical mechanics is viewed as statistical inference based on incomplete knowledge, but not as physical theory. All the results can be obtained by the Principle of ME (i.e., MEM). Specifically, the probability distribution is determined by maximizing entropy under the constraints of known knowledge, then all the thermodynamic quantities are calculated. It should be pointed out that "entropy" in the Principle of ME refers to the information-theoretic entropy S_I.

Thus, we have the following mathematical problem (discrete case):

$$\text{Maximization}: \quad S_I = -\sum_i p_i \ln(p_i/m_i) \,, \tag{5.4.1}$$

$$\text{Constraints}: \quad \sum_i p_i = 1 \,, \quad \text{(normalization condition)} \,, \tag{5.4.2a}$$

$$< f_r(x) >= \sum_i p_i f_r(x_i) \,, \quad r = 1, \ldots, m \,; \tag{5.4.2b}$$

Determine the ME solution $p_i, i = 1, 2, \ldots$.

Here $f_r(x), r = 1, \ldots, m$ are known functions; p_i is the probability that the microscopic quantity x assumes the value x_i; the averages (expectations) $< f_r(x) >, r = 1, \ldots, m$ are the measured macroscopic quantities; $m_i, i = 1, \ldots$ are the measure in the entropy expression, which represent degeneracies, as we will see later. What we need to solve now is also a moment problem (with multiple constraints of the first moments as shown later).

Introduce the Lagrange multipliers $\lambda_0, \lambda_r, r = 1, \ldots, m$, form the objective function, and let its partial derivatives with respect to p_i be zero; it is easy to show the solution to be

$$p_i = m_i \exp\left[-\lambda_0 - \sum_{r=1}^m \lambda_r f_r(x_i)\right] \,. \tag{5.4.3}$$

To prove that these p_i really give the maximum value of S_I, we need the inequality

$$-\sum_i q_i \ln(q_i/m_i) \le -\sum_i q_i \ln(p_i/m_i) \,, \tag{5.4.4}$$

(the equality holds if and only if $q_i = p_i$), where $\{q_i\}$ and $\{p_i\}$ are arbitrary probability distributions. Inequality (5.4.4) is equivalent to

$$\sum_i q_i \ln(q_i/p_i) \ge 0 \,,$$

which can be shown using the inequality $x \ln x \ge x - 1$ ($x \ge 0$) with $x = q_i/p_i$.

Now suppose that $\{p_i\}$ is a solution given by (5.4.3) while $\{q_i\}$ is an arbitrary solution. (A solution means a probability distribution satisfying the constraints (5.4.2a,b).) Then,

$$\begin{aligned}
S_{Iq} &= -\sum_i q_i \ln(q_i/m_i) \\
&\leq -\sum_i q_i \ln(p_i/m_i) \\
&= -\sum_i q_i \left[-\lambda_0 - \sum_{r=1}^m \lambda_r f_r(x_i) \right] \\
&= \lambda_0 \sum_i q_i + \sum_{r=1}^m \lambda_r \sum_i q_i f_r(x_i) \\
&= \lambda_0 + \sum_{r=1}^m \lambda_r < f_r(x) > .
\end{aligned}$$

On the other hand,

$$\begin{aligned}
S_{Ip} &= -\sum_i p_i \ln(p_i/m_i) \\
&= -\sum_i p_i \left[-\lambda_0 - \sum_{r=1}^m \lambda_r f_r(x_i) \right] \\
&= \lambda_0 + \sum_{r=1}^m \lambda_r < f_r(x) > .
\end{aligned}$$

So we conclude that

$$S_{Iq} \leq S_{Ip} . \tag{5.4.5}$$

The maximum value of information-theoretic entropy is

$$S_{I\,\mathrm{max}} = \lambda_0 + \sum_{r=1}^m \lambda_r < f_r(x) > . \tag{5.4.6}$$

Define the partition function by

$$Z(\lambda_r, r = 1, \ldots, m) = \sum_i m_i \exp\left[-\sum_{r=1}^m \lambda_r f_r(x_i) \right] . \tag{5.4.7}$$

Then, from the normalization condition and (5.4.3), we have

$$p_i = \frac{m_i}{Z} \exp\left[-\sum_{r=1}^m \lambda_r f_r(x_i) \right] , \tag{5.4.8}$$

$$\lambda_0 = \ln Z . \tag{5.4.9}$$

From the mathematical point of view, the partition function $Z = e^{\lambda_0}$ is a divider for normalizing p_i. On the other hand, from the physical point of view, it contains the information about the distribution of particles in x. All the thermodynamic quantities can be expressed in terms of the partition function. Some useful formulae are presented in the next subsection.

5.4.2 Useful Formulae

1. The maximum value of information-theoretic entropy is

$$S_{I\,\mathrm{max}} = \ln Z + \sum_{r=1}^{m} \lambda_r < f_r(x) > ,$$ (5.4.10)

which is obtained from combining (5.4.9, 5.4.6).

   2. The formulae for the averages are

$$< f_r(x) >= -\frac{\partial}{\partial \lambda_r} \ln Z , \quad r = 1, \ldots, m ,$$ (5.4.11)

which are equations for determining $\lambda_1, \ldots, \lambda_m$. For $r \geq 2$, usually they are solved necessarily by numerical method.

Proof. From (5.4.7, 5.4.8), it follows that

$$\begin{aligned}
\frac{\partial \ln Z}{\partial \lambda_r} &= \frac{1}{Z}\frac{\partial Z}{\partial \lambda_r} \\
&= \frac{1}{Z}\sum_i m_i \exp\left[-\sum_{r=1}^{m}\lambda_r f_r(x_i)\right][-f_r(x_i)] \\
&= -\sum_i p_i f_r(x_i) \\
&= -<f_r(x)> .
\end{aligned}$$

   3. The formulae for variance are

$$\Delta^2 f_r(x) = \frac{\partial^2}{\partial \lambda_r^2}\ln Z , \quad r = 1, \ldots, m .$$ (5.4.12)

Proof.

$$\begin{aligned}
\Delta^2 f_r(x) &= <f_r^2(x)> - <f_r(x)>^2 \\
&= \sum_i p_i f_r^2(x_i) - \left(-\frac{\partial}{\partial \lambda_r}\ln Z\right)^2 .
\end{aligned}$$

In the last line,

$$\text{the first term} = \frac{1}{Z}\frac{\partial^2 Z}{\partial \lambda_r^2}$$

since

$$\begin{aligned}
\frac{1}{Z}\frac{\partial^2 Z}{\partial \lambda_r^2} &= \frac{1}{Z}\sum_i m_i \exp\left[-\sum_{r=1}^{m}\lambda_r f_r(x_i)\right] f_r^2(x_i) \\
&= \sum_i p_i f_r^2(x_i) ;
\end{aligned}$$

$$\text{the second term} \;=\; -\left(\frac{1}{Z}\frac{\partial Z}{\partial \lambda_r}\right)^2$$

$$=\; -\frac{1}{Z^2}\left(\frac{\partial Z}{\partial \lambda_r}\right)^2 .$$

Hence,

$$\Delta^2 f_r(x) \;=\; \frac{1}{Z}\frac{\partial^2 Z}{\partial \lambda_r^2} - \frac{1}{Z^2}\left(\frac{\partial Z}{\partial \lambda_r}\right)^2$$

$$=\; \frac{\partial^2}{\partial \lambda_r^2}\ln Z .$$

Q.E.D.

To sum up, the average of $f_r(x)$ is equal to the negative of the first partial derivative of $\ln Z$ with respect to λ_r while the variance is equal to the second partial derivative of $\ln Z$ with respect to λ_r.

4. If an f_r contains a parameter α_k (e.g., temperature T or volume V), then

$$< \frac{\partial f_r}{\partial \alpha_k} >= -\frac{1}{\lambda_r}\frac{\partial}{\partial \alpha_k}\ln Z . \tag{5.4.13}$$

Proof. In this case the partition function also contains the parameter α_k. The left- and right-hand sides of (5.4.13) are, respectively,

$$\text{LHS} \;=\; \sum_i p_i \frac{\partial f_r}{\partial \alpha_k} ,$$

$$\text{RHS} \;=\; -\frac{1}{\lambda_r}\cdot\frac{1}{Z}\frac{\partial Z}{\partial \alpha_k}$$

$$=\; -\frac{1}{\lambda_r}\cdot\frac{1}{Z}\frac{\partial}{\partial \alpha_k}\sum_i m_i \exp\left[-\sum_{\substack{l=1\\l\neq r}}^{m}\lambda_l f_l(x_i) - \lambda_r f_r(x_i;\alpha_k)\right]$$

$$=\; -\frac{1}{\lambda_r}\cdot\frac{1}{Z}\sum_i m_i \exp\left[-\sum_{l=1}^{m}\lambda_l f_l(x_i)\right]\left(-\lambda_r\frac{\partial f_r}{\partial \alpha_k}\right)$$

$$=\; \sum_i p_i \frac{\partial f_r}{\partial \alpha_k} .$$

So (5.4.13) results.

5. Finally, the average of any given function $g(x)$ is

$$< g(x) >= \sum_i p_i g(x_i) .$$

In the preceding derivations, all (macroscopic) quantities were assumed to be independent of both spatial coordinates and time at the very beginning.

That is to say, the system is in (thermodynamical) equilibrium (without an external field). If a system is in a nonequilibrium steady state, then some or all quantities are dependent on the spatial coordinates but all are independent of time. If a system is not in a steady state, then some or all quantities are dependent on spatial coordinates and/or time. In the last two cases, the ME problems and solutions are formally similar to those for the equilibrium case. However, the situation is far more difficult in both concept and mathematical manipulation. We will restrict our attention to the equilibrium case, to which the remaining three sections of this chapter will be devoted. Readers are, if necessary, referred to standard textbooks, e.g., [5.12, 5.13], for the prerequisite knowledge.

5.5 Distributions of Particles Among Energy Levels

This section is concerned with three distributions of particles among energy levels in a quantum system of particles. The corresponding distribution in a classical system of particles, the Boltzmann distribution, is derived as a limit of the result for a quantum system.

Consider an isolated (having no energy or substance exchange with environment) quantum system in equilibrium consisting of near-independent and identical particles. Given the total number of particles, N, and the total energy of the system, E, determine the distribution of particles among energy levels, i.e., the number of particles, n_i, having energy ε_i, $i = 1, 2, \ldots$. There are three distributions of particles, $\{n_i\}$, according to the three statistical laws particles obey. In applying each statistics, we will briefly introduce the method used in conventional statistical mechanics first, and then use the Principle of ME. Our major goal is to show that the Principle of ME and the conventional statistical mechanics method give exactly the same results, but the former is simpler in the sense having fewer physical assumptions with easier mathematical manipulation. In addition, we will establish a relationship between the information-theoretic entropy and thermodynamic or statistical-mechanics entropy.

5.5.1 Boltzmann Distribution

The situation here is that particles are identical but distinguishable (e.g., localized), and hence Maxwell-Boltzmann statistics applies.

1. Conventional Statistical Mechanics Method

A fundamental physical assumption in conventional statistical mechanics is *equiprobabilities* assumption: If an isolated system is in equilibrium, it is found with equal probability in each one of its accessible quantum states. Hence, a macrostate corresponding to the largest number of quantum states is the most probable. The corresponding distribution of particles $\{n_i\}$ is considered to be real.

Suppose G_i distinct quantum states have the same energy ε_i, i.e., G_i is the degeneracy of energy level ε_i. It can be derived by permutation and combination formulae that the number of quantum states corresponding to the distribution of particles $\{n_i\}$ is

$$W(\{n_i\}) = \frac{N!}{\prod_i n_i!} \prod_i G_i^{n_i} ,$$

(N is the total number of particles). Utilizing Stirling's formula, $\ln m! \approx m(\ln m - 1)$ $(m \gg 1)$, we have

$$\ln W \approx N \ln N - \sum_i n_i \ln n_i + \sum_i n_i \ln G_i . \tag{5.5.1}$$

Under the constraints

$$\sum_i n_i = N , \tag{5.5.2a}$$

$$\sum_i n_i \varepsilon_i = E , \tag{5.5.2b}$$

maximizing $\ln W$ yields the most probable distribution $\{n_i\}$:

$$n_i = N \frac{G_i}{Z} \mathrm{e}^{-\beta \varepsilon_i} , \tag{5.5.3a}$$

which is equivalent to

$$\frac{n_i}{N} = \frac{G_i}{Z} \mathrm{e}^{-\beta \varepsilon_i} , \tag{5.5.3b}$$

where Z is the partition function,

$$Z = \sum_i G_i \mathrm{e}^{-\beta \varepsilon_i} . \tag{5.5.4}$$

Equations (5.5.3a,b) are the Boltzmann distribution of particles among energy levels. The determination of the constant β will be mentioned later.

2. Principle of Maximum Entropy

In (5.4.1, 5.4.2a,b), p_i is interpreted as the probability that a particle is found to be in energy level ε_i (i.e., in any one of the G_i quantum states), m_i as G_i (the degeneracy of ε_i), $f(x) = x$, $(r = 1)$, x is the energy ε of a particle and its average is $\bar{\varepsilon} = E/N$. Then, the constraints (5.4.2a,b) are of the specific forms

$$\sum_i p_i = 1 , \tag{5.5.5a}$$

$$\sum_i p_i \varepsilon_i = \bar{\varepsilon} \, . \tag{5.5.5b}$$

The ME solution can be written out immediately after (5.4.8):

$$p_i = \frac{G_i}{Z} e^{-\lambda \varepsilon_i} . \tag{5.5.6}$$

The partition function is, following (5.4.7),

$$Z = \sum_i G_i e^{-\lambda \varepsilon_i} , \tag{5.5.7}$$

which is the same form as (5.5.4).

In the thermodynamic limit ($N \to \infty$ and the number density of particles remains constant), the proportion n_i/N is equal to the probability p_i in a sense (the law of large numbers), and (5.5.6) is equivalent to (5.5.3b). The constraints (5.5.5a,b, 5.5.2a,b) are also equivalent. Equation (5.5.1) becomes

$$\begin{aligned}
\ln W &= -N \sum_i (n_i/N) \ln[(n_i/N)/G_i] \\
&= -N \sum_i p_i \ln(p_i/G_i) \\
&= N S_I \, ,
\end{aligned}$$

where S_I is the information-theoretic entropy per particle, $N S_I$ is the information-theoretic entropy of the system. For a system in equilibrium, both $\ln W$ and $N S_I$ attain the maximum values. On the other hand, according to the Boltzmann relation, for the thermodynamic entropy S, it holds that

$$S = k \ln W \, ,$$

(k is the Boltzmann constant). Combining the last two expressions yields the relation

$$S = k S_{I\,\mathrm{max}} \, , \tag{5.5.8}$$

where $S_{I\,\mathrm{max}}$ is the information-theoretic entropy of the system. Equation (5.5.8) is a basic relation in Predictive Statistical Mechanics. From the preceding discussion we see that maximizing $\ln W$ and maximizing S_I are equivalent in mathematics. This is why the two methods give the same results. But the solution by the Principle of ME is simpler by virtue of its existing results.

The multiplier λ in (5.5.7) is determined in the same way as determining β in (5.5.4) in conventional statistical mechanics. The result is $\lambda = \beta = 1/kT$, where k is the Boltzmann constant and T is the absolute temperature of the system. It should be pointed out that λ cannot be related to T only by the constraints (5.5.5a,b). Thermodynamical formulae must be used. Once the partition function has been obtained, various thermodynamic quantities can then be calculated.

In a classical system, even nonlocalized particles are distinguishable. The mechanical states of particles are represented by points in a (six-dimensional orthogonal) phase space. The number of phase cells (i.e., the number of microstates) in a volume element of phase space, $\mathrm{d}\boldsymbol{q}\mathrm{d}\boldsymbol{p}$ ($\boldsymbol{q}$ and $\boldsymbol{p}$ represent position and momentum, respectively) is

$$G_i = \mathrm{d}\tau = \mathrm{d}\boldsymbol{q}\mathrm{d}\boldsymbol{p}/h^3 \ ,$$

(h is the Plank constant). The particles in a volume element have the same energy ε. The number of them is denoted by $n(\varepsilon)$. It follows from (5.5.3b) that

$$\frac{n(\varepsilon)}{N} = \frac{1}{Z}\mathrm{e}^{-\beta\varepsilon}\mathrm{d}\tau \ .$$

This is the Boltzmann distribution of particles among energies in a classical system. The partition function, after (5.5.4), is

$$Z = \int \mathrm{e}^{-\beta\varepsilon}\mathrm{d}\tau \ .$$

5.5.2 Fermi-Dirac and Bose-Einstein Distributions

The situation here is that particles are identical and indistinguishable (non-localized), and hence Fermi-Dirac or Bose-Einstein statistics applies.

1. Conventional Statistical Mechanics Method

The method used here also relies on the assumption of equiprobabilities. Under the constraints of the total number of particles and the total energy, (5.5.2a,b), the most probable distribution $\{n_i\}$ is determined by maximizing the corresponding number of quantum states. Being different from the situation in deriving the Boltzmann distribution, particles here are not only identical but also indistinguishable. In addition, for a system consisting of Fermions *Pauli's exclusion principle* takes effect: A single quantum state can contain at most one particle.

Using permutation and combination formulae, the number of quantum states corresponding to the distribution $\{n_i\}$ of particles is found to be

$$W_{\mathrm{FD}} \ = \ \prod_i \frac{G_i!}{n_i!(G_i - n_i)!} \ , \quad \text{(Fermion system)} \ ,$$

$$W_{\mathrm{BE}} \ = \ \prod_i \frac{(n_i + G_i - 1)!}{n_i!(G_i - 1)!} \ , \quad \text{(Boson system)} \ .$$

Utilizing Stirling's formula, the above two formulae give

$$\ln W_{\mathrm{FD}} \approx \sum_i -n_i \ln n_i + G_i \ln G_i - (G_i - n_i) \ln(G_i - n_i) \,,$$

$$(\text{Fermion system}) \,, \qquad (5.5.9)$$

$$\ln W_{\mathrm{BE}} \approx \sum_i -n_i \ln n_i - G_i \ln G_i + (G_i + n_i) \ln(G_i + n_i) \,,$$

$$(\text{Boson system}) \,. \qquad (5.5.10)$$

Then, after mathematical manipulation similar to that in deriving the Boltzmann distribution, it turns out that the most probable distributions $\{n_i\}$ are

$$n_i = \frac{G_i}{e^{\alpha + \beta \varepsilon_i} + 1} \,, \quad (\text{Fermi-Dirac distribution}) \,, \qquad (5.5.11)$$

$$n_i = \frac{G_i}{e^{\alpha + \beta \varepsilon_i} - 1} \,, \quad (\text{Bose-Einstein distribution}) \,. \qquad (5.5.12)$$

It is well known that for $e^{\alpha} \gg 1$ (nondegeneracy case) the above two distributions approach the Boltzmann distribution.

2. Principle of Maximum Entropy

Since particles are identical and indistinguishable, and in particular Pauli's exclusion principle must be obeyed for a Fermion system, the measure m_i in the entropy definition cannot be simply interpreted as G_i. The formulation of the Principle of ME here is somewhat different from that in the previous subsection.

Let p_{ijn} be the probability that the jth quantum state of energy level ε_i contains n particles. Then, the average number of particles in energy level ε_i is

$$<n_i> = \sum_{j=1}^{G_i} \sum_n p_{ijn} n \,. \qquad (5.5.13)$$

When p_{ijn} is the ME solution, $\{<n_i>\}$ calculated by the above formula is the ME distribution of particles among energy levels. In the thermodynamic limit, the average value $<n_i>$ is equal to the exact value n_i.

Since particles are identical, indistinguishable and near-independent, the information-theoretic entropy of the system is

$$S_I = -\sum_i \sum_{j=1}^{G_i} \sum_n p_{ijn} \ln p_{ijn} \,. \qquad (5.5.14)$$

The constraints are

$$\sum_n p_{ijn} = 1 \,, \quad i = 1, 2, \ldots, \quad j = 1, \ldots, G_i \,, \quad (\text{normalization}) \,, \qquad (5.5.15)$$

$$\sum_i \sum_{j=1}^{G_i} \sum_n p_{ijn} n = N \,, \quad (\text{given the total number of particles}) \,, \qquad (5.5.16)$$

$$\sum_i \sum_{j=1}^{G_i} \sum_n p_{ijn} n \varepsilon_i = E \,, \quad \text{(given the energy of the system)}\,. \qquad (5.5.17)$$

From the above expressions, the ME solution is found, after some mathematical manipulation, to be

$$p_{ijn} = e^{-(\lambda_1 + \lambda_2 \varepsilon_i)n} \Big/ \sum_n e^{-(\lambda_1 + \lambda_2 \varepsilon_i)n} \,, \qquad (5.5.18)$$

where λ_1 and λ_2 are the Lagrange multipliers for (5.5.16, 5.5.17), respectively.

(1) *Fermi-Dirac Distribution.* According to Pauli's exclusion principle, $n = 0, 1$. The ME solution $\{< n_i >\}$ is obtained by substituting (5.5.18) in (5.5.13):

$$\begin{aligned}
< n_i > \; &= \; \sum_{j=1}^{G_i} \sum_{n=0}^{1} n e^{-(\lambda_1 + \lambda_2 \varepsilon_i)n} \Big/ \sum_{n=0}^{1} e^{-(\lambda_1 + \lambda_2 \varepsilon_i)n} \\
&= \; \frac{G_i e^{-(\lambda_1 + \lambda_2 \varepsilon_i)}}{1 + e^{-(\lambda_1 + \lambda_2 \varepsilon_i)}} \\
&= \; \frac{G_i}{e^{\lambda_1 + \lambda_2 \varepsilon_i} + 1} \,.
\end{aligned} \qquad (5.5.19)$$

Substituting (5.5.18) ($n = 0, 1$ in the summation in the denominator) in (5.5.14), after some mathematical manipulation, yields the maximum information-theoretic entropy,

$$\begin{aligned}
S_{I\,\mathrm{max}} \; = \; &- \sum_i < n_i > \ln < n_i > - G_i \ln G_i \\
&+ (G_i - < n_i >) \ln(G_i - < n_i >) \,.
\end{aligned} \qquad (5.5.20)$$

Comparing (5.5.19) with (5.5.11), we see that in the thermodynamic limit the conventional statistical mechanics method and the Principle of ME give the same results. Furthermore, comparing (5.5.20) with (5.5.9), we see that the maximum information-theoretic entropy and $\ln W_{FD}$ are equal. From the Boltzmann relation,

$$S = k \ln W \,,$$

it follows that

$$S = k S_{I\,\mathrm{max}} \,, \qquad (5.5.21)$$

where S is the thermodynamic entropy.

(2) *Bose-Einstein Distribution.* Here Pauli's exclusion principle is left out of consideration, and hence $n = 0, 1 \ldots$. The ME distribution $\{< n_i >\}$ is obtained by substituting (5.5.18) in (5.5.13):

$$< n_i > = \sum_{j=1}^{G_i} \sum_{n=0}^{\infty} n e^{-(\lambda_1 + \lambda_2 \varepsilon_i)n} \Big/ \sum_{n=0}^{\infty} e^{-(\lambda_1 + \lambda_2 \varepsilon_i)n} \,.$$

Utilizing the formulae

$$\sum_{n=0}^{\infty} e^{-an} = 1/(1 - e^{-a}), \quad \sum_{n=0}^{\infty} n e^{-an} = e^{-a}/(1 - e^{-a})^2 , \quad (a > 0),$$

the above formula becomes

$$< n_i > = \frac{G_i}{e^{\lambda_1 + \lambda_2 \varepsilon_i} - 1} . \tag{5.5.22}$$

Substituting (5.5.18) ($n = 0, 1 \ldots$ in the summation in the denominator) in (5.5.14), after some mathematical manipulation, yields the maximum information-theoretic entropy,

$$\begin{aligned} S_{I\,\mathrm{max}} &= -\sum_i < n_i > \ln < n_i > + G_i \ln G_i \\ &\quad -(G_i + < n_i >) \ln(G_i + < n_i >) . \end{aligned} \tag{5.5.23}$$

By comparison of (5.5.22) and (5.5.12), we see that in the thermodynamic limit, the conventional statistical mechanics method and the Principle of ME give the same results. Moreover, by comparison of (5.5.23) and (5.5.10), we see that the maximum information-theoretic entropy and $\ln W_{\mathrm{BE}}$ are equal. From the Boltzmann relation, it follows that

$$S = k S_{I\,\mathrm{max}} , \tag{5.5.24}$$

where S is the thermodynamic entropy.

Finally, the constants λ_1 (α) and λ_2 (β) are determined by thermodynamical formulae. The results are

$$\begin{aligned} \lambda_1 &= \alpha = -\mu/kT , \\ \lambda_2 &= \beta = 1/kT , \end{aligned}$$

where μ is the chemical potential.

5.6 Classical Statistical Ensembles

Consider a system consisting of N identical particles. The mechanical state of each particle is described by specifying its position $\boldsymbol{q}$ and momentum $\boldsymbol{p}$. $\boldsymbol{q}$ and $\boldsymbol{p}$ have three components each, so the mechanical state of a system is described by specifying $(\boldsymbol{q}_1, \ldots, \boldsymbol{q}_N ; \boldsymbol{p}_1, \ldots, \boldsymbol{p}_N)$, which is represented by a point in a $6N$-dimensional phase space, or commonly called the Γ-space. So a system is represented by a point moving with time in the Γ-space. The $6N$ coordinates of the Γ-space are denoted by (q, p) for short.

A statistical ensemble is a very large collection of imaginary independent systems, each has the same structure and is under the same macroscopic conditions as the system being studied. So an ensemble is represented by a very large collection of points in the Γ-space. Let the density $f(q, p, t)$ be the probability that at time t a point representing a system of the ensemble is found to be in a unit volume element near the point (q, p) in the Γ-space. Consequently,

$$\mathrm{d}W = f(q, p, t)\mathrm{d}q\mathrm{d}p$$

is the probability that at time t a point representing a system is found to be in the volume element $\mathrm{d}q\mathrm{d}p$ near the point (q, p). The number of phase cells contained in the volume element $\mathrm{d}q\mathrm{d}p$ is

$$\mathrm{d}\Gamma = \frac{\mathrm{d}q\mathrm{d}p}{N! h^{3N}} \, , \tag{5.6.1}$$

(h is the Plank constant). $\mathrm{d}\Gamma$ is a dimensionless volume element.

Change the form of $\mathrm{d}W$:

$$\begin{aligned}
\mathrm{d}W &= f(q, p, t)\mathrm{d}q\mathrm{d}p \\
&= f(q, p, t)N! h^{3N}\mathrm{d}\Gamma \\
&= \rho(q, p, t)\mathrm{d}\Gamma \, , \tag{5.6.2}
\end{aligned}$$

where the density $\rho(q, p, t)$ represents the probability that at time t a point representing a system is found to be in a phase cell near the point (q, p). The normalization condition is

$$\int \rho(q, p, t)\mathrm{d}\Gamma = 1 \, . \tag{5.6.3}$$

Given the density $\rho(q, p, t)$, the ensemble average of any mechanical quantity $A(q, p)$ can be calculated by

$$< A > = \int A(q, p)\rho(q, p, t)\mathrm{d}\Gamma \, . \tag{5.6.4}$$

The information-theoretic entropy is defined by

$$S_I = -\int \rho(q, p, t) \ln \rho(q, p, t)\mathrm{d}\Gamma \, , \tag{5.6.5}$$

which is the same as Gibb's entropy. (S_I for grand canonical ensembles is defined somewhat differently, as will be seen later.) By Liouville's theorem it can be shown that S_I of an isolated system does not vary with time even in an irreversible process. This is a conceptual difficulty encountered when the Principle of ME is applied to irreversible processes.

In studying ensembles of equilibrium systems, none of the quantities introduced above contains time t explicitly, so we have $f(q, p)$, $\rho(q, p)$, etc.

We will study three kinds of classical ensembles. They are: microcanonical ensemble, i.e., ensemble of isolated systems (having no energy or substance exchange with environment); canonical ensemble, i.e., ensemble of closed systems (having energy but no substance exchange with environment); and grand canonical ensemble, i.e., ensemble of open systems (having energy and substance exchanges with environment). Our task is to determine the density $\rho(q, p)$. As before, the conventional statistical mechanics method is first briefly introduced, and then the Principle of ME is used.

5.6.1 Microcanonical Ensemble

1. Conventional Statistical Mechanics Method

Let isolated systems in an ensemble be characterized by macroscopic parameters volume V, the number of particles N, and energy E (with accuracy $\Delta E/E \ll 1$). Based on the assumption of equiprobabilities, the density of microcanonical distribution is found to be

$$\rho(q,p) = \begin{cases} \Omega^{-1}(E,N,V), & (E \le H(q,p) \le E + \Delta E), \\ 0, & (H(q,p) < E, \ H(q,p) > E + \Delta E) \end{cases} \tag{5.6.6}$$

($H(q,p)$ is the Hamiltonian, i.e., energy of the system). Ω is the total number of microstates (phase cells) having energies lying between E and $E + \Delta E$, i.e., the total number of points in the layer $E \sim E + \Delta E$ in the Γ-space,

$$\Omega(E,N,V) = \int_{E \le H(q,p) \le E + \Delta E} \mathrm{d}\Gamma. \tag{5.6.7}$$

In this way $\rho(q,p)$ satisfies the normalization condition automatically.

In the classical mechanics limit, $\Delta E \to 0$ and the density becomes

$$\rho(q,p) = \Omega^{-1}(E,N,V)\delta(H(q,p) - E).$$

Thus, the microstates of systems are distributed only on the hypersurface $H(q,p) = E$. Accordingly, the total number of microstates is

$$\Omega(E,N,V) = \int \delta(H(q,p) - E)\mathrm{d}\Gamma,$$

where the integration is carried out only on the hypersurface $H(q,p) = E$.

2. Principle of Maximum Entropy

$$\text{Maximization}: \quad S_I = -\int \rho(q,p) \ln \rho(q,p)\mathrm{d}\Gamma,$$

$$\text{Constraints}: \quad \int_{E \le H(q,p) \le E + \Delta E} \rho(q,p)\mathrm{d}\Gamma = 1, \quad (\text{normalization}).$$

It is easy to show the ME solution to be the same as (5.6.6). But here the assumption of equiprobabilities is not explicitly invoked.

The maximum information-theoretic entropy is

$$S_{I\,\max} = \ln \Omega. \tag{5.6.8}$$

According to the Boltzmann relation, the thermodynamic entropy

$$S = k \ln \Omega;$$

so that

$$S = kS_{I\,\max}. \tag{5.6.9}$$

5.6.2 Canonical Ensemble

1. Conventional Statistical Mechanics Method

Let closed systems in an ensemble be characterized by macroscopic parameters volume V, the number of particles N, and temperature T (energy H varies but its average value is fixed). The system being studied and the environment are together considered as an isolated system. Based on the assumption of equiprobabilities, after rather lengthy derivation, it turns out that the density of canonical distribution is

$$\rho(q,p) = \frac{1}{Z}e^{-\beta H(q,p)} \, , \tag{5.6.10}$$

where the partition function

$$Z(T,N,V) = \int e^{-\beta H(q,p)}\mathrm{d}\Gamma \, . \tag{5.6.11}$$

Later, we will see the determination of the constant β.

2. Principle of Maximum Entropy

$$\text{Maximization :} \qquad S_I = -\int \rho(q,p)\ln\rho(q,p)\mathrm{d}\Gamma \, , \tag{5.6.12}$$

$$\text{Constraints :} \qquad \int \rho(q,p)\mathrm{d}\Gamma = 1 \, , \tag{5.6.13a}$$

$$< H >= \int H(q,p)\rho(q,p)\mathrm{d}\Gamma \, ,$$

$$\text{(given average energy) ;} \quad \text{(5.6.13b)}$$

Determine the ME solution $\rho(q,p)$.

By the Lagrange multiplier method, the solution is

$$\rho(q,p) = \frac{1}{Z}e^{-\lambda H(q,p)} \, , \tag{5.6.14}$$

where the partition function

$$Z(T,N,V) = \int e^{-\lambda H(q,p)}\mathrm{d}\Gamma \, . \tag{5.6.15}$$

Based on the following inequality similar to (5.4.4):

$$-\int \rho'\ln\rho'\mathrm{d}\Gamma \leq -\int \rho'\ln\rho\mathrm{d}\Gamma \, , \tag{5.6.16}$$

(the equality holds if and only if $\rho' = \rho$), where ρ' and ρ are two arbitrary densities, utilizing the constraints (5.6.13a,b), and proceeding as in proving (5.4.5) in Sect. 5.4, it can be shown that $\rho(q,p)$ in (5.6.14) is really the ME solution.

Comparing (5.6.14) with (5.6.10), and (5.6.15) with (5.6.11), we see that the conventional statistical mechanics method and the Principle of ME give the same results.

Substituting (5.6.14) in (5.6.12) yields the maximum information-theoretic entropy,

$$
\begin{aligned}
S_{I\,\max} &= -\int \rho(-\ln Z - \lambda H)\mathrm{d}\Gamma \\
&= \ln Z + \lambda < H > .
\end{aligned}
\tag{5.6.17}
$$

On the other hand, the Helmholtz free energy is

$$
F = U - TS = -kT \ln Z ,
$$

where S is the thermodynamic entropy, U is the internal entropy of the system. Put it in the form

$$
S = k \ln Z + \frac{U}{T}
$$

which is compared with (5.6.17), it follows that

$$
\lambda = \beta = 1/kT ,
\tag{5.6.18}
$$

$$
S = kS_{I\,\max} .
\tag{5.6.19}
$$

5.6.3 Grand Canonical Ensemble

1. Conventional Statistical Mechanics Method

Let open systems in an ensemble be characterized by macroscopic parameters volume V, chemical potential μ (the number of particles N varies but its average value is fixed), and temperature T (energy H varies but its average value is fixed). The system being studied and the environment are together considered as an isolated system. On the basis of the assumption of equiprobabilities, through tedious derivation, it turns out finally that the density of grand canonical distribution is

$$
\rho_N(q,p) = \frac{1}{Z}\mathrm{e}^{-\alpha N - \beta H(q,p)} ,
\tag{5.6.20}
$$

which is the probability that a point representing a system consisting of N particles is found to be in a phase cell near the point (q,p) in the $6N$-dimensional Γ-space. The grand partition function is

$$
Z(T,\mu,V) = \sum_{N=0}^{\infty} \int \mathrm{e}^{-\alpha N - \beta H(q,p)}\mathrm{d}\Gamma .
\tag{5.6.21}
$$

Later, we will see the determination of the constants α and β .

2. Principle of Maximum Entropy

$$\text{Maximization}: \quad S_I = \sum_{N=0}^{\infty} - \int \rho_N(q,p) \ln \rho_N(q,p) \mathrm{d}\Gamma \ , \qquad (5.6.22)$$

$$\text{Constraints}: \quad \sum_{N=0}^{\infty} \int \rho_N(q,p) \mathrm{d}\Gamma = 1 \ , \qquad (5.6.23a)$$

$$<N> = \sum_{N=0}^{\infty} N \int \rho_N(q,p) \mathrm{d}\Gamma \ ,$$

$$\text{(given the average number of particles)} \ , \quad (5.6.23b)$$

$$<H> = \sum_{N=0}^{\infty} \int H(q,p) \rho_N(q,p) \mathrm{d}\Gamma \ ,$$

$$\text{(given average energy)} \ ; \qquad (5.6.23c)$$

Determine the ME solution $\rho_N(q,p)$.

The solution obtained by the Lagrange multiplier method is

$$\rho_N(q,p) = \frac{1}{Z} e^{-\lambda_1 N - \lambda_2 H(q,p)} \ , \qquad (5.6.24)$$

where the grand partition function

$$Z(T,\mu,V) = \sum_{N=0}^{\infty} \int e^{-\lambda_1 N - \lambda_2 H(q,p)} \mathrm{d}\Gamma \ . \qquad (5.6.25)$$

Utilizing the inequality

$$-\sum_{N=0}^{\infty} \int \rho'_N \ln \rho'_N \mathrm{d}\Gamma \leq -\sum_{N=0}^{\infty} \int \rho'_N \ln \rho_N \mathrm{d}\Gamma \ , \qquad (5.6.26)$$

(the equality holds if and only if $\rho'_N = \rho_N$) and the constraints (5.6.23a–c), it can be shown that $\rho_N(q,p)$ in (5.6.24) is really the ME solution. (See the proof for (5.4.5) in Sect. 5.4.)

Comparing (5.6.24) with (5.6.20), and (5.6.25) with (5.6.21), we see that the conventional statistical mechanics method and the Principle of ME give the same results.

Substituting (5.6.24) in (5.6.22) yields the maximum information-theoretic entropy,

$$\begin{aligned} S_{I\,\mathrm{max}} &= -\sum_{N=0}^{\infty} \int \rho_N(-\ln Z - \lambda_1 N - \lambda_2 H) \mathrm{d}\Gamma \\ &= \ln Z + \lambda_1 <N> + \lambda_2 <H> \ . \end{aligned} \qquad (5.6.27)$$

On the other hand, the grand potential is

$$J = U - TS - \mu N = -kT \ln Z \ ,$$

which is put in the form

$$S = k \ln Z - \frac{\mu}{T} N + \frac{U}{T} \ .$$

Comparing it with (5.6.27) yields

$$\lambda_1 = \alpha = -\mu/kT \ , \quad \lambda_2 = \beta = 1/kT \ , \tag{5.6.28}$$

$$S = kS_{I\,\text{max}} \ . \tag{5.6.29}$$

5.7 Quantum Statistical Ensembles

In this section an expression of information-theoretic entropy for a quantum statistical ensemble is derived first, then distributions for microcanonical, canonical, and grand canonical quantum ensembles are derived by the Principle of ME. This section is parallel with the preceding section in contents but the brief introduction of the conventional statistical mechanics method is left out since the description would be just a repetition of that in the preceding section and the results would be the same as those from the Principle of ME in this section.

The expressions of information-theoretic entropy we have used up to now are all of the form $-\sum p_i \ln p_i$ or its modified versions, where p_i's are scalars. In quantum mechanics, mechanical quantities are represented by operators, which are matrices in matrix mechanics. In a quantum statistical ensemble the distribution of systems among quantum states is represented by a density matrix ρ. On the analogy of $-\sum p_i \ln p_i$, the expression of information-theoretic entropy now can be written as

$$S_I = -\text{Tr}\{\rho \ln \rho\} \ ,$$

where Tr stands for the trace of a matrix. This expression can be derived from the definition of information-theoretic entropy.

Let accessible (pure or compound) states to systems in an ensemble be represented by wave functions ψ_i, $i = 1, 2, \ldots$, which are vectors:

$$\psi_i = \begin{pmatrix} a_{1i} \\ a_{2i} \\ \vdots \end{pmatrix} \ , \quad \sum_n |a_{ni}|^2 = 1 \ . \tag{5.7.1}$$

(Only the discrete case is considered for simplicity.) $\psi_i, i = 1, 2, \ldots$ form an orthonormal and complete system:

$$|(\psi_i, \psi_j)|^2 = \sum_m a_{mi}^* a_{mj} = \delta_{ij} \ . \tag{5.7.2}$$

For a system in quantum state ψ_i, the expectation of mechanical quantity F is

$$< F >_i = (\psi_i, F\psi_i) = \sum_m \sum_n a^*_{mi} F_{mn} a_{ni} \ . \tag{5.7.3}$$

Suppose in an ensemble the probability that a system is found to be in quantum state ψ_i is w_i. The ensemble average of $< F >_i$ is

$$< F > = \sum_i w_i < F >_i = \sum_m \sum_n \sum_i w_i a^*_{mi} a_{ni} F_{mn} \ .$$

Define a density matrix $\rho = (\rho_{nm})$ by

$$\rho_{nm} = \sum_i w_i a^*_{mi} a_{ni} \ . \tag{5.7.4}$$

Then,

$$\begin{aligned}
< F > &= \sum_m \sum_n \rho_{nm} F_{mn} \\
&= \mathrm{Tr}\{F\rho\} \ .
\end{aligned} \tag{5.7.5}$$

Given the density matrix ρ, the observational value (i.e., ensemble average) of the mechanical quantity F can calculated.

Since ψ_i and ψ_j $(i \neq j)$ are orthogonal, and hence the events that a system is found to be in state ψ_i and in state ψ_j are mutually exclusive, the information-theoretic entropy of the ensemble is

$$S_I = -\sum_i w_i \ln w_i \ . \tag{5.7.6}$$

Now we show that w_i is an eigenvalue of ρ and ψ_i is the corresponding eigenvector, $i = 1, 2, \ldots$, i.e.,

$$\rho\psi_i = w_i \psi_i \ .$$

From (5.7.1, 5.7.4), it follows that the nth components of the column vector $\rho\psi_i$ is

$$\begin{aligned}
(\rho\psi_i)_n &= \sum_m \sum_l w_l a^*_{ml} a_{nl} \cdot a_{mi} \\
&= \sum_l w_l a_{nl} \sum_m a^*_{ml} a_{mi} \\
&= \sum_l w_l a_{nl} \delta_{li} \\
&= w_i a_{ni} \ ,
\end{aligned}$$

which is just the nth component of the column vector $w_i\psi_i$. Note that (5.7.2) was utilized in the above.   Q.E.D.

A function of eigenvalue w_i of the matrix ρ, $f(w_i)$, is an eigenvalue of the function of the matrix ρ, $f(\rho)$. So we have

$$(\rho \ln \rho)\psi_i = (w_i \ln w_i)\psi_i \,,$$

i.e., $\rho \ln \rho \sim W$ ($\sim$ stands for similar), where W is a diagonal matrix with elements $w_{ii} = w_i \ln w_i$. Consequently, from (5.7.6) it follows that

$$
\begin{aligned}
S_I &= -\mathrm{Tr}\{W\} \\
    &= -\mathrm{Tr}\{\rho \ln \rho\} \,.
\end{aligned}
\tag{5.7.7}
$$

The density matrix satisfies the normalization condition

$$\mathrm{Tr}\{\rho\} = 1 \tag{5.7.8}$$

since, utilizing (5.7.4, 5.7.1),

$$\mathrm{Tr}\{\rho\} = \sum_n \rho_{nn} = \sum_i w_i \sum_n a_{ni}^* a_{ni} = \sum_i w_i = 1 \,.$$

Let ρ' be a density matrix corresponding to a distribution $\{w_i'\}$. From the inequality

$$-\sum_i w_i' \ln w_i' \le -\sum_i w_i' \ln w \,,$$

it follows immediately that

$$-\mathrm{Tr}\{\rho' \ln \rho'\} \le -\mathrm{Tr}\{\rho' \ln \rho\} \,, \tag{5.7.9}$$

(the equality holds if and only if $\rho' = \rho$), where ρ' and ρ are two arbitrary density matrices.

In the rest of this section we derive, by the Principle of ME, density matrices of distribution of systems among energy levels (and, additionally, the numbers of particles for a grand canonical ensemble) for three kinds of quantum ensembles.

5.7.1 Microcanonical Ensemble

$$\text{Maximization}: \quad S_I = -\mathrm{Tr}\{\rho \ln \rho\} \,, \tag{5.7.10}$$

$$\text{Constraints}: \quad \mathrm{Tr}\{\rho\} = 1 \,.$$

It is easy to find the ME solution ρ (a diagonal matrix):

$$
\rho_{nn} =
\begin{cases}
\Omega^{-1}(E, N, V) \,, & (E \le H(q,p) \le E + \Delta E) \,, \\
0 \,, & (H(q,p) < E, \; H(q,p) > E + \Delta E) \,.
\end{cases}
\tag{5.7.11}
$$

Here Ω is the total number of quantum states having energies between E and $E + \Delta E$ ($\Delta E/E \ll 1$). Equation (5.7.11) gives the density of microcanonical distribution for a (microcanonical) quantum ensemble. This is the same as the result from the assumption of equiprobabilities.

Substituting (5.7.11) in (5.7.10) yields the maximum information-theoretic entropy,

$$S_{I\,\max} = -\sum_{n=1}^{\Omega} \Omega^{-1} \ln \Omega^{-1} = \ln \Omega \; .$$

From the Boltzmann relation, the relation between the thermodynamic entropy S and $S_{I\,\max}$ follows immediately:

$$S = kS_{I\,\max} \; . \tag{5.7.12}$$

5.7.2 Canonical Ensemble

$$\text{Maximization}: \quad S_I = -\text{Tr}\{\rho \ln \rho\} \; , \tag{5.7.13}$$

$$\text{Constraints}: \quad \text{Tr}\{\rho\} = 1 \; , \tag{5.7.14a}$$

$$<H> = \text{Tr}\{H\rho\} \; , \quad \text{(given average energy)}\; ; \tag{5.7.14b}$$

Determine the ME solution ρ.

Introduce the Lagrange multipliers α and λ (scalars), form the objective functional

$$G = -\text{Tr}\{\rho \ln \rho\} - \alpha(\text{Tr}\{\rho\} - 1) - \lambda(\text{Tr}\{H\rho\} - <H>) \; ,$$

and let the variation of G with respect to ρ be zero, i.e.,

$$0 = \delta G = \text{Tr}\{[-(I + \ln \rho) - \alpha I - \lambda H]\delta\rho\} \; ,$$

where I is the identity matrix. Since $\delta\rho$ is arbitrary, we obtain

$$\rho = e^{-\alpha I - \lambda H} = e^{-\alpha}e^{-\lambda H} \; .$$

Note that α has be replaced by $(\alpha + 1)$. Substituting the above expression in the normalization condition (5.7.14a) yields

$$1 = \text{Tr}\{e^{-\alpha}e^{-\lambda H}\} = e^{-\alpha}\text{Tr}\{e^{-\lambda H}\} \; .$$

Define the partition function

$$Z(T, N, V) = e^{\alpha} = \text{Tr}\{e^{-\lambda H}\} \; . \tag{5.7.15}$$

Then, the density matrix can be put in the form

$$\rho = \frac{1}{Z}e^{-\lambda H} \; . \tag{5.7.16}$$

Now we show that ρ in (5.7.16) is really the ME solution. Let ρ' be an arbitrary solution. From (5.7.9) we obtain

$$\begin{aligned} S_{I\rho'} &= -\text{Tr}\{\rho' \ln \rho'\} \\ &\leq -\text{Tr}\{\rho' \ln \rho\} \\ &= -\text{Tr}\{\rho'(-I \ln Z - \lambda H)\} \\ &= \ln Z \, \text{Tr}\{\rho'\} + \lambda\text{Tr}\{H\rho'\} \\ &= \ln Z + \lambda <H> \; . \end{aligned}$$

Note that constraints (5.7.14a,b) were utilized in the above (ρ' is a solution). On the other hand,

$$
\begin{aligned}
S_{I\rho} &= -\text{Tr}\{\rho \ln \rho\} \\
&= -\text{Tr}\{\rho(-I \ln Z - \lambda H)\} \\
&= \ln Z + \lambda < H > .
\end{aligned}
$$

Comparing the two expressions above, we see

$$
S_{I\rho'} \leq S_{I\rho} . \tag{5.7.17}
$$

The maximum information-theoretic entropy is

$$
S_{I\,\text{max}} = \ln Z + \lambda < H > . \tag{5.7.18}
$$

The density matrix ρ in (5.7.16) is the density of canonical distribution for a (canonical) quantum ensemble.

Utilizing the formula for the Helmholtz free energy as in Sect. 5.6.2, we get

$$
\begin{aligned}
\lambda &= 1/kT , \tag{5.7.19} \\
S &= kS_{I\,\text{max}} . \tag{5.7.20}
\end{aligned}
$$

5.7.3 Grand Canonical Ensemble

$$
\text{Maximization}: \quad S_I = -\text{Tr}\{\rho \ln \rho\} , \tag{5.7.21}
$$

$$
\text{Constraints}: \quad \text{Tr}\{\rho\} = 1 , \tag{5.7.22a}
$$

$$
< N >= \text{Tr}\{N\rho\} ,
$$
$$
\text{(given the average number of particles)} , \tag{5.7.22b}
$$

$$
< H >= \text{Tr}\{H\rho\} , \quad \text{(given average energy)} ; \tag{5.7.22c}
$$

Determine the ME solution ρ.

Form the objective functional

$$
\begin{aligned}
G &= -\text{Tr}\{\rho \ln \rho\} - \alpha(\text{Tr}\{\rho\} - 1) - \lambda_1(\text{Tr}\{N\rho\} - < N >) \\
&\quad -\lambda_2(\text{Tr}\{H\rho\} - < H >) .
\end{aligned}
$$

Let

$$
0 = \delta G = \text{Tr}\{[-(I + \ln \rho) - \alpha I - \lambda_1 N - \lambda_2 H]\delta\rho\} .
$$

Then,

$$
\rho = e^{-\alpha I - \lambda_1 N - \lambda_2 H} = e^{-\alpha} e^{-\lambda_1 N - \lambda_2 H} .
$$

From the normalization condition (5.7.22a), it follows that

$$1 = e^{-\alpha} \text{Tr}\{e^{-\lambda_1 N - \lambda_2 H}\} .$$

Defining the grand partition function,

$$Z(T, \mu, V) = e^{\alpha} = \text{Tr}\{e^{-\lambda_1 N - \lambda_2 H}\} , \tag{5.7.23}$$

then the density matrix can be written as

$$\rho = \frac{1}{Z} e^{-\lambda_1 N - \lambda_2 H} . \tag{5.7.24}$$

In the same way as proving (5.7.17) in the preceding subsection, we can show that (5.7.24) is really the ME solution. The maximum information-theoretic entropy is

$$S_{I\,\text{max}} = \ln Z + \lambda_1 < N > + \lambda_2 < H > . \tag{5.7.25}$$

The density matrix ρ in (5.7.24) is the density of grand canonical distribution for a (grand canonical) quantum ensemble. Utilizing the formula for grand potential as in Sect. 5.6.3, the Lagrange multipliers are determined to be

$$\lambda_1 = -\mu/kT , \quad \lambda_2 = 1/kT , \tag{5.7.26}$$

and the relation between the thermodynamic entropy S and the maximum information-theoretic entropy is found to be

$$S = kS_{I\,\text{max}} . \tag{5.7.27}$$

Appendices

A. Cepstral Analysis

In this appendix for Chap. 3 we introduce the basic concept of the cepstral analysis system (CAS), the input/output relationship; the properties of the complex cepstrum; the I/O relationship when the input is a causal and minimum-phase sequence, and the relationship between the complex and real cepstra for an (additionally) real-valued sequence. They will play an important role in MEM2 [A.1, A.2].

A.1 Cepstral Analysis System

Given a sequence $y(n)$, define a sequence $\hat{y}(n)$ by

$$\hat{y}(n) = \text{IZT}[\, \log \text{ZT}[y(n)]\,]\,, \quad n = \ldots, -1, 0, 1, \ldots, \tag{A.1}$$

where ZT (IZT) denotes the (inverse) Z-transform, and log the natural logarithm. $\hat{y}(n)$ is called the *complex cepstrum* of $y(n)$. The restrictive "complex" is used since the logarithm is generally a complex one.

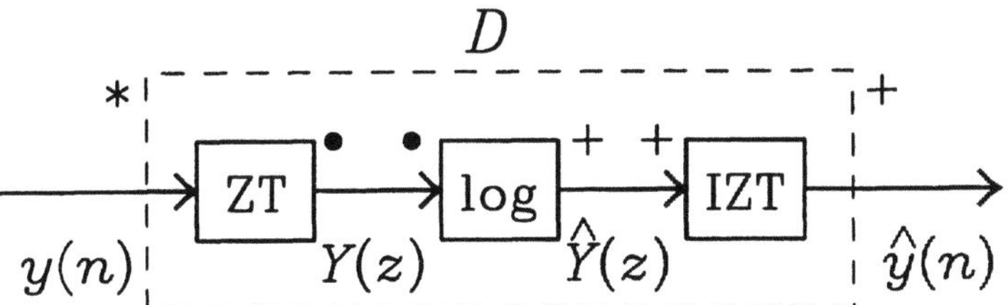

Fig. A.1. Z-transformation realization for the complex cepstral analysis system D. Complex $\log Y(z) = \log |Y(z)| + \mathrm{i}\arg[Y(z)]$

The transformation from $y(n)$ to $\hat{y}(n)$ is depicted in Fig. A.1. This is called the Z-transformation realization for the complex CAS.

In order that the logarithmic function can be defined, $Y(z) = \text{ZT}[y(n)]$ must be analytic and nonzero in an annular region of the complex z-plane. In this regard, particularly the argument (phase) $\arg[Y(z)]$ must be continuous but cannot simply take the principal value. Since we are usually interested only in stable signals, the annular region should contain the unit circle. In such a case the complex CAS is usually realized by FT, i.e.,

$$\hat{y}(n) = \text{IFT}[\,\log \text{FT}[y(n)]\,]\,, \tag{A.2}$$

which can also be depicted by Fig. A.1 with the Z-transformation replaced by FT.

For the FT realization, if only the real part of the logarithm is of interest, then define

$$\hat{y}_R(n) = \text{IFT}[\,\log |\text{FT}[y(n)]|\,]\,. \tag{A.3}$$

$\hat{y}_R(n)$ is called the *real cepstrum* of $y(n)$.

The complex CAS, D, is also called the *characteristic system* in a homomorphic system for convolution, which means that this system D performs homomorphic transformation between the input $y(n)$ and output $\hat{y}(n)$, the "addition" at the input being convolution while the "addition" at the output being ordinary addition.

Let

$$y(n) = y_1(n) * y_2(n) = \sum_{k=-\infty}^{\infty} y_1(k) \cdot y_2(n - k)\,, \tag{A.4}$$

then

$$\begin{aligned}
Y(z) &= Y_1(z) \cdot Y_2(z)\,, \\
\hat{Y}(z) &= \hat{Y}_1(z) + \hat{Y}_2(z)\,, \\
\hat{y}(n) &= \hat{y}_1(n) + \hat{y}_2(n)\,.
\end{aligned}$$

Thus, the two sequences of convolution at the input are transformed to those of addition at the output:

$$D[y_1(n) * y_2(n)] = D[y_1(n)] + D[y_2(n)]\,.$$

It may be the case that separating $\hat{y}_1(n)$ and $\hat{y}_2(n)$ at the output is easier than separating $y_1(n)$ and $y_2(n)$ at the input; so we can take $\hat{y}_1(n)$ or $\hat{y}_2(n)$ and then recover $y_1(n)$ or $y_2(n)$ by the inverse system D^{-1}. This kind of deconvolution method has important applications in areas such as speech and seismic signal processings. As a matter of fact, it was first in speech signal processing that the concept "cepstrum" was introduced. (*Cepstrum* paraphrases the word *spectrum*. The others, *quefrency* – *frequency*, *repiod* – *period*, etc. are not in common use.)

A.2 I/O Relationship

The derivative of $\hat{Y}(z)$ with respect to z is

$$\hat{Y}'(z) = \frac{\mathrm{d}}{\mathrm{d}z}[\log Y(z)] = \frac{Y'(z)}{Y(z)}\,,$$

which is multiplied by $zY(z)$ to yield

$$zY'(z) = [z\hat{Y}'(z)] \cdot Y(z) \ .$$

Then, performing IZT on the two sides and utilizing

$$\begin{aligned}
Z[ny(n)] &= -zY'(z) \ , \\
Z[y_1(n) * y_2(n)] &= Y_1(z) \cdot Y_2(z) \ ,
\end{aligned}$$

it follows that

$$-n y(n) = [-n\hat{y}(n)] * y(n) \ ,$$

that is,

$$n y(n) = \sum_{k=-\infty}^{\infty} k\hat{y}(k)y(n-k) \ .$$

Dividing by n on the two sides finally yields

$$y(n) = \sum_{k=-\infty}^{\infty} \left(\frac{k}{n}\right) \hat{y}(k)y(n-k) \ , \quad n \neq 0 \ . \tag{A.5}$$

Equation (A.5) is the relationship between the input $y(n)$ and the output $\hat{y}(n)$ for the complex CAS. Here the only requirement to $Y(z)$ is that it should be analytic and nonzero in an annular region containing the unit circle. When the CAS is realized by FT, (A.5) can also be derived, utilizing properties of FT, in much the same way as above, cf. [3.6, Sect. 2.5], As a matter of fact, this just means letting $z = e^{i\omega}$.

Equation (A.5) is an implicit relationship between $y(n)$ and $\hat{y}(n)$, representing a noncausal system. Any algorithm based on (A.5) must be an iterative one. Nevertheless, under certain conditions (A.5) can be simplified to a recursion formula which can be used in computation (noniterative).

A.3 Properties of the Complex Cepstrum

The general properties of the complex cepstrum are listed below.

Property 1. If $y(n)$ is a causal and minimum-phase sequence, then the complex cepstrum is causal:

$$\hat{y}(n) = 0 \quad \text{for} \quad n < 0 \ .$$

Property 2. If $y(n)$ is an anticausal and maximum-phase sequence, then the complex cepstrum is anticausal:

$$\hat{y}(n) = 0 \quad \text{for} \quad n > 0 \ .$$

Property 3. If $y(n)$ is of finite duration, then the complex cepstrum $\hat{y}(n)$ will nevertheless have infinite duration; conversely, if $\hat{y}(n)$ is of finite duration, then $y(n)$ will nevertheless have infinite duration.

If the input $y(n)$ has a rational ZT, then the Properties 1 and 2 can be stated more specifically, and Property $4'$ can be added. Now the ZT of $y(n)$ is

$$Y(z) = Az^R \frac{\displaystyle\prod_{k=1}^{m_i}(1 - a_k z^{-1}) \prod_{k=1}^{m_o}(1 - b_k z)}{\displaystyle\prod_{k=1}^{p_i}(1 - c_k z^{-1}) \prod_{k=1}^{p_o}(1 - d_k z)} \,, \tag{A.6}$$

where $|a_k|$, $|b_k|$, $|c_k|$, $|d_k| < 1$, so that a_k and c_k are zeros and poles inside the unit circle, respectively; while b_k and d_k are zeros and poles outside the unit circle, respectively. The factor z^R can be left out and A can be assumed positive by appropriate measurement.

Expanding $\hat{Y}(z) = \log Y(z)$ and utilizing

$$\log(1 - \alpha z^{-1}) \;=\; -\sum_{n=1}^{\infty} \frac{\alpha^n}{n} z^{-n} \,, \quad |z| > |\alpha| \,,$$

$$\log(1 - \beta z^{+1}) \;=\; -\sum_{n=1}^{\infty} \frac{\beta^n}{n} z^{+n} \,, \quad |z| < |\beta^{-1}| \,,$$

from (A.6) it follows that the IZT of $\hat{Y}(z)$, $\hat{y}(n)$, is

$$\hat{y}(n) = \begin{cases} \log A \,, & n = 0 \,, \\[2mm] -\displaystyle\sum_{k=1}^{m_i} \frac{a_k^n}{n} + \sum_{k=1}^{p_i} \frac{c_k^n}{n} \,, & n > 0 \,, \\[2mm] \displaystyle\sum_{k=1}^{m_o} \frac{b_k^{-n}}{n} - \sum_{k=1}^{p_o} \frac{d_k^{-n}}{n} \,, & n < 0 \,. \end{cases} \tag{A.7}$$

From (A.7) we observe the following properties of the complex cepstrum in the particular case where the ZT of the input $y(n)$ is a rational function:

Property $1'$. If $y(n)$ is a causal and minimum-phase sequence (no poles or zeros outside the unit circle, i.e., $b_k = d_k = 0$), then the complex cepstrum is causal and satisfies

$$\hat{y}(n) = \begin{cases} \log A \,, & n = 0 \,, \\[2mm] -\displaystyle\sum_{k=1}^{m_i} \frac{a_k^n}{n} + \sum_{k=1}^{p_i} \frac{c_k^n}{n} \,, & n > 0 \,, \\[2mm] 0 \,, & n < 0 \,. \end{cases} \tag{A.8}$$

Property $2'$. If $y(n)$ is an anticausal and maximum-phase sequence (no poles or zeros inside the unit circle, i.e., $a_k = c_k = 0$), then the complex cepstrum is anticausal and satisfies

$$\hat{y}(n) = \begin{cases} \log A \,, & n = 0 \,, \\ \displaystyle\sum_{k=1}^{m_o} \frac{b_k^{-n}}{n} - \sum_{k=1}^{p_o} \frac{d_k^{-n}}{n} \,, & n < 0 \,, \\ 0 \,, & n > 0 \,. \end{cases} \tag{A.9}$$

Property 4′. The complex cepstrum decays at least as fast as $1/n$. Specifically,

$$|\hat{y}(n)| < C \left| \frac{\alpha^n}{n} \right| \,, \quad -\infty < n < \infty \,, \tag{A.10}$$

where C is a constant and $\alpha = \max\{|a_k|, |b_k|, |c_k|, |d_k|\}$.

A.4 I/O Relationship for Minimum-Phase Input

When the input $y(n)$ is a causal and minimum-phase sequence, $y(n) = 0$ for $n < 0$, $\hat{y}(n) = 0$ for $n < 0$ (Property 2). The I/O relationship is

$$y(n) = \begin{cases} 0 \,, & n < 0 \,, \\ \exp[\hat{y}(0)] \,, & n = 0 \,, \\ \displaystyle\hat{y}(n)y(0) + \sum_{k=0}^{n-1} \left(\frac{k}{n}\right) \hat{y}(k)y(n - k) \,, & n > 0 \,. \end{cases} \tag{A.11}$$

The bottom line for $n > 0$ comes from (A.5), and the middle line for $n = 0$ can be proved. Equation (A.11) is equivalent to

$$\hat{y}(n) = \begin{cases} 0 \,, & n < 0 \,, \\ \log y(0) \,, & n = 0 \,, \\ \displaystyle\frac{y(n)}{y(0)} - \sum_{k=0}^{n-1} \left(\frac{k}{n}\right) \hat{y}(k)\frac{y(n - k)}{y(0)} \,, & n > 0 \,. \end{cases} \tag{A.12}$$

Formulae (A.11, A.12) each represent a causal system. These recursion formulae can be used in computation (noniterative).

Relationships corresponding to (A.11, A.12) for an anticausal and maximum-phase input $y(n)$ are not given here since we will not use them.

If $y(n)$ is real-valued in addition to causal and of minimum-phase, the complex cepstrum $\hat{y}(n)$ and real cepstrum $C(n) = \hat{y}_R(n)$ satisfy the following relationship:

$$C(n) = \begin{cases} 1/2\hat{y}(n) \,, & n > 0 \,, \\ \hat{y}(0) \,, & n = 0 \,, \\ 1/2\hat{y}(-n) \,, & n < 0 \,, \end{cases} \tag{A.13}$$

which is shown in Fig. A.2. Note that the real CAS only has the FT realization while the complex cepstrum $\hat{y}(n)$ is independent of the realization of the complex CAS.

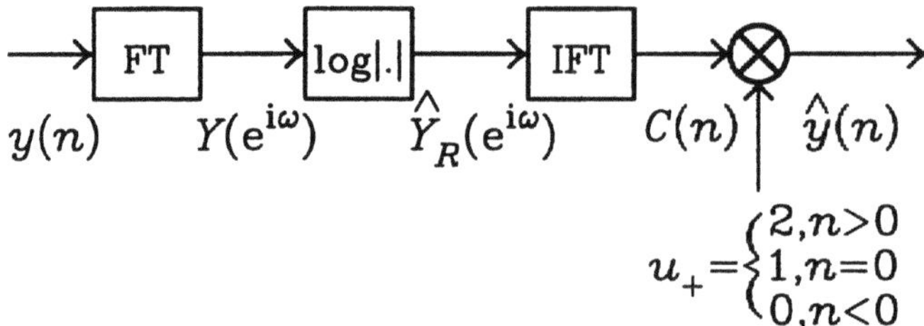

Fig. A.2. The complex CAS is realized by the real CAS when the input is a real, causal and minimum-phase sequence

Formula (A.13) can be proved as follows: Put $Y(e^{i\omega})$ in the polar form while $\hat{Y}(e^{i\omega})$ in the Cartesian form, then

$$
\begin{aligned}
Y(e^{i\omega}) &= |Y(e^{i\omega})|e^{i\,\arg[Y(e^{i\omega})]} \,, \\
\hat{Y}(e^{i\omega}) &= \hat{Y}_R(e^{i\omega}) + i\hat{Y}_I(e^{i\omega}) \,.
\end{aligned}
\tag{A.14}
$$

Since $y(n)$ is a real sequence,

$$\hat{Y}_R(e^{i\omega}) = \log|Y(e^{i\omega})| \text{ is an even function in } \omega,$$

$$\hat{Y}_I(e^{i\omega}) = \arg[Y(e^{i\omega})] \text{ is an odd function in } \omega.$$

$\hat{y}(n)$ is also a real sequence, which can be decomposed into an even part and an odd part, i.e.,

$$\hat{y}(n) = \hat{y}_e(n) + \hat{y}_o(n) \,, \tag{A.15}$$

where

$$\hat{y}_e(n) = \frac{1}{2}[\hat{y}(n) + \hat{y}(-n)] \,, \tag{A.16a}$$

$$\hat{y}_o(n) = \frac{1}{2}[\hat{y}(n) - \hat{y}(-n)] \,. \tag{A.16b}$$

The properties of FT state that the FT of a real and even sequence is a real and even function, while the FT of a real and odd sequence is an imaginary and odd function. Comparing (A.15) with (A.14) we see that the FT of $\hat{y}_e(n)$ is $\hat{Y}_R(e^{i\omega})$. That is to say, $\hat{y}_e(n)$ is just the real cepstrum of $y(n)$, $C(n)$:

$$
\begin{aligned}
\hat{y}_e(n) &= \text{IFT}[\,\hat{Y}_R(e^{i\omega})\,] = \text{IFT}[\,\log|Y(e^{i\omega})|\,] \\
&= \text{IFT}[\,\log|\text{FT}[y(n)]|\,] \,.
\end{aligned}
$$

From (A.16a), noticing $y(n) = 0$ for $n < 0$, it follows that

$$
\begin{aligned}
C(n) = \hat{y}_e(n) &= \tfrac{1}{2}[\hat{y}(n) + 0] && \text{for} \quad n > 0 \,, \\
C(n) = \hat{y}_e(n) &= \tfrac{1}{2}[\hat{y}(0) + \hat{y}(0)] && \text{for} \quad n = 0 \,, \\
C(n) = \hat{y}_e(n) &= \tfrac{1}{2}[0 + \hat{y}(-n)] && \text{for} \quad n < 0 \,,
\end{aligned}
$$

and (A.13) results. The conditions are: The input $y(n)$ is a real, causal and minimum-phase sequence.   Q.E.D.

Finally, we would like to say a few words about the technical terms "complex cepstrum" and "real cepstrum". Here the restrictive "complex" and "real" refer merely to the logarithm, but have nothing to do with the values (complex or real) of sequences. The complex and real cepstra of a real sequence are both real, otherwise the two sides of (A.5), (A.3), etc. do not match. For the same reason, the complex and real cepstra of a complex sequence are both complex. Particularly, if $y(n)$ is an ACF sequence and hence is positive definite, then $Y(e^{i\omega})$ is positive, and therefore the complex and real cepstra of $y(n)$ are identical, and $\hat{y}(n)$ is Hermitian when the FT realization is used. By "cepstrum" we usually mean real cepstrum with a few exceptions. In this regard context will help.

B. Image Restoration

In this appendix for Chap. 3 we introduce the basic concept and relationships in image formation and restoration, and the relationship between image restoration and spectral estimation. They are preliminary to the study of MEM2 in image restoration.

B.1 Image Formation

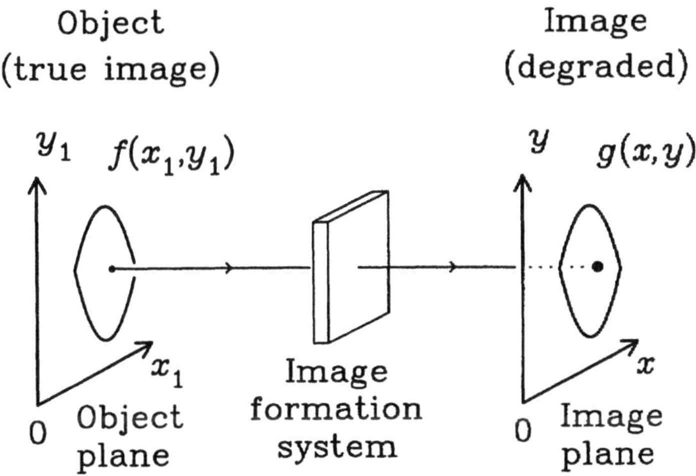

Fig. B.1. Schematic diagram of image formation (spatial domain)

The formation of a 2-D image is schematically depicted in Fig. B.1 [A.3]. The radiations emitted from an object (or true image) on the object plane on the left go through the image formation system in the middle, then reach the image plane on the right and form an image. The distribution of radiation intensity of the object is represented by a continuous function $f(x_1, y_1)$, while the intensity distribution of the image is represented by a continuous function $g(x, y)$. The formed image is generally different from the object in one way or another since the image formation system is almost always imperfect. So sometimes $g(x, y)$ is explicitly called the *degraded* image.

Ignoring nonlinear effect, the image formation system can be modeled by a linear system in Fig. B.2, which consists of a 2-D linear filter and an adder in cascade.

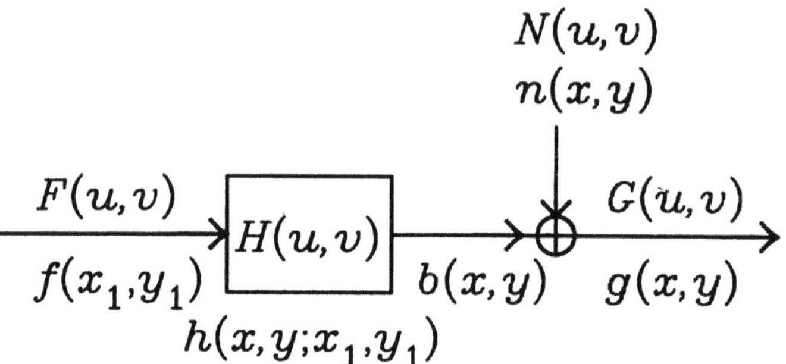

Fig. B.2. A linear model of the image formation system

1. Spatial Domain Representation

The output from the filter is

$$b(x, y) = \int\!\!\int_{-\infty}^{+\infty} h(x, y;\ x_1, y_1) f(x_1, y_1) \mathrm{d}x_1 \mathrm{d}y_1 \ , \tag{B.1}$$

where $h(x, y;\ x_1, y_1)$ is called the *point spread function* (PSF), which is, in fact, the impulse response of the filter. The term PSF means that if the object contains only one point at (p, q), $f(x_1, y_1) = \delta_{x_1-p, y_1-q}$, then the output is

$$\begin{aligned} b_0(x, y) &= \int\!\!\int_{-\infty}^{+\infty} h(x, y;\ x_1, y_1) \delta_{x_1-p, y_1-q} \mathrm{d}x_1 \mathrm{d}y_1 \\ &= h(x, y;\ p, q) \ . \end{aligned}$$

For a perfect image formation system, $h(x, y;\ p, q) = \delta_{x-p, y-q}$, the image of a point of the object is also a point. Usually, each point of the object will spread to form an area of finite size on the image plane, resulting in a blurred image of the whole object.

When only additive noise is under consideration in practice, the noise is represented by $n(x, y)$ input to the adder. The degraded image, which is the final output from the image formation system, is

$$g(x, y) = b(x, y) + n(x, y) \ . \tag{B.2}$$

So, we may say in words: "degraded" = "blurred" + "noisy".

When the image formation system is space-invariant, or homogeneous, the PSF depends only on the differences of the corresponding coordinates, then (B.1) becomes

$$b(x,y) = \int\int_{-\infty}^{+\infty} h(x-x_1, y-y_1)f(x_1,y_1)\mathrm{d}x_1\mathrm{d}y_1$$
$$= h(x,y) * f(x,y) , \tag{B.3}$$

where $*$ denotes 2-D convolution. In this case a displacement of the object will only result in a displacement of the image but not a change of image configuration. The situation here is similar to that in a time-invariant system.

2. Spatial Frequency Domain Representation

From now on we only deal with homogeneous linear systems. Substitute $b(x,y)$ of (B.3) in (B.2) and perform FT on the two sides; then

$$G(u,v) = H(u,v) \cdot F(u,v) + N(u,v) , \tag{B.4}$$

where

u, v are the spatial frequencies;
$G(u,v)$ is the visibility, i.e., the FT, of the degraded image;
$H(u,v)$ the transfer function of the system, i.e., the FT of $h(x,y)$;
$F(u,v)$ the visibility of the true image (object);
$N(u,v)$ the FT of the noise $n(x,y)$.

It should be pointed out that depending on data acquisition, the noise on the (u,v)-plane, $N(u,v)$, may be unrelated to the noise on the (x,y)-plane, $n(x,y)$.

B.2 Image Restoration

Image restoration is the following problem:

Given the PSF h (or the transfer function H), the statistical properties of the noise n (or N), and/or some prior knowledge about the true image f (or its visibility F) (e.g., $f \geq 0$, f is space limited), estimate the true image f (or F) from the measured degraded image g (or its visibility G).

In image restoration, our goal is to make the processed image be as close to the true image as possible in intensity distribution. The closeness can be objectively defined. In a related technique, image enhancement, on the other hand, the goal is to process an image so that some features of the image manifest themselves best for image analysis. The criterion for "best" is essentially subjective. An enhanced image may well be quite different from the original (true) image in intensity distribution.

Image restoration is essentially a kind of deconvolution operation. It can be carried out in the spatial domain, based on (B.2, B.3), or in the spatial frequency domain, based on (B.4), or alternatively in the two domains.

A comprehensive introduction of image restoration methods is beyond the scope of this appendix. What we like to do is to show that the direct inversion method is not feasible in general.

In the spatial frequency domain, direct inversion means dividing the two sides of (B.4) by $H(u,v)$, resulting in

$$F(u,v) = \frac{G(u,v)}{H(u,v)} - \frac{N(u,v)}{H(u,v)} \ . \tag{B.5}$$

The estimated visibility is

$$\hat{F}(u,v) = \frac{G(u,v)}{H(u,v)} \ . \tag{B.6}$$

Then, the estimated image $\hat{f}(x,y)$ is the IFT of $\hat{F}(u,v)$.

Problems with the direct inversion method are:

(1) Normally $H(u,v)$ has zeros; at these points the division operation is meaningless.

(2) For the points having very small $|H(u,v)|$, although the division can be done theoretically, the ignored noise will be enlarged to an intolerable extent.

Therefore, in general, the direct inversion method is not applicable to image restoration. An exception might be the case where the signal-to-noise ratio is fairly high, and the zero or near zero values of $H(u,v)$ are reasonably altered.

B.3 Relationship Between Image Restoration and Spectral Estimation

From the above description we see that a homogeneous image formation system is equivalent to a 2-D space-invariant linear filter (noise may be added) in spectral analysis. An analogy between image restoration and spectral estimation of time series can be established as shown in Table B.1.

Table B.1. Analogy between image restoration and spectral estimation

	Given data		To be estimated
Time series spectral estimation	Partial ACF R	FT $\rightleftharpoons$ IFT	Spectrum S
Image restoration	Partial visibility G	IFT $\rightleftharpoons$ FT	True image f

Note the FT direction. In this regard, convention is followed in the above table. The difference between spectral estimation and image restoration in the

FT direction is not essential; it makes only a little difference to mathematical formulae and practical computation.

In principle, spectral estimation and image restoration may be thought to be equivalent. However, we should keep in mind their differences as stated in the following:

1. Time series is 1-D while image signal is usually 2-D. This difference in dimensionality may sometimes result in an insurmountable obstacle when one attempts to use spectral analysis methods for image signal processing.

On the other hand, for the simplicity and convenience of notation and description, sometimes image signal is put in the 1-D form. This is true especially for image signal of discrete form. In doing this, pixels of a 2-D image are arranged in such a way that the end of a column (or row) is joined to the beginning of the next column (or row) (lexicographic ordering) to form a long (1-D) column (or row) vector. For a 2-D image composed of $N \times N$ pixels, the image vectors and noise vector each have N^2 components,

$$\boldsymbol{f} = (f_1 \ldots, f_{N^2})^{\mathrm{T}} , \qquad \boldsymbol{g} = (g_1, \ldots, g_{N^2})^{\mathrm{T}} ,$$
$$\boldsymbol{b} = (b_1, \ldots, b_{N^2})^{\mathrm{T}} , \qquad \boldsymbol{n} = (n_1, \ldots, n_{N^2})^{\mathrm{T}} ,$$

and the PSF is represented by an $N^2 \times N^2$ matrix H. The image formation is represented by

$$\boldsymbol{g} = \boldsymbol{b} + \boldsymbol{n} = H\boldsymbol{f} + \boldsymbol{n} . \tag{B.7}$$

2. In spectral estimation, the measured data are usually the time series but not its ACF. In contrast, in image restoration the data are almost always the visibility (indirect imaging) or degraded image (direct imaging); they form an FT pair.

3. In spectral estimation of time series, in most cases noise is weak and hence is not under consideration. However, noise in image formation systems is usually not negligible, so the noise in data is taken into account in almost all the algorithms for image restoration.

References

Chapter 1

1.1 M. W. Zemansky: *Heat and Thermodynamics* (McGraw–Hill, New York, NY 1981)

1.2 G. Silviu: *Information Theory with Applications* (McGraw-Hill, New York, NY 1977)

1.3 E. T. Jaynes: Phys. Rev. **106**, 620–630 (1957); ibid. **108**, 171–190 (1957)

1.4 J. P. Burg: "Maximum entropy spectral analysis"; paper presented at the 37th Annu. Int'l Meet. Soc. Explor. Geophys. (Oklahoma City, OK 1967)

1.5 B. R. Frieden: J. Opt. Soc. Am. **62**, 511–518 (1972)

1.6 N. Wu: "The maximum entropy method and its application in radio astronomy"; Ph.D. thesis, Sydney University, Sydney, Australia (1985)

Chapter 2

2.1 S. Haykin (ed.): *Nonlinear Methods of Spectral Analysis*, 2nd edn., Topics Appl. Phys., Vol. 34 (Springer, Berlin, Heidelberg 1983) Chap. 2

2.2 J. Skilling, S. F. Gull: "Algorithms and applications", in *Maximum–Entropy and Bayesian Methods in Inverse Problems*, ed. by C. R. Smith, W. T. Grandy (Reidel, Dordrecht, Netherlands 1985) pp. 83–132

2.3 J. A. Edward, M. M. Fitelson: IEEE Trans. IT-**19**, 232–234 (1973)

2.4 J. P. Burg: "Maximum entropy spectral analysis"; paper presented at the 37th Annu. Int'l Meet. Soc. Explor. Geophys. (Oklahoma City, OK 1967)

2.5 A. Van den Bos: IEEE Trans. IT-**17**, 493–494 (1971)

2.6 M. M. Komesaroff, I. Lerche: "Extending the Fourier transform – the positivity constraint", in *Image Formation from Coherence Functions in Astronomy*, ed. by C. van Schooneveld (Reidel, Dordrecht, Netherlands 1979) pp. 241–247

2.7 J. E. Shore, R. W. Johnson: IEEE Trans. IT-**36**, 26–37 (1980)

2.8 R. W. Johnson, J. E. Shore: IEEE Trans. ASSP-**32**, 129–137 (1984)

2.9 X. Zhuang, L. Chen, S. Chen: IEEE Trans. ASSP-**41**, 1730–1734 (1993)

2.10 T. J. Cornwell: "The use of Bayesian statistics in image estimation from interferometer data", in *Image Formation from Coherence Functions in Astronomy*, ed. by C. van Schooneveld (Reidel, Dordrecht, Netherlands 1979) pp. 227–234

2.11 C. L. Byrne, R. M. Fitzgerald: IEEE Trans. ASSP-31, 734–736 (1983)

2.12 T. J. Ulrych, T. N. Bishop: Rev. Geophysics and Space Phys. **13**, 183–200 (1975)

2.13 S. Kay: *Modern Spectral Estimation: Theory and Application* (Prentice-Hall, Englewood Cliffs, NJ 1988) p. 213
2.14 J. S. Lim, N. A. Malik: IEEE Trans. ASSP-**29**, 401–413 (1981)
2.15 S. J. Wernecke, L. R. D'Addario: IEEE Trans. C-**26**, 351–364 (1977)
2.16 J. G. Ables: Astron. Astrophys. Suppl. **15**, 383–393 (1974)
2.17 J. P. Burg: "Maximum entropy spectral analysis"; Ph.D. dissertation, Stanford University, Stanford, CA (1975)
2.18 H. J. Newton: *TIMESLAB: A Time Series Analysis Laboratory* (Wadsworth, Belmont, CA 1988)
2.19 L. Marple: IEEE Trans. ASSP-**28**, 441–454 (1980)
2.20 I. Barrodale, R. E. Eriskson: Geophys. **45**, 420–446 (1980)
2.21 D. M. Lin, Y. H. Mao: Scientia Sinica, Series A, China **XXVII**, 196–212 (1984)
2.22 S. M. Kay, S. L. Marple: IEEE Proc. **69**, 1380–1419 (1981)
2.23 H. Akaike: Ann. Inst. Statist. Math. **22**, 203–217 (1970)
2.24 H. Akaike: IEEE Trans. AC-**19**, 716–723 (1974)
2.25 H. Akaike: Biometrica **66**, 237–242 (1979)
2.26 E. Parzen: IEEE Trans. AC-**19**, 723–730 (1974)
2.27 M. Wax: IEEE Trans. ASSP-**36**, 581–588 (1988)
2.28 S. K. Katsikas, S. D. Likothanassis, D. G. Lainiotis: IEEE Trans. ASSP-**38**, 872–876 (1990)
2.29 P. M. Djurić, S. M. Kay: IEEE Trans. SP-**40**, 2829–2833 (1992)

Chapter 3

3.1 B. R. Frieden: J. Opt. Soc. Am. **62**, 511–518 (1972)
3.2 T. S. Huang (ed.): *Picture Processing and Digital Filtering*, 2nd edn., Topics Appl. Phys., Vol. 6 (Springer, Berlin, Heidelberg 1979) Sect. 5.17
3.3 S. F. Gull, J. Skilling: "The entropy of an image", in *Maximum–Entropy and Bayesian Methods in Inverse Problems*, ed. by C. R. Smith, W. T. Grandy (Reidel, Dordrecht, Netherlands 1985) pp. 287–301
3.4 T. J. Cornwell, K. F. Evans: Astron. Astrophys. **143**, 77–83 (1985)
3.5 S. F. Gull, J. Skilling: "The maximum entropy method", in *Indirect Imaging*, ed. by J. A. Roberts (Cambridge University Press, Cambridge, UK 1984) pp. 267–279
3.6 N. Wu: "The maximum entropy method and its application in radio astronomy"; Ph.D. thesis, Sydney University, Sydney, Australia (1985)
3.7 N. Wu: IEEE Trans. ASSP-**31**, 486–491 (1983)
3.8 L. Zhang, Z. Wu: AMSE Rev. **11**, 23–41 (1989)
3.9 C. Nadeu, M. Bertran-Salvans, J. Solé: Signal Processing, **10**, 7–18 (1986)
3.10 Q. Cheng: Kexue Tongbao, China **30**, 436–440 (1985)
3.11 H. J. Trussell: IEEE Trans. ASSP-**28**, 114–117 (1980)
3.12 N. Wu: IEEE Trans. ASSP-**36**, 294–296 (1988)
3.13 R. Willingale: Mon. Not. R. Astr. Soc. **194**, 359–364 (1981)
3.14 S. F. Gull, J. G. Daniell: Nature **272**, 686–690 (1978)
3.15 N. Wu: Astron. Astrophys. **139**, 555–557 (1984)
3.16 M. A. Delsuc: "A new maximum entropy processing algorithm, with applications to nuclear magnetic resonance experiments ", in *Maximum Entropy and Bayesian Methods*, ed. by J. Skilling (Kluwer, Dordrecht, Netherlands 1989) pp. 285–290
3.17 J. Skilling, R. K. Bryan: Mon. Not. R. Astr. Soc. **211**, 111–124 (1984)
3.18 X. Zhuang, E. Østevold, R. M. Haralick: IEEE Trans. ASSP-**35**, 208–218 (1987)

3.19 J. Skilling: "The axiom of maximum entropy", in *Maximum–Entropy and Bayesian Methods in Science and Engineering* (vol. 1), ed. by G. J. Erickson, C. R. Smith (Kluwer, Dordrecht, Netherlands 1989) pp. 173–187

3.20 N. Weir: "Applications of maximum entropy techniques to HST data", in *Proc. 3rd ESO/ST-ECF Data Analysis Workshop*, ed. by P. J. Grosbol, R. H. Warmels (ESO, Garching, Germany 1991) pp. 115–129

3.21 S. F. Gull, J. Skilling: *Quantified Maximum Entropy MemSys5 Users' Manual* (Maximum Entropy Data Consultants Ltd. of England, Cambridge, UK 1991)

3.22 S. F. Gull: "Developments in maximum entropy data analysis", in *Maximum Entropy and Bayesian Methods*, ed. by J. Skilling (Kluwer, Dordrecht, Netherlands 1989) pp. 53–71

3.23 N. Weir, S. Djorgovski: "MEM: new techniques, application, and photometry", in *The Restoration of HST Images and Spectra*, ed. by R. L. White, R. J. Allen (STScI, Baltimore, MD 1991) pp. 31–38

3.24 N. Wu: "MEM package for image restoration in IRAF", in *Astronomical Data Analysis Software and Systems II*, ed. by R. J. Hanisch, R. J. V. Brissenden, J. Barnes (Boston, MA 1992) pp. 520–523

3.25 N. Wu: "MEM task for image restoration in IRAF", in *Astronomical Data Analysis Software and Systems IV*, ed. by R. A. Shaw, H. E. Payne, J. J. E. Hayes (STScI, Baltimore, MD 1994) pp. 305–308

3.26 N. Wu: "Model updating in the MEM algorithm", in *The Restoration of HST Images and Spectra II*, ed. by R. J. Hanisch, R. L. White (STScI, Baltimore, MD 1994) pp. 58–63

3.27 P. K. Piña, R. C. Putter: Publications of Astronomical Society of the Pacific, **105**, 630–637 (1993)

3.28 R. C. Putter, P. K. Piña: "Pixon–based image reconstruction"; paper presented at the 13th Int'l Workshop on Maximum Entropy and Bayesian Methods (Santa Barbara, CA 1993)

3.29 J. Nuñez, J. Llacer: Publications of Astronomical Society of the Pacific, **105**, 1192–1208 (1993)

3.30 J. Nuñez, J. Llacer: "HST image restoration with variable resolution", in *The Restoration of HST Images and Spectra II*, ed. by R. J. Hanisch, R. L. White (STScI, Baltimore, MD 1994) pp. 123–130

3.31 J. Starck, E. Pantin: "Deconvolution by the multiscale maximum entropy method", in *Astronomical Data Analysis Software and Systems V*, ed. by G. H. Jacoby, J. Barnes (Tucson, AZ 1995) pp. 191–194

3.32 J. Skilling, A. W. Strong, K. Bennett: Mon. Not. R. Astr. Soc. **187**, 145–152 (1979)

3.33 D. Ward-Thompson, D. S. Berry, E. I. Robson: Mon. Not. R. Astr. Soc. **257**, 180–186 (1992)

3.34 G. Heidbreder: "Maximum entropy applications in radar", in *Maximum Entropy and Bayesian Methods*, ed. by W. T. Grandy, L. H. Schick (Kluwer, Dordrecht, Netherlands 1991) pp. 127–136

3.35 B. Borden: IEEE Trans. SP-**40**, 969–973 (1992)

3.36 R. K. Shevgaonkar: Astron. Astrophys. **176**, 159–170 (1987)

3.37 G. Minerbo: Computer Graphics and Image Processing **10**, 48–68 (1979)

Chapter 4

4.1 J. A. Högbom: "The introduction of a priori knowledge in certain processing algorithms", in *Image Formation from Coherence Functions in Astronomy*, ed. by C. van Schooneveld (Reidel, Dordrecht, Netherlands 1979) pp. 237–239

4.2 R. Nityananda, R. Narayan: J. Astrophys. Astr. **3**, 419–450 (1982)

4.3 R. Narayan, R. Nityanada: "Maximum entropy – flexibility versus fundamentalism", in *Indirect Imaging*, ed. by J. A. Roberts (Cambridge University Press, Cambridge, UK 1984) pp. 282–290

4.4 J. W. Woods: IEEE Trans. IT-**22**, 552–559 (1976)

4.5 X. Zhuang, R. M. Haralik, Y. Zhao: IEEE Trans. SP-**39**, 1478–1480 (1991)

4.6 J. Skilling, R. K. Bryan: Mon. Not. R. Astr. Soc. **211**, 111–124 (1984)

4.7 J. E. Shore, R. W. Johnson: IEEE Trans. IT-**36**, 26–37 (1980)

4.8 N. Wu: "Bayesian inference, MEM and consistency"; paper presented at the 4th Valencia Int'l Meet. on Bayesian Statistics (Peñiscola, Spain 1991)

4.9 A. Baggeroer: IEEE Trans. IT-**22**, 534–545 (1976)

4.10 H. Akaike: Ann. Inst. Statist. Math. **21**, 407–419 (1969)

4.11 D. Burshtein, E. Weistein: IEEE Trans. ASSP-**35**, 504–510 (1987)

4.12 S. Haykin (ed.): *Nonlinear Methods of Spectral Analysis*, 2nd edn., Topics Appl. Phys., Vol. 34 (Springer, Berlin, Heidelberg 1983) Sect. 2.9

4.13 R. Kromer: "Asymptotic properties of the autoregressive spectral estimator"; Ph.D. dissertation, Stanford University, Stanford, CA (1970)

4.14 R. Lacoss: Geophys. **36**, 661–675 (1971)

4.15 D. M. Lin, Y. H. Mao: Scientia Sinica, Series A, China **XXVII**, 196–212 (1984)

4.16 W. Chen, G. Stegen: J. Geophys. Res. **79**, 3019–3022 (1974)

4.17 N. Malik, J. Lim: IEEE Trans. ASSP-**30**, 788–797 (1982)

4.18 C. Nadeu, M. Bertran-Salvans, J. Solé: Signal Processing, **10**, 7–18 (1986)

4.19 N. Wu: "The maximum entropy method and its application in radio astronomy"; Ph.D. thesis, Sydney University, Sydney, Australia (1985)

4.20 N. Wu: IEEE Trans. ASSP-**35**, 1355–1358 (1987)

4.21 L. Marple: "Resolution of conventional Fourier, autoregressive, and special ARMA methods of spectrum analysis", in *Proc. IEEE Int'l Conf. on ASSP* (Hartford, CT 1977) pp. 74–77

4.22 R. Bhandari: Astron. Astrophys. **70**, 331–333 (1978)

4.23 P. F. Fougere, E. J. Zawalick, H. R. Radoski: Phys. Earth Planet. Interiors **12**, 201–207 (1976)

4.24 R. W. Herring: IEEE Trans. ASSP-**28**, 692–701 (1980)

4.25 N. Wu: "On the resolvability of power spectrum estimates in the maximum entropy method"; paper presented at the 1st Beijing Postdoctoral Annual Conf. (Beijing, China 1988)

4.26 P. F. Fougere: J. Geophys. Res. **82**, 1051–1054 (1977)

4.27 D. N. Swingler: IEEE Trans. ASSP-**28**, 257–259 (1980)

4.28 M. Kaveh, G. A. Lippert: IEEE Trans. ASSP-**31**, 438–444 (1983)

4.29 M. K. Ibrahim: IEEE Trans. ASSP-**35**, 1476–1477 (1987)

4.30 S. M. Kay: IEEE Trans. ASSP-**26**, 467–469 (1978)

Chapter 5

5.1 L. R. Mead, N. Papanicolaou: J. Math. Phys. **25**, 2404–2417 (1984)

5.2 D. V. Widder: *The Laplace Transform* (Princeton University Press, Princeton, NJ 1946)

5.3 S. Ciulli, M. Mounsif, N. Gorman, T. D. Spearman: J. Math. Phys. **31**, 1717–1719 (1991)

5.4 S. Kopeć, G. L. Bretthorst: "Entropy, moments and probability theory", in *Maximum Entropy and Bayesian Methods*, ed. by A. Mohammad-Djafari, G. Demoments (Kluwer, Dordrecht, Netherlands 1993) pp. 25–30

5.5 J. G. Wheeler, R. G Gordon: J. Chem. Phys. **51**, 5566–5583 (1969)

5.6 L. R. Mead: J. Math. Phys. **27**, 2903–2907 (1986)

5.7 S. Kopeć: J. Math. Phys. **32**, 1269–1272 (1991)

5.8 S. Kopeć: "On application of MaxEnt to solving Fredholm integral equations", in *Maximum Entropy and Bayesian Methods*, ed. by A. Mohammad-Djafari, G. Demoments (Kluwer, Dordrecht, Netherlands 1993) pp. 63–66

5.9 P. Hick, G. Stevens: Astron. Astrophys. **172**, 350–358 (1987)

5.10 E. T. Jaynes: Phys. Rev. **106**, 620–630 (1957); ibid. **108**, 171–190 (1957)

5.11 E. T. Jaynes: "Predictive statistical mechanics"; in *Frontiers of Nonequilibrium Statistical Physics*, Proc. NATO Adv. Study Institute, ed. by G. T. Moore, M. U. Scully (Pelnum, New York, NY 1986) pp. 33–56

5.12 K. Huang: *Statistical Mechanics* (Wiley, New York, NY 1987)

5.13 F. Reif: *Statistical Physics*, Berkeley Physics Course, vol. 5 (McGraw–Hill, New York, NY 1965)

Appendices

A.1 A. V. Oppenheim, R. W. Schafer: *Digital Signal Processing* (Prentice–Hall, Englewood Cliffs, NJ 1975) Chap. 10

A.2 J. M. Tribolet: *Seismic Applications of Homomorphic Signal Processing* (Prentice–Hall, Englewood Cliffs, NJ 1979)

A.3 B. R. Hunt: IEEE Trans. C-**26**, 219–229 (1977)

Index

Springer Series in Information Sciences

Editors: Thomas S. Huang   Teuvo Kohonen   Manfred R. Schroeder
Managing Editor: H. K. V. Lotsch

If you have any concerns about our products,
you can contact us on
ProductSafety@springernature.com

In case Publisher is established outside the EU,
the EU authorized representative is:
Springer Nature Customer Service Center GmbH
Europaplatz 3, 69115 Heidelberg, Germany

Printed by Libri Plureos GmbH
in Hamburg, Germany